Natural Product Inspired Scaffolds

Mohit Kumar • Deepika Kathuria •
Ajay Sharma
Editors

Natural Product Inspired Scaffolds

Applications in Tissue Engineering

Springer

Editors
Mohit Kumar
Department of Mechanical Engineering
Chandigarh University
Mohali, Punjab, India

Deepika Kathuria
Chandigarh College of Pharmacy
Chandigarh Group of Colleges
Landran, Punjab, India

Ajay Sharma
Department of Biomolecular Sciences,
School of Pharmacy
University of Mississippi
Oxford, USA

ISBN 978-981-97-3110-7 ISBN 978-981-97-3111-4 (eBook)
https://doi.org/10.1007/978-981-97-3111-4

This Springer imprint is published by the registered company Springer Nature Singapore Pte Ltd.
The registered company address is: 152 Beach Road, #21-01/04 Gateway East, Singapore 189721, Singapore

Dr Mohit Kumar
The book is dedicated to my Daughter "Paavni"

Dr Deepika Kathuria
The book is dedicated to my dearest son "Nyle Verma"

Dr Ajay Sharma
The book is dedicated to my wife "Garima Bhardwaj"

Preface

Tissue engineering is one of the burgeoning fields mostly used in the replacement of damaged tissue. The key players of this process are scaffolds, cells, and growth factors. A scaffold is a base material designed to mimic the microarchitecture and mechanical properties of the mature tissue to be regenerated. To date various materials have been utilized for the development of scaffolds; however, the study for the search of natural scaffolds has been accelerated owing to their biological as well as mechanical functional similarities with the in vivo tissues. The main categories of natural biomaterials are protein-based biomaterials, polysaccharide-based biomaterials, glycosaminoglycan-derived biomaterials, and tissue/organ-derived biomaterials. Based on their use, the scaffold needs to be chosen; for instance, stiff scaffolds are required with tuned porosity for the bone. The key features of natural scaffolds include biocompatibility, biodegradability, non-genotoxicity, non-teratogenicity, high surface area, porosity, pre-existing vascular networks, and safety. Thus the use of natural scaffolds is highly recommendable. Various advanced techniques have been developed for the fabrication of scaffolds including conventional and rapid prototyping (for the development of 3D porous scaffold). The incorporation of plant extracts in these bioscaffolds has also been studied widely.

This book explores various aspects of natural biomaterials used for bioscaffolds, fabrication techniques, their properties, and synthesis, among other topics. It discusses about the biomedical applications of natural bioscaffolds in tissue engineering, such as in bone, skin, and nerve tissue regeneration, as well as in plastic surgery, wound healing, and breast reconstruction. It also deals with the biomedical applications of natural gums-based scaffolds, including xanthan, gellan, arabic, guar, and ghatti gum. The book also discusses the toxicological and regulatory aspects of bioscaffolds.

Mohali, Punjab, India — Mohit Kumar
Landran, Punjab, India — Deepika Kathuria
Oxford, USA — Ajay Sharma

Contents

Editors and Contributors

About the Editors

Mohit Kumar is an Assistant Professor in the Department of Mechanical Engineering at the University Institute of Engineering, Gharaun, Mohali. He completed his graduation (B. Tech) from Kurukshetra University, Kurukshetra, Master of Technology in Mechanical Engineering (spl. Welding and Fabrication) from Sant Longowal Institute of Engineering and Technology (SLIET, Deemed University), and his PhD in Mechanical Engineering (polymer composites) from Thapar Institute of Engineering and Technology. Overall, he has received fellowships totaling over 18 lakhs INR from 2014 to March 2021, including the MHRD fellowship in M.Tech and DAE-BRNS, BARC fellowship for a sponsored project. He was awarded Gold Medal for his academic excellence in M.Tech. He has published over 20 peer-reviewed papers and book chapters (research articles—16, book chapters—4) in reputed international/national journals and books. Additionally, he is an editorial review board member of two international journals. His research activities have focused on mechanical characterizations of polymer composite materials, the aging of FRP composites, and FEM. At present, his area of interest is on natural composites derived from natural polymers, with applications in tissue engineering.

Deepika Kathuria is an Assistant Professor at Chandigarh Colleges of Pharmacy, Chandigarh Group of Colleges, Landran, Punjab, India. She completed her master's in pharmaceuticals and doctorate in pharmaceutical sciences from the National Institute of Pharmaceutical Education and Research, Mohali, India. She has published over 30 peer-reviewed papers in reputed international journals. She has various honors/fellowships/awards to her credit, i.e., Award of INSPIRE fellowship (JRF) from DST, Award of INSPIRE fellowship (SRF) from DST, Bristol-Mayer Squibb fellowship (merit-cum-topper scholarship during M.Pharm), Newton Bhabha Placement, DST-exchange program. She does research in organic chemistry, medicinal chemistry, analytical chemistry, natural products, and CADD. She has worked with medicinal plants, pharmaceuticals, and cosmeceuticals, allowing her to bring a multidisciplinary perspective to this topic.

Ajay Sharma is a Natural Product Chemist with a strong organic chemistry, photochemistry, biochemistry, and ethnobotany background. Presently, he is working as a post doctoral fellow at Department of Biomolecular Sciences, School of Pharmacy, University of Mississippi, Oxford, USA. He worked as an Assistant Professor at the University Centre for Research and Development, Department of Chemistry, Chandigarh University, Mohali, India. He completed his graduation (BSc Medical) from Govt. College Hoshiarpur (affiliated to Panjab University, Chandigarh), master's degree in Organic Chemistry (Natural Products) from Punjab Agricultural University, Ludhiana, and his PhD (Organic Chemistry—Natural Products) from SLIET (Sant Longowal Institute of Engineering & Technology, Deemed University), Punjab. He has published over 45 peer-reviewed papers and book chapters in reputed international/national journals and books. Presently, he has five edited books in production with various international publishers. He has various honors/fellowship/awards to his credit, i.e., Role of Honor from PAU Ludhiana (2011), merit-cum-topper scholarship during MSc (PAU Ludhiana) (2009-10), MHRD Fellowship Award for PhD (2012), Award of INSPIRE fellowship (JRF, 2013) from DST, Award of INSPIRE fellowship (SRF, 2015) from DST, First Prize in Poster Award at National Conference (NICS-2016), and Best Poster Award in TSFS (2016). He does research in organic chemistry, natural products chemistry, photochemistry, and food chemistry. His research activities have focused on the isolation and characterization of bioactive extracts from traditional medicinal plants and evaluation of their biological potential. Currently, his research group at Chandigarh University is investigating the field of environmentally friendly synthesis of metal nanoparticles and carbon nanodots using plant extracts/plant-based biomass followed by assessing their biological, catalytic, and sensing potential. To date, he has successfully guided six MSc dissertation students.

Contributors

Ashish Baldi University College of Pharmacy, Guru Kashi University, Talwandi Sabo, Bathinda, Punjab, India

Rinki Bharali Department of Botany, Madhab Choudhury College, Barpeta, Assam, India

Romesh Borgohain Department of Chemistry, Birjhora Mahavidyalaya, Bongaigaon, Assam, India

Munmi Borkataky Department of Life Sciences, Dibrugarh University, Dibrugarh, Assam, India

Tridip Boruah Department of Botany, Madhab Choudhury College, Barpeta, Assam, India

Shilpa Chandel Faculty of Pharmaceutical Sciences, ICFAI University, Kalujhanda, Baddi, Himachal Pradesh, India

Twinkle Chetia Department of Botany, Cotton University, Guwahati, Assam, India

Rosana de Carvalho Cruz Department of Biochemistry and Immunology, Institute of Biological Sciences, Belo Horizonte, MG, Brazil

Bioengineering Laboratory, Department of Mechanical Engineering, Universidade Federal de Minas Gerais (UFMG), Belo Horizonte, MG, Brazil

Babita Dhiman Electronics and Communication Engineering, UIC, Chandigarh University, Mohali, Punjab, India

Sanjay Dhoble Department of Physics, R.T.M. Nagpur University, Nagpur, India

Swati Gajbhiye Department of Physics, R.T.M. Nagpur University, Nagpur, India

Khushbu Gumber Department of Chemistry, Chandigarh University, Gharaun, Mohali, Punjab, India

Mitesh Ikar Department of Chemical Engineering, Vishwakarma Institute of Technology, Pune, Maharashtra, India

Naser Kalhor Department of Mesenchymal Stem Cells, Academic Center for Education, Culture and Research, Qom, Iran

Gurpreet Kaur Department of Zoology, Akal University, Talwandi Sabo, Punjab, India

Navjot Kaur Amar Shaheed Baba Ajit Singh Jujhar Singh Memorial College of Pharmacy, Bela, Ropar, Punjab, India

Pawan Kumar Department of Pharmaceutical Science and Technology, Maharaja Ranjit Singh Punjab Technical University, Bathinda, Punjab, India

Ram Kumar University College of Pharmacy, Guru Kashi University, Talwandi Sabo, Bathinda, Punjab, India

Sanjeev Kumar University Institute of Pharma Sciences, Chandigarh University, Gharuan, Mohali, Punjab, India

Akshita Mahajan Department of Chemistry, Chandigarh University, Gharaun, Mohali, Punjab, India

Hanie Mahaki Vascular and Endovascular Surgery Research Center, Mashhad University of Medical Sciences, Mashhad, Iran

Hamed Manoochehri The Persian Gulf Marine Biotechnology Research Center, The Persian Gulf Biomedical Sciences Research Institute, Bushehr University of Medical Sciences, Bushehr, Iran

Ratul Nath Department of Life Sciences, Dibrugarh University, Dibrugarh, Assam, India

Navneesh Department of Pharmaceutical Chemistry, ISF College of Pharmacy, Moga, Punjab, India

Parijat Pandey Department of Pharmaceutical Sciences, Gurugram University, Gurugram, Haryana, India

Bharat Parashar Faculty of Pharmaceutical Sciences, ICFAI University, Kalujhanda, Baddi, Himachal Pradesh, India

Morjina Parveen Department of Botany, Madhab Choudhury College, Barpeta, Assam, India

Susuma Paul Department of Botany, Madhab Choudhury College, Barpeta, Assam, India

Aaditi Pendse Department of Ayurveda, KLE, Belagavi, Karnataka, India

Himu Roy Department of Botany, Cotton University, Guwahati, Assam, India

Sunil Sable Department of Chemical Engineering, Vishwakarma Institute of Technology, Pune, Maharashtra, India

Sapna Saini Janta College of Pharmacy, Butana, Sonepat, Haryana, India

Reihaneh Seyedebrahimi Cellular and Molecular Research Center, Qom University of Medical Sciences, Qom, Iran

Anatomy Department, Faculty of Medicine, Qom University of Medical Sciences, Qom, Iran

Rahul Kumar Sharma Amar Shaheed Baba Ajit Singh Jujhar Singh Memorial College of Pharmacy, Bela, Ropar, Punjab, India

Ravinder Sharma Faculty of Pharmaceutical Sciences, ICFAI University, Kalujhanda, Baddi, Himachal pradesh, India

Shailesh Sharma Amar Shaheed Baba Ajit Singh Jujhar Singh Memorial College of Pharmacy, Bela, Ropar, Punjab, India

UCRD, Chandigarh University, Gharuan, Mohali, Punjab, India

Mohsen Sheykhhasan Cellular and Molecular Research Center, Qom University of Medical Sciences, Qom, Iran

Baljeet Singh Amar Shaheed Baba Ajit Singh Jujhar Singh Memorial College of Pharmacy, Bela, Ropar, Punjab, India

Simrandeep Singh UCRD, Chandigarh University, Gharuan, Mohali, Punjab, India

Darshita Singha Department of Life Sciences, Dibrugarh University, Dibrugarh, Assam, India

Sushant Faculty of Pharmaceutical Sciences, The ICFAI University, Baddi, Himachal Pradesh, India

Hamid Tanzadehpanah Antimicrobial Resistance Research Center, Mashhad University of Medical Sciences, Mashhad, Iran

Desmond Tobin The Charles Institute of Dermatology, School of Medicine, The Conway Institute, University College Dublin, Belfeld, Dublin 4, Ireland

Kiran Yadav Faculty of Pharmaceutical Sciences, ICFAI University, Kalujhanda, Baddi, Himachal Pradesh, India

Piao Yang Department of Molecular Genetics, College of Arts and Sciences, The Ohio State University, Columbus, OH, USA

Amirhossein Ahmadieh Yazdi Stem Cell Biology Research Center, Yazd Reproductive Sciences Institute, Shahid Sadoughi University of Medical Sciences, Yazd, Iran

Introduction to Natural Product-Based Tissue Engineering

1

Darshita Sinha, Ratul Nath, and Munmi Borkataky

Abstract

Tissue engineering has emerged as a promising avenue for developing regenerative therapies and functional tissue substitutes. In this context, the integration of natural product-based scaffolds has garnered considerable attention due to their inherent biocompatibility and bioactivity. These scaffolds offer a unique opportunity to create biomimetic environments that support cell adhesion, proliferation, and differentiation, ultimately facilitating tissue regeneration. This chapter presents an overview of the utilization of natural product-derived compounds in the design and fabrication of tissue engineering. Natural products, sourced from plants, animals, and microorganisms, contain a diverse array of bioactive molecules that can influence various cellular responses critical for tissue repair. The biocompatibility of these compounds ensures minimal adverse effects on the host organism and aligns with the growing focus on sustainable and eco-friendly medical approaches. The chapter also highlights key advantages of natural product-based scaffolds, including their ability to mimic the extracellular matrix composition and structure, their potential to promote angiogenesis and neurogenesis, and their modulation of immune responses. Examples of prominent natural products, such as chitosan, collagen, and plant-derived extracts, are discussed in the context of their applications in tissue engineering. Furthermore, the chapter underscores the interdisciplinary nature of this field, requiring collaboration between materials scientists, biologists, and engineers to develop scaffolds that effectively harness the regenerative potential of natural products. The integration of natural product-based scaffolds represents a promising strategy in tissue engineering, offering a versatile platform to enhance tissue regeneration. As research in this domain advances, the synergy between natural product bioactiv-

D. Sinha · R. Nath (✉) · M. Borkataky
Department of Life Sciences, Dibrugarh University, Dibrugarh, Assam, India
e-mail: ratulnath@dibru.ac.in

M. Kumar et al. (eds.), *Natural Product Inspired Scaffolds*,
https://doi.org/10.1007/978-981-97-3111-4_1

ity and scaffold design holds the potential to revolutionize regenerative medicine, paving the way for innovative therapies and treatments in the realm of tissue engineering and beyond.

Keywords
Tissue engineering · Regeneration · Bioscaffold · Natural product · Medicines

1.1 Introduction

Tissue engineering is an interdisciplinary science that combines engineering, biology, and medical concepts to construct functioning and viable biological tissues or organs (Socci et al. 2023). It is extremely important in regenerative medicine, which is concerned with repairing or replacing damaged or tissues and organs that are diseased in the body. Unlike traditional treatments, which generally focus on symptoms, tissue engineering aims to repair the structure and function of damaged tissues, offering a potential cure rather than just symptom management (Li et al. 2022).

The history of tissue engineering can be traced back to the mid-twentieth century, with significant developments and milestones occurring over the decades that followed. The field emerged from a convergence of expertise in biology, engineering, and material science, with the goal of creating functional biological tissues to replace or repair damaged or diseased organs within the human body. In the 1970s, pioneers like Robert Langer and Joseph and Jay Vacanti made early contributions to the field, exploring the possibilities of tissue regeneration and the concept of growing replacement organs in the laboratory (Saltzman 2004). The 1980s and 1990s marked a significant turning point for tissue engineering. The National Science Foundation (NSF) and the National Institutes of Health (NIH) began to invest in research initiatives in tissue engineering, providing essential funding and support. The landmark publication Langer and Vacanti's *Tissue Engineering* in 1993 helped define the field, and this era saw the advent of cell-based therapies and the use of biomaterials to support tissue growth (Langer and Vacanti 1993). During the early 2000s, researchers made significant strides in tissue engineering, leveraging advances in stem cell research, nanotechnology, and bioprinting (Sivaraman et al. 2019). The discovery and utilization of induced pluripotent stem cells (iPSCs) brought forth a new source of cells for tissue regeneration (Zhang and Yang-De 2015). Bioprinting technologies gained traction, enabling the precise layering of cells and biomaterials to create 3D structures mimicking native tissues (Ashammakhi et al. 2019). In the present day and beyond, tissue engineering is poised to reach new heights. Continued research in stem cell biology, gene editing technologies like CRISPR-Cas9, tissue design, and development promise to accelerate progress (Prasad et al. 2021). Personalized medicine and the ability to tailor tissue engineering solutions to an individual's genetic makeup and specific needs are becoming central themes, indicating a bright future for regenerative medicine.

The significance of tissue engineering in regenerative medicine stems from its ability to address key healthcare issues such as organ transplant shortages, chronic

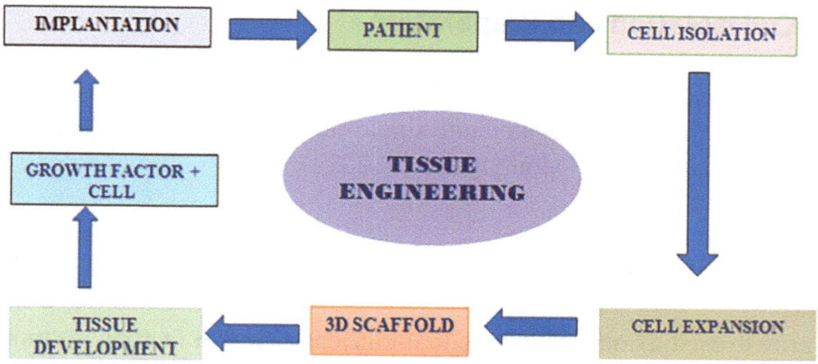

Fig. 1.1 Schematic of the scaffold-based tissue engineering approach

degenerative diseases, and severe tissue injuries (Blair et al. 2017). Scientists and researchers can use tissue engineering techniques to create custom-made tissues and organs that closely mimic natural tissues, lowering the risk of immune rejection and enhancing patient outcomes.

Moreover, tissue engineering has the potential to revolutionize personalized medicine. It enables the development of patient-specific tissues, which may lower the likelihood of adverse responses and improve therapy efficacy (Ikada 2006). The field also opens new avenues for studying disease mechanisms, drug testing, and the development of more effective therapies (Low and Tagle 2017). Tissue engineering holds the promise of revolutionizing the way we approach healthcare as our understanding of biology, biomaterials, and biomanufacturing processes advances, providing hope for those with chronic illnesses, organ failure, or catastrophic traumas. The continuous research and development in the sector have the potential to alter medical practice and greatly improve the quality of life for countless patients around the world.

Using natural products as a foundation for tissue engineering approaches entails utilizing the intrinsic qualities of substances obtained from live organisms to create scaffolds, matrices, or bioactive components that aid in tissue regeneration and growth. These natural compounds function as biocompatible and bioactive materials that promote cellular connections and tissue integration (Kharaziha et al. 2021).

By harnessing the unique properties of natural products, tissue engineering approaches can enhance the success and efficacy of tissue regeneration, potentially leading to more effective treatments for various diseases, injuries, and degenerative conditions. Figure 1.1 represents the schematic diagram of the scaffold-based tissue engineering approach.

1.2 Methods Used in Tissue Engineering

There are several common methods used in tissue culture:

1.2.1 Cell Isolation and Preparation

Cells are obtained from a tissue or organism through enzymatic or mechanical disaggregation and then disaggregated into single cells or small clusters using enzymes or mechanical dissociation (Montanari et al. 2022).

1.2.2 Cell Culture Media Preparation

Create growth media containing essential nutrients, growth factors, hormones, and supplements for cell growth, and customize them based on specific cell type and culture requirements (Valk et al. 2010).

1.2.3 Sterilization and Aseptic Technique

Maintain a sterile environment by using aseptic techniques to prevent the contamination of the culture, including proper cleaning, the use of laminar flow hoods, and autoclaving equipment and media (Rediguieri et al. 2016).

1.2.4 Cell Cultivation

The process involves seeding isolated cells into appropriate culture vessels and monitoring their growth conditions, including temperature, humidity, pH, and gas levels (Carrier et al. 1999).

1.2.5 Subculturing and Passaging

Cells are periodically transferred to new culture vessels with fresh media to maintain viability and activity and then split to lower densities for growth and prevent contact inhibition (Tan et al. 2012).

1.2.6 Cryopreservation

To ensure cell viability, cells are maintained at low temperatures below −80 °C and then used cryoprotectants and controlled freezing rates to minimize damage during the freezing process (Costa et al. 2017; Urbani et al. 2017).

1.2.7 Cell Line Authentication

Verify the identity and characteristics of cell lines using techniques like DNA profiling, isoenzyme analysis, or short tandem repeat (STR) analysis to ensure accurate research results (Stacey et al. 2007).

1.2.8 Differentiation and Specialized Culture

Cell differentiation can be induced through specialized culture conditions, differentiation-inducing agents, or genetic modifications, enhancing the understanding of biological processes and mimicking specific tissue environments (Hwang et al. 2008).

1.2.9 Cloning and Selection

Techniques like limiting dilution and cloning rings are used to isolate and grow clonal populations of cells from a heterogeneous culture, selectively cultivating cells with desired characteristics (Guilak et al. 2006).

1.2.10 Microscopy and Characterization

Utilize microscopy to observe cell morphology, growth patterns, and interactions in culture, and characterize cell behavior, gene expression, and protein production using techniques like immunostaining, PCR, Western blotting, and flow cytometry (Chan et al. 2021).

1.2.11 Transfection and Genetic Manipulation

Techniques like transfection, electroporation, or viral transduction are used to introduce foreign genes, nucleic acids, or proteins into cells for genetic studies, protein expression, or gene editing (Fajrial et al. 2020).

These methods collectively enable researchers to study cell behavior, disease mechanisms, drug testing, and various biological processes in a controlled laboratory environment. There are various factors which affect the tissue engineering which are provided in Fig. 1.2.

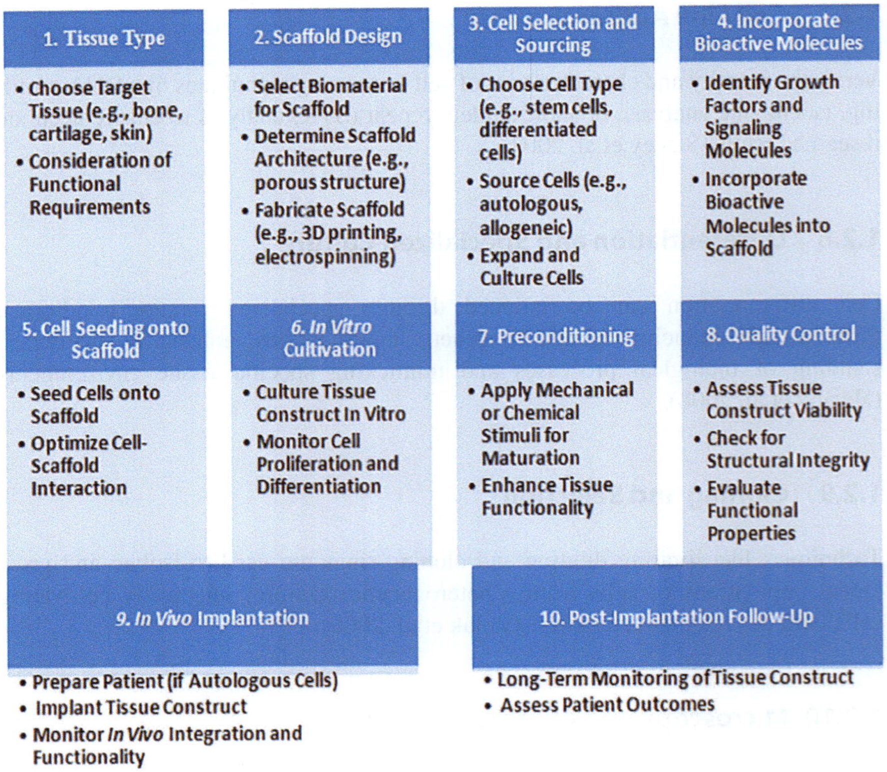

Fig. 1.2 This flowchart outlines the key steps in tissue engineering, from choosing the tissue to postimplantation monitoring. Researchers and engineers consider various factors, techniques, and decisions in each step. Specifics can vary based on factors like tissue type and biomaterials used

1.3 Natural Products Used in Tissue Engineering

Natural products play a crucial role in tissue engineering due to their biocompatibility, bioactivity, and potential to enhance tissue regeneration. Table 1.1 shows some natural products commonly used in tissue engineering:

These natural products contribute to the design and development of biomimetic scaffolds, hydrogels, and matrices that can support cell growth, tissue regeneration, and ultimately aid in the advancement of tissue engineering approaches.

1.4 Key Reasons for Using Natural Products in Tissue Engineering

Using natural products in tissue engineering offers several advantages, aligning with the goal of creating biocompatible and functional tissues. Here are key reasons for incorporating natural products in tissue engineering:

Table 1.1 Natural products with their applications

S. no.	Natural products	Application	References
1.	Collagen	Collagen is a biocompatible scaffold, used in tissue engineering and delivery systems for soft tissue applications like the skin. Its mechanical, degradation, and water-uptake properties make it suitable for customization. Collagen is mainly isolated from animal tissues, which may be unsafe due to potential viral contamination	Dahl et al. (2007)
2.	Gelatin	Derived from collagen, gelatin is used to create hydrogels for tissue scaffolds, drug delivery systems, and cell encapsulation due to its biocompatibility and tunable properties	Huang and Fu (2010)
3.	Hyaluronic acid (HA)	A glycosaminoglycan found in the extracellular matrix, used for its hydrating and lubricating properties and as a component in scaffolds for cartilage and skin tissue engineering	Bae et al. (2011)
4.	Chitosan	Derived from chitin, chitosan is a biopolymer with antimicrobial properties and is used in scaffolds for tissue engineering, particularly in wound healing and cartilage regeneration	Islam et al. (2020)
5.	Alginate	Derived from seaweed, alginate is commonly used in hydrogels for cell encapsulation and tissue engineering applications due to its biocompatibility and gel-forming properties	Sahoo and Biswal (2021)
6.	Silk fibroin	Derived from silk, silk fibroin is used in various forms (e.g., hydrogels, films) for tissue engineering due to its biocompatibility, biodegradability, and mechanical strength	Sun et al. (2021)
7.	Laminin	A glycoprotein present in the extracellular matrix, used to enhance cell adhesion, migration, and differentiation in tissue engineering applications	Hidalgo-Bastida and Cartmell (2010)
8.	Fibrin	Derived from fibrinogen, fibrin is used as a scaffold material for tissue engineering, particularly for wound healing and tissue regeneration, due to its biocompatibility and ability to support cell adhesion	Noori et al. (2017)
9.	Elastin	A protein providing elasticity to tissues, used in tissue engineering to mimic the elasticity of native tissues like the blood vessels and skin	Daamen et al. (2007)
10.	Matrigel	A basement membrane matrix derived from Engelbreth-Holm-Swarm (EHS) mouse sarcoma cells, used as a 3D matrix for cell culture and tissue engineering applications	Aisenbrey and Murphy (2020)

(continued)

Table 1.1 (continued)

S. no.	Natural products	Application	References
11.	Decellularized extracellular matrix (ECM)	ECM derived from tissues (e.g., the heart, the liver) after removal of cellular components, used to provide a natural and tissue-specific scaffold for tissue engineering	Zhang et al. (2022)
12.	Chondroitin sulfate	A glycosaminoglycan found in cartilage and connective tissues, used in tissue engineering for cartilage repair and regeneration	Yang et al. (2020)
13.	Keratin	A protein found in the hair, nails, and skin, used in tissue engineering for wound healing, skin regeneration, and as a scaffold material	Xu et al. (2013)
14.	Dextran	A polysaccharide, commonly used to develop hydrogels for tissue engineering and drug delivery due to its biocompatibility and tunable properties	Sood et al. (2021)
15.	Glycosaminoglycans (GAGs)	Various GAGs (e.g., heparin, dermatan sulfate) are used in tissue engineering for their roles in cell signaling, growth factor binding, and ECM formation	Sodhi and Panitch (2020)
16.	Agarose	A polysaccharide derived from seaweed, used to create hydrogels for cell encapsulation and tissue engineering applications due to its biocompatibility and gel-forming properties	Beaumont et al. (2021)
17.	Pectin	A polysaccharide found in plant cell walls, used in tissue engineering for its biocompatibility and potential applications in hydrogels and scaffolds	Bacakova et al. (2014)
18.	Lignin	A complex aromatic polymer found in plant cell walls, used in tissue engineering for its potential as a biocompatible and bioactive scaffold material	Witzler et al. (2018)
19.	Fucoidan	A sulfated polysaccharide derived from brown seaweed, used for its anti-inflammatory and regenerative properties in tissue engineering	Senthilkumar et al. (2017)
20.	Growth factors (e.g., BMP, TGF-β, FGF, VEGF)	Natural signaling molecules that play a vital role in cell proliferation, differentiation, and tissue development, often used to enhance tissue engineering outcomes by promoting specific cell behaviors	Chen et al. (2010)
21.	Demineralized bone matrix (DBM)	A product derived from bone tissue after removal of minerals, used in bone tissue engineering to provide a scaffold rich in growth factors and proteins that promote bone regeneration	Gruskin et al. (2012)

(continued)

Table 1.1 (continued)

S. no.	Natural products	Application	References
22.	Shark cartilage	Used as a source of chondroitin sulfate and other glycosaminoglycans, often incorporated into scaffolds for cartilage tissue engineering and wound healing applications	Yang et al. (2020)
23.	Osteonectin	A glycoprotein found in the bone, used in bone tissue engineering to promote cell adhesion, migration, and tissue regeneration	Sivakumar et al. (2023)
24.	Sericin	A protein derived from silk, used for its potential in promoting cell proliferation, adhesion, and wound healing in tissue engineering, especially for skin and dermal applications	Arango et al. (2021)
25.	Grape seed extract	Contains bioactive compounds like proanthocyanidins that possess antioxidant properties and potential for promoting tissue regeneration and wound healing in tissue engineering	Marrazzo and O'Leary (2020)
26.	*Moringa oleifera* extract	Rich in bioactive compounds such as flavonoids and polyphenols, used for its anti-inflammatory and antioxidant properties in tissue engineering, particularly for skin and wound healing applications	Rubio-Elizalde et al. (2019)
27.	*Centella asiatica* extract	Contains triterpenoids and flavonoids, used for its potential in promoting collagen synthesis, angiogenesis, and wound healing in tissue engineering, especially for skin and dermal applications	Hussin et al. (2020)
28.	Ginseng extract	Contains bioactive compounds like ginsenosides that have been explored for their potential in promoting cell proliferation, angiogenesis, and wound healing in tissue engineering, especially in skin and dermal applications	Sharma et al. (2021)
29.	Astaxanthin	A carotenoid antioxidant found in microalgae and seafood, used for its anti-inflammatory and antioxidant properties in tissue engineering, particularly for skin, eye, and muscle tissue regeneration	Chou et al. (2020)
30.	Soy protein	A protein derived from soybeans, used to create scaffolds in tissue engineering due to its biocompatibility and potential applications in promoting cell adhesion and tissue regeneration	Jahangirian et al. (2019)

(continued)

Table 1.1 (continued)

S. no.	Natural products	Application	References
31.	Pine bark extract (Pycnogenol)	Rich in flavonoids and polyphenols, used for its antioxidant and anti-inflammatory properties in tissue engineering, particularly for skin, vascular, and joint tissue regeneration	Secim-Karakaya et al. (2021)
32.	Curcumin	A compound found in turmeric, known for its anti-inflammatory and antioxidant properties, often used in tissue engineering for its potential in reducing inflammation and promoting tissue healing	Ahangari et al. (2019)
33.	Bee propolis	A resinous substance collected by bees, used for its potential antimicrobial and anti-inflammatory properties in tissue engineering, particularly in wound healing and oral tissue applications	Dumitru et al. (2022)

1. *Biocompatibility*: Natural products that include extracellular matrix components and proteins are inherently biocompatible. They mimic the native microenvironment of cells, ultimately promoting cell adhesion, migration, and tissue integration (Naahidi et al. 2017).
2. *Cell signaling and bioactivity*: Natural products often contain bioactive molecules, growth factors, and signaling cues that regulate cell behavior and tissue development. These bioactive components can enhance cell proliferation, differentiation, and tissue maturation (Rao et al. 2018).
3. *Structural mimicry*: Natural materials, like collagen and fibrin, possess structures similar to those found in native tissues. This structural mimicry supports the formation of organized tissues with appropriate mechanical properties (Vieira et al. 2017; Singh et al. 2022).
4. *Reduced immunogenicity*: Natural products derived from the body's own tissues or from non-immunogenic sources are less likely to elicit an immune response. This is crucial for preventing rejection and inflammation when the engineered tissue is implanted (Raghavan et al. 2012).
5. *Promotion of angiogenesis*: Some natural products, such as certain polysaccharides and growth factors, can stimulate angiogenesis (the formation of new blood vessels). This is essential for ensuring an adequate blood supply to the engineered tissue (Matsuda et al. 2013).
6. *Ease of degradation and remodeling*: Natural biomaterials often exhibit controlled degradation over time, allowing for remodeling and replacement by native tissue. This property is important for dynamic tissues that undergo structural changes during development or regeneration (Lam et al. 2008; Kumar et al. 2020, 2021a, b, c).
7. *Support for cell adhesion and migration*: Natural products provide a substrate that facilitates cell adhesion and migration, critical for cell colonization and

tissue formation. Integrins and other cell adhesion molecules interact effectively with natural materials (Ayala et al. 2011).

8. *Versatility in source and composition*: Natural products can be sourced from a variety of tissues, plants, or animals, providing versatility in composition. This allows researchers to tailor biomaterial properties to specific tissue engineering applications (Khan and Ahmad 2013).
9. *Environmental and sustainability considerations*: Utilizing natural products aligns with sustainability goals, as they can be sourced from renewable materials. This is particularly relevant as the field of tissue engineering explores environmentally friendly alternatives (Williams 2019; Kumar et al. 2021a, b, c).
10. *Clinical translation and approval*: Natural products may have an advantage in the regulatory approval process, as they are often derived from materials with a history of safe use in the body. This can expedite the translation of tissue-engineered products to clinical applications (Dlaska et al. 2015).

1.5 Advantages of Tissue Engineering

Regenerative potential: Tissue engineering holds promise for regenerating damaged or diseased tissues and organs, allowing patients to regain lost function and improve their quality of life. Regenerative medicine and tissue engineering can allow the improvement of patients' quality of life through availing novel treatment options (Dzobo et al. 2019).

1. Reduced dependency on donor organs: Tissue engineering can potentially address the shortage of donor organs by creating lab-grown organs or tissues, reducing the need for organ transplantation and the associated waiting times and risks (Giwa et al. 2017).
2. Personalized medicine: Tissue engineering allows for the development of patient-specific tissues, enabling personalized treatment approaches and reducing the risk of immune rejection (Sarkar et al. 2023).
3. In vitro drug testing and screening: Engineered tissues can be used for drug testing, providing a more accurate representation of human responses to drugs compared to traditional cell cultures or animal models (Marei et al. 2022).
4. Understanding disease mechanisms: Creating disease models using engineered tissues helps researchers understand disease mechanisms at a cellular and tissue level, aiding in drug discovery and development (Huang et al. 2020).
5. Biocompatibility and biodegradability: Biomaterials used in tissue engineering can be designed to be biocompatible and biodegradable, ensuring they integrate well with the host tissue and are eventually replaced by natural tissue (Eldeeb et al. 2022).
6. Minimal risk of disease transmission: Unlike donor organs, engineered tissues reduce the risk of disease transmission since they are created in a controlled environment and from known cell sources (Sohn et al. 2020).

7. Facilitating research and education: Engineered tissues serve as valuable tools for research and education, allowing scientists and students to study tissue behavior, disease progression, and potential treatments (Villasante and Vunjak-Novakovic 2015).

1.6 Disadvantages of Tissue Engineering

1. Complexity and challenges in engineering: Creating functional, complex tissues or organs with the required cellular architecture and functionality remains a significant technical challenge (Atala et al. 2012).
2. Biological and mechanical properties: Achieving the optimal biological and mechanical properties of engineered tissues to mimic natural tissues is a persistent challenge in tissue engineering (Li et al. 2021a, b).
3. Immunological response: Despite efforts to create patient-specific tissues, there can still be an immunological response upon transplantation, requiring immune suppression and posing potential risks (Bilodeau et al. 2020).
4. Cost and scalability: Tissue engineering processes can be expensive due to the cost of materials, specialized equipment, and skilled labor. Scaling up production to meet widespread demand is also a concern (Wu et al. 2020).
5. Ethical considerations: Ethical considerations, particularly regarding the use of stem cells and the creation of organs/tissues, remain a significant concern, necessitating careful regulation and public discussion (Peng et al. 2020).
6. Lack of standardization and regulations: Standardization of tissue engineering techniques and regulatory frameworks to ensure safety, efficacy, and consistency in tissue production and transplantation is an ongoing challenge (Arabi et al. 2020).
7. Technological limitations: Current technology may not yet allow for the creation of highly complex organs with intricate vasculature and neural networks, limiting the types of tissues that can be effectively engineered (Atala et al. 2012).

   It's important to note that ongoing research and technological advancements may address some of these disadvantages and further enhance the potential of tissue engineering in the future.

1.7 Case Studies

Here are four case studies of tissue engineering:

1.7.1 Bladder Tissue Engineering

In this study, seven patients with end-stage bladder disease were treated with cystoplasty using autologous engineered bladder tissues as an alternative to using gastrointestinal segments. Bladder biopsies were taken from each patient, and urothelial and muscle cells were cultured and seeded onto biodegradable bladder-shaped scaffolds. After about 7 weeks, the engineered bladder constructs were used for reconstruction, with or without an omental wrap. The follow-up period ranged from 22 to 61 months, and the results showed significant improvements in bladder function, with the greatest benefits observed in composite engineered bladders with an omental wrap. The engineered bladders exhibited structural adequacy and maintained normal physiological functions without adverse effects. The study concludes that autologous engineered bladder tissues can be successfully used in cystoplasty, offering a promising alternative to traditional approaches (Atala et al. 2006).

1.7.2 In Situ Bone Tissue Engineering

In addressing donor-site morbidity in autologous bone grafts, the study successfully applied tissue engineering techniques with the arteriovenous loop model to generate vascularized large bone grafts. Two cases involving large bone defects were presented, one in the radius and one in the tibia, where arteriovenous loops served as vascular axes placed in the defects. Bone generation was achieved using cancellous bone and fibrin glue in one case, and β-tricalcium phosphate/hydroxyapatite, fibrin glue, and autotransplanted bone marrow aspirate in the other. Postoperative examinations after 36 and 72 months revealed patent arteriovenous loops and completely healed bone defects, demonstrating the successful axially vascularized in situ tissue-engineered bone generation in large bone defects without significant donor-site defects (Horch et al. 2014).

1.7.3 Bone Marrow Cell

In this study on cardiovascular surgery, the use of foreign materials is recognized for its potential drawbacks, including unwanted side effects and limited growth potential. To address these issues, the researchers developed a tissue-engineering technique employing bone marrow cells (BMCs) for clinical treatments. The process involved harvesting saphenous vein samples from patients, mincing them, culturing the cells, and seeding them onto a biodegradable scaffold. Initially performed in May 1999, this method was later refined using aspirated BMCs seeded onto the scaffold on the day of surgery. This modified technique was applied to 22 patients from November 2001. The results demonstrated no morbidity issues such as thrombogenic complications, stenosis, or obstruction of tissue-engineered autografts. Importantly, there were no mortalities associated with these techniques. The findings suggest that using BMCs seeded onto a biodegradable scaffold for creating tissue-engineered vascular autografts (TEVAs) is a promising and ideal strategy. The study provides strong evidence supporting the justification and validity of the protocol for clinical trials in tissue engineering (Matsumura et al. 2003).

1.7.4 Heart Valve Tissue Engineering

A study explores transcatheter valve implantation as an alternative to surgery, particularly for high-risk elderly patients. Bioprosthetic transcatheter valves currently face degeneration issues, limiting their suitability for younger patients. The study introduces tissue-engineered heart valves (TEHVs) with self-repair capacity as a next-generation solution. In a sheep model, transcatheter implantation of homologous TEHVs displayed successful functionality for up to 24 weeks. However, issues such as mild central regurgitation and compromised leaflet coaptation were observed over time, emphasizing the need for further refinement in tissue engineering for sustained efficacy (Driessen-Mol et al. 2014).

1.8 Future Prospects

Tissue engineering, a multidisciplinary field at the intersection of biology, engineering, and material science, holds immense promise for revolutionizing healthcare in the future. As we move forward, the prospects for tissue engineering appear increasingly bright and transformative. One of the key advancements lies in personalized medicine, where tissues and organs can be engineered to match an individual's specific genetic makeup, reducing the risk of immune rejection and enhancing overall effectiveness of treatments.

Furthermore, the integration of cutting-edge technologies, such as 3D bioprinting and gene editing, allows for precise control over tissue structure and function. Three-dimensional bioprinting enables the fabrication of complex, functional tissues and organs by layering bioinks containing cells and supportive biomaterials. CRISPR-Cas9 gene editing facilitates the modification of cells at the genetic level, potentially correcting genetic disorders and optimizing tissue compatibility.

Moreover, the development of smart biomaterials and bioactive scaffolds can provide cues to guide cell behavior and tissue regeneration. These materials can mimic the native extracellular matrix, aiding in the seamless integration of engineered tissues within the body. Additionally, advancements in nanotechnology hold promise for delivering therapeutic agents at a cellular or molecular level, enhancing tissue engineering's potential for targeted and effective treatments.

In conclusion, the future of tissue engineering is undoubtedly promising, paving the way for a new era of healthcare. By harnessing the power of personalized medicine, innovative technologies, and advanced biomaterials, we are poised to address critical healthcare challenges and improve the quality of life for countless individuals. As research continues to evolve and technologies progress, tissue engineering will likely play an increasingly pivotal role in reshaping the landscape of medicine, offering hope for patients in need of regenerative solutions.

References

Ahangari N, Kargozar S, Ghayour-Mobarhan M, Baino F, Pasdar A, Sahebkar A et al (2019) Curcumin in tissue engineering: a traditional remedy for modern medicine. Biofactors 45(2):135–151

Aisenbrey EA, Murphy WL (2020) Synthetic alternatives to Matrigel. Nat Rev Mater 5(7):539–551

Arabi M, Mohamadi-Jahani F, Alavi-Moghadam S, Abedi M (2020) Standards and regulatory frameworks (for cell- and tissue-based products). In: Learning materials in biosciences, pp 89–97

Arango MC, Montoya Y, Peresin MS, Bustamante J, Álvarez-López C (2021) Silk sericin as a biomaterial for tissue engineering: a review. Int J Polym Mater Polym Biomater 70(16):1115–1129

Ashammakhi N, Ahadian S, Xu C, Montazerian H, Ko H, Nasiri R, De Barros NR, Khademhosseini A (2019) Bioinks and bioprinting technologies to make heterogeneous and biomimetic tissue constructs. Mater Today Bio 1:100008

Atala A, Bauer SB, Soker S, Yoo JJ, Retik AB (2006) Tissue-engineered autologous bladders for patients needing cystoplasty. Lancet 367(9518):1241–1246

Atala A, Kasper FK, Mikos AG (2012) Engineering complex tissues. Sci Transl Med 4(160):160rv12

Ayala R, Zhang C, Yang D, Hwang Y, Aung A, Shroff SS, Arce FT, Lal R, Arya G, Varghese S (2011) Engineering the cell–material interface for controlling stem cell adhesion, migration, and differentiation. Biomaterials 32(15):3700–3711

Bacakova L, Novotná K, Parizek M (2014) Polysaccharides as cell carriers for tissue engineering: the use of cellulose in vascular wall reconstruction. Physiol Res 63:S29

Bae MS, Yang DH, Lee JB, Heo DN, Kwon Y, Youn IC, Choi K, Hong JH, Kim GT, Choi YS, Hwang EH, Kwon IK (2011) Photo-cured hyaluronic acid-based hydrogels containing simvastatin as a bone tissue regeneration scaffold. Biomaterials 32(32):8161–8171.

Beaumont M, Tran R, Vera G, Niedrist D, Rousset A, Pierre R, Forget A (2021) Hydrogel-forming algae polysaccharides: from seaweed to biomedical applications. Biomacromolecules 22(3):1027–1052

Bilodeau C, Goltsis O, Rogers IM, Post M (2020) Limitations of recellularized biological scaffolds for human transplantation. J Tissue Eng Regen Med 14(3):521–538

Blair NF, Frith TJR, Barbaric I (2017) Regenerative medicine: advances from developmental to degenerative diseases. Adv Exp Med Biol 1007:225–239

Carrier RL, Papadaki M, Rupnick M, Schoen FJ, Bursac N, Langer R, Vunjak-Novakovic G (1999) Cardiac tissue engineering: cell seeding, cultivation parameters, and tissue construct characterization. Biotechnol Bioeng 64(5):580–589

Chan YH, Lee YC, Hung CY, Yang PJ, Lai PC, Feng SW (2021) Three-dimensional spheroid culture enhances multipotent differentiation and stemness capacities of human dental pulp-derived mesenchymal stem cells by modulating MAPK and NF-kB signaling pathways. Stem Cell Rev Rep 17(5):1810–1826

Chen FM, Zhang M, Wu ZF (2010) Toward delivery of multiple growth factors in tissue engineering. Biomaterials 31(24):6279–6308

Chou HY, Ma DL, Leung CH, Chiu CC, Hour TC, Wang HMD (2020) Purified astaxanthin from *Haematococcus pluvialis* promotes tissue regeneration by reducing oxidative stress and the secretion of collagen in vitro and in vivo. Oxid Med Cell Longev 2020:13

Costa RS, De Souza FMS, Senhorini JA, Veríssimo-Silveira R. Ninhaus-Silveira A. (2017) Effects of cryoprotectants and low temperatures on hatching and abnormal embryo development of Prochilodus lineatus (Characiformes: Prochilodontidae). Neotropical Ichthyology/Neotropical Ichthyology 15(3)

Dahl SLM, Vaughn ME, Niklason LE. (2007) An ultrastructural analysis of collagen in tissue engineered arteries. Annals of Biomedical Engineering 35(10):1749–1755

Daamen W, Veerkamp J, Vanhest J, Vankuppevelt T. (2007) Elastin as a biomaterial for tissue engineering. Biomaterials 28(30):4378–4398

Dlaska CE., Andersson G, Brittberg M, Suedkamp NP, Raschke MJ, Schuetz MA. (2015). Clinical Translation in Tissue Engineering—The Surgeon's View. Current Molecular Biology Reports, 1(2):61–70
Driessen-Mol A, Emmert MY, Dijkman PE, Frese L, Sanders B, Weber B et al (2014) Transcatheter implantation of homologous "off-the-shelf" tissue-engineered heart valves with self-repair capacity: long-term functionality and rapid in vivo remodeling in sheep. J Am Coll Cardiol 63(13):1320–1329
Dumitru CD, Neacsu IA, Grumezescu AM, Andronescu E (2022) Bee-derived products: chemical composition and applications in skin tissue engineering. Pharmaceutics 14(4):750
Dzobo K, Motaung KS, Adesida AB (2019) Recent trends in decellularized extracellular matrix BioINKs for 3D printing: an updated review. Int J Mol Sci 20(18):4628
Eldeeb AE, Salah S, Elkasabgy NA (2022) Biomaterials for tissue engineering applications and current updates in the field: a comprehensive review. AAPS PharmSciTech 23(7):267
Fajrial AK, He QQ, Wirusanti NI, Slansky JE, Ding X (2020) A review of emerging physical transfection methods for CRISPR/Cas9-mediated gene editing. Theranostics 10(12):5532
Giwa S, Lewis J, Alvarez LM, Langer R, Roth AE, Church GM, Markmann JF, Sachs DH, Chandraker A, Wertheim JA, Rothblatt M, Boyden ES, Eidbo EE, Lee WPA, Pomahac B, Brandacher G, Weinstock DM, Elliott GD, Nelson DP, Toner M (2017) The promise of organ and tissue preservation to transform medicine. Nat Biotechnol 35(6):530–542
Gruskin E, Doll BA, Futrell FW, Schmitz JP, Hollinger JO (2012) Demineralized bone matrix in bone repair: history and use. Adv Drug Deliv Rev 64(12):1063–1077
Guilak F, Lott KE, Awad HA, Cao Q, Hicok KC, Fermor B, Gimble JM (2006) Clonal analysis of the differentiation potential of human adipose-derived adult stem cells. J Cell Physiol 206(1):229–237
Hidalgo-Bastida LA, Cartmell SH (2010) Mesenchymal stem cells, osteoblasts and extracellular matrix proteins: enhancing cell adhesion and differentiation for bone tissue engineering. Tissue Eng Part B Rev 16(4):405–412
Horch RE, Beier JP, Kneser U, Arkudas A (2014) Successful human long-term application of *in situ* bone tissue engineering. J Cell Mol Med 18(7):1478–1485
Huang D, Gibeley SB, Xu C, Xiao Y, Celik O, Ginsberg HN, Leong KW (2020) Engineering liver microtissues for disease modeling and regenerative medicine. Adv Funct Mater 30(44):1909553
Huang S, Fu X. (2010). Naturally derived materials-based cell and drug delivery systems in skin regeneration. Journal of Controlled Release 142(2):149–159
Hussin HM, Lawi MM, Haflah NHM, Kassim AYM, Idrus RBH, Lokanathan Y (2020) Centella asiatica (L.)-neurodifferentiated mesenchymal stem cells promote the regeneration of peripheral nerve. Tissue Eng Regen Med 17:237–251
Hwang NS, Varghese S, Elisseeff J. (2008) Controlled differentiation of stem cells. Advanced Drug Delivery Reviews 60(2):199–214.
Ikada Y (2006) Challenges in tissue engineering. J R Soc Interface 3(10):589–601
Islam MM, Shahruzzaman M, Biswas S, Sakib MN, Rashid TU. (2020) Chitosan based bioactive materials in tissue engineering applications-A review. Bioactive Materials 5(1):164–183
Jahangirian H, Azizi S, Rafiee-Moghaddam R, Baratvand B, Webster TJ (2019) Status of plant protein-based green scaffolds for regenerative medicine applications. Biomol Ther 9(10):619
Khan F, Ahmad SR (2013) Polysaccharides and their derivatives for versatile tissue engineering application. Macromol Biosci 13(4):395–421
Kharaziha M, Baidya A, Annabi N (2021) Rational design of immunomodulatory hydrogels for chronic wound healing. Adv Mater 33(39):2100176
Kumar M, Saini JS, Bhunia H (2020) Performance of mechanical joints prepared from carbon-fiber-reinforced polymer nanocomposites under accelerated environmental aging. J Mater Eng Perform 29:7511–7525
Kumar M, Saini JS, Bhunia H, Chowdhury SR (2021a) Behavior of mechanical joints prepared from EB cured CFRP nanocomposites subjected to hygrothermal aging under bolt preloads. Appl Compos Mater 28:271–296
Kumar M, Saini JS, Bhunia H, Chowdhury SR (2021b) Aging of bolted joints prepared from electron-beam-cured multiwalled carbon nanotube-based nanocomposites with variable torques. Polym Compos 42(8):4082–4104

Kumar M, Saini JS, Bhunia H (2021c) Investigations on MWCNT embedded carbon/epoxy composite joints subjected to hygrothermal aging under bolt preloads. Fibers Polym 22(7):1957–1975
Lam C, Hutmacher DW, Schantz J, Woodruff MA, Teoh SH (2008) Evaluation of polycaprolactone scaffold degradation for 6 months *in vitro* and *in vivo*. J Biomed Mater Res A 90A(3):906–919
Langer R, Vacanti JP (1993) Tissue engineering. Science 260(5110):920–926
Li Y, Liu Y, Li R, Bai H, Zhu Z, Zhu L et al (2021a) Collagen-based biomaterials for bone tissue engineering. Mater Des 210:110049
Li Z, Xiang S, Li EN, Fritch MR, Alexander PG, Lin H, Tuan RS (2021b) Tissue engineering for musculoskeletal regeneration and disease modeling. In: Organotypic models in drug development, pp 235–268
Li L, Zhang X, Zhou J, Zhang L, Xue J, Tao W (2022) Non-invasive thermal therapy for tissue engineering and regenerative medicine. Small 18(36):2107705
Low LA, Tagle DA (2017) Tissue chips–innovative tools for drug development and disease modeling. Lab Chip 17(18):3026–3036
Marrazzo P, O'Leary C (2020) Repositioning natural antioxidants for therapeutic applications in tissue engineering. Bioengineering 7(3):104
Marei I, Samaan TA, Al-Quradaghi MA, Farah AA, Mahmud SH, Ding H. Triggle CR (2022) 3d tissue-engineered vascular drug screening platforms: promise and considerations. Frontiers in cardiovascular medicine 9
Matsuda K, Falkenberg KJ, Woods AA, Choi YS, Morrison WA, Dilley RJ (2013) Adipose-derived stem cells promote angiogenesis and tissue formation for *in vivo* tissue engineering. Tissue Eng A 19(11–12):1327–1335
Matsumura G, Hibino N, Ikada Y, Kurosawa H, Shinoka T (2003) Successful application of tissue engineered vascular autografts: clinical experience. Biomaterials 24(13):2303–2308
Montanari M, Burattini S, Ciacci C, Ambrogini P, Carloni S, Balduini, W, Lopez, D, Panza G, Papa S, & Canonico B (2022) Automated—Mechanical procedure compared to gentle enzymatic tissue dissociation in cell function studies Biomolecules 12(5):701
Naahidi S, Jafari M, Logan M, Wang Y, Yuan Y, Bae H, Dixon B, Chen P (2017) Biocompatibility of hydrogel-based scaffolds for tissue engineering applications. Biotechnol Adv 35(5):530–544
Noori A, Ashrafi SJ, Vaez-Ghaemi R, Hatamian-Zaremi A, Webster TJ (2017) A review of fibrin and fibrin composites for bone tissue engineering. Int J Nanomedicine 12:4937–4961
Peng Y, Huang X, Zhou Q (2020) Ethical and policy considerations for human embryo and stem cell research in China. Cell Stem Cell 27(4):511–514
Prasad KC, George A, Ravi NS, Mohankumar KM (2021) CRISPR/Cas based gene editing: marking a new era in medical science. Mol Biol Rep 48:4879
Raghavan SS, Woon CY, Kraus A, Megerle K, Choi M, Pridgen BC, Pham H, Chang J (2012) Human flexor tendon tissue engineering: decellularization of human flexor tendons reduces immunogenicity *in vivo*. Tissue Eng A 18(7–8):796–805
Rao S, Harini B, Shadamarshan RPK, Balagangadharan K, Selvamurugan N (2018) Natural and synthetic polymers/bioceramics/bioactive compounds-mediated cell signalling in bone tissue engineering. Int J Biol Macromol 110:88–96
Rediguieri CF, Sassonia RC, Dua K, Kikuchi IS, Pinto TDJA (2016) Impact of sterilization methods on electrospun scaffolds for tissue engineering. Eur Polym J 82:181–195
Rubio-Elizalde I, Bernáldez-Sarabia J, Moreno-Ulloa A, Vilanova C, Juárez P, Licea-Navarro A, Castro-Ceseña AB (2019) Scaffolds based on alginate-PEG methyl ether methacrylate-Moringa oleifera-Aloe vera for wound healing applications. Carbohydr Polym 206:455–467
Sahoo DR, Biswal T (2021) Alginate and its application to tissue engineering. SN Applied Sciences/SN Applied Sciences 3(1).
Saltzman WM (2004) Tissue engineering: engineering principles for the design of replacement organs and tissues. Oxford University Press, New York
Sarkar N, Bhumiratana S, Geris L, Papantoniou I, Grayson W (2023) Bioreactors for engineering patient-specific tissue grafts. Nat Rev Bioeng 1(5):361–377
Secim-Karakaya P, Saglam-Metiner P, Yesil-Celiktas O (2021) Antimicrobial and wound healing properties of cotton fabrics functionalized with oil-in-water emulsions containing Pinus brutia bark extract and Pycnogenol® for biomedical applications. Cytotechnology 73:423–431

Senthilkumar K, Ramajayam G, Venkatesan J, Kim SK, Ahn BC (2017) Biomedical applications of fucoidan, seaweed polysaccharides. In: Seaweed polysaccharides. Elsevier, Amsterdam, pp 269–281

Sharma A, Khanna S, Kaur G, Singh I (2021) Medicinal plants and their components for wound healing applications. Future Journal of Pharmaceutical Sciences 7(1)

Singh K, Kumar M, Saini JS (2022) Metal inserts for performance improvement in glass fiber-reinforced polymer nanocomposite joints under accelerated aging conditions. Polym Compos 43(3):1448–1461

Sivaraman S, Sinha A, Lim K, Kim J, Rao RD, Jensen H (2019) Nanotechnology-based stem cell tissue engineering with a focus on regeneration of cardiovascular systems. Springer eBooks, pp 1–67

Sivakumar PM, Yetisgin AA, Demir E, Sahin SB, Cetinel S. (2023) Polysaccharide-bioceramic composites for bone tissue engineering: A review. International Journal of Biological Macromolecules 250:126237.

Socci MC, Rodríguez G, Oliva E, Fushimi S, Takabatake K, Nagatsuka H et al (2023) Polymeric materials, advances and applications in tissue engineering: a review. Bioengineering 10(2):218

Sodhi H, Panitch A (2020) Glycosaminoglycans in tissue engineering: a review. Biomol Ther 11(1):29

Sohn S, Van Buskirk M, Buckenmeyer MJ, Londoño R, Faulk DM (2020) Whole organ engineering: approaches, challenges, and future directions. Appl Sci 10(12):4277

Sood A, Gupta A, Agrawal G (2021) Recent advances in polysaccharides based biomaterials for drug delivery and tissue engineering applications. Carbohydr Polym Technol Appl 2:100067

Stacey GN, Byrne E, Hawkins JR (2007) DNA fingerprinting and characterization of animal cell lines. In: Animal cell biotechnology: methods and protocols. Humana Press, Totowa, NJ, pp 123–145

Sun W, Gregory DA, Tomeh MA, Zhao X (2021) Silk fibroin as a functional biomaterial for tissue engineering. Int J Mol Sci 22(3):1499

Tan Q, Lui PPY, Rui YF (2012) Effect of in vitro passaging on the stem cell-related properties of tendon-derived stem cells—implications in tissue engineering. Stem Cells Dev 21(5):790–800

Urbani L, Maghsoudlou P, Milan A, Menikou M, Hagen CK, Totonelli G, Camilli, C, Eaton S, Burns A, Olivo A, De Coppi P. (2017). Long-term cryopreservation of decellularised oesophagi for tissue engineering clinical application. PloS One 12(6):e0179341.

Van Der Valk J, Brunner D, De Smet K, Svenningsen, ÅF, Honegger P, Knudsen L, Lindl T, Noraberg J, Price A, Scarino M, Gstraunthaler G (2010) Optimization of chemically defined cell culture media – Replacing fetal bovine serum in mammalian in vitro methods. Toxicology in Vitro24(4):1053–1063

Vieira, S., Vial, S., Reis, R. L., & Oliveira, J. M. (2017). Nanoparticles for bone tissue engineering. Biotechnology Progress, 33(3), 590–611.

Villasante A, Vunjak-Novakovic G (2015) Tissue-engineered models of human tumors for cancer research. Expert Opin Drug Discov 10(3):257–268

Williams D (2019) Challenges with the development of biomaterials for sustainable tissue engineering. Front Bioeng Biotechnol 7:127

Witzler M, Alzagameem A, Bergs M, El Khaldi-Hansen B, Klein SE, Hielscher D et al (2018) Lignin-derived biomaterials for drug release and tissue engineering. Molecules 23(8):1885

Wu Q, Liu J, Wang X, Feng L, Wu J, Zhu X, Wen W, Gong X (2020) Organ-on-a-chip: recent breakthroughs and future prospects. Biomed Eng Online 19(1):9

Xu S, Sang L, Zhang Y, Wang X, Li X (2013) Biological evaluation of human hair keratin scaffolds for skin wound repair and regeneration. Mater Sci Eng C 33(2):648–655

Yang J, Shen M, Wen H, Luo Y, Huang R, Rong L, Xie J (2020) Recent advance in delivery system and tissue engineering applications of chondroitin sulfate. Carbohydr Polym 230:115650

Zhang X, Yang-De Z (2015) Tissue engineering applications of three-dimensional bioprinting. Cell Biochem Biophys 72(3):777–782

Zhang X, Chen X, Hong H, Hu R, Liu J, Liu C (2022) Decellularized extracellular matrix scaffolds: recent trends and emerging strategies in tissue engineering. Bioact Mater 10:15–31

2 Fabrication Strategies for the Synthesis of Natural Product-Based Scaffold

Rosana de Carvalho Cruz

Abstract

A scaffold can be defined as a structure to which cells and growth factors can be added, with a view to replacing diseased tissue. The construction of a scaffold provides mechanical support and is based on premises such as cells, administration of signals and/or growth factors, and type of extracellular matrix required. Biopolymers are natural structures that are excellent candidates for the construction of scaffolds since they provide characteristics of biocompatibility, antigenicity, and different structure. Natural polymers frequently used in the construction include polysaccharides, alginate, chitosan, proteins, collagen, and gelatin. In tissue engineering, the type of material and the technique that be applied in the construction will be considered, considering the type of tissue that will be replaced. Porosity, biocompatibility, chemical stability, and structural performance are also parameters of significant importance in the development of scaffolds. Different techniques can be used to manufacture scaffolds using natural products and they are classified into conventional and rapid prototyping. The choice is related to the objective you want to achieve and the type of tissue for which the scaffold was developed. Much research still needs to be carried out to build a scaffold of entirely natural products that have broader applications. Research needs to be conducted responsibly, safely, and without compromising results as this line is directly linked to the future of biotechnology.

R. de Carvalho Cruz (✉)
Department of Biochemistry and Immunology, Institute of Biological Sciences, Universidade Federal de Minas Gerais (UFMG), Belo Horizonte, MG, Brazil

Bioengineering Laboratory, Department of Mechanical Engineering, Universidade Federal de Minas Gerais (UFMG), Belo Horizonte, MG, Brazil

M. Kumar et al. (eds.), *Natural Product Inspired Scaffolds*,
https://doi.org/10.1007/978-981-97-3111-4_2

Keywords
Scaffold · Natural products · Manufacturing strategies · Tissue engineering

2.1 Introduction

Damage to or failures of organs are increasingly common conditions in a world population that is growing increasingly older day by day, has poor eating habits, a sedentary lifestyle, and different classifiable types of stress.

A scaffold can be defined as a structure to which cells and growth factors can be added, with a view to replacing diseased tissue. Tissue engineering is a multidisciplinary branch that seeks to build scaffolds for different clinical applicability with the aim of reducing the number of transplants or surgeries, or even for use as means of vehicles for drug delivery. The construction of a scaffold provides mechanical support and is based on premises such as cells, administration of signals and/or growth factors, and type of extracellular matrix (ECM) required. It is a temporary structure and must, therefore, accomplish its role when necessary. Moreover, it must suffer degradation when its function is no longer required. Different parameters are considered for its construction. Among them are features such as biodegradability, mechanical and functional stability, cost and particularly the characteristic of not promoting an immunological response that leads to its rejection. Topological characteristics, hydrodynamic impact, and structural corrosion are also important variables to consider when 3D constructions are being contemplated. The manufacturing method must be considered so that it will not interfere with parameters such as pore structures (if there are any), their degradation, and anisotropy, which could compromise the performance and function of an ECM. Scaffolds can be prepared using natural or synthetic polymers. The choice is related to the parameters required for their structure and performance of the function for which they were developed (Costa et al. 2017; Bakhtiari et al. 2023; Krishani et al. 2023).

Polymers are chemical substances with diverse structures and compositions, which can be used in the construction of ECM. Natural polymers that are frequently used in the construction of ECMs, for example, include polysaccharides, alginate, chitosan, proteins, collagen, and gelatin. Once again, the choice depends on the objective to be achieved and the tissue in which the scaffold will act since there are advantages and disadvantages in every choice. When the use of natural products versus synthetic products is compared, these differences can be seen more clearly. In general, synthetic polymers are good candidates due to their mechanical properties and versatility when different techniques are used for working with them, and generally, at a lower cost. Nevertheless, they frequently fail to mimic the "in vivo" physiological conditions. The choice of a polymer for use as a biomaterial must comply with criteria involving morphology, compatibility, pore size and porosity, bioactivity, 3D geometry, antigenicity, and its capacity for degradation. This must occur without compromising the newly formed cells, as generally occurs with natural polymers (Reddy et al. 2021; Rosellini and Cascone 2023).

For each polymer it is possible to choose construction strategies and techniques such as cross-linking, microfluidics, 3D printing, and electrospinning. All of these techniques have variations that allow them to be adapted to the construction of different scaffolds for different applications (Rosellini and Cascone 2023). Moreover, some techniques involve processes such as decellularization to produce an ECM, followed by cell repopulation. They can be used for different applications, for example, such as heart muscles, valves and blood vessels, the skin, or even neural tissue. Biomaterials produced in this way have the advantage of providing a microenvironment close to that of the original tissue. Thus, they have a greater chance of promoting good future regeneration. Furthermore, during the construction of a decellularized extracellular matrix (dECM), immunogenic cells and molecules are removed, thereby minimizing the ability to activate an immune response. The goal is to develop a scaffold that is functional and has both mechanical and biological properties. Techniques for producing a dECM involve chemical, physical, and enzymatic treatments that are often combined to ensure efficacy in obtaining the dECM (Brown et al. 2022; Zhang et al. 2022).

Scaffolds made from natural products can be used in different applications. A good example of their use is in the correction of bone defects, high-energy trauma, tumor resection, and skeletal injuries. Natural structures of animal origin, such as regions of bone trabeculae, can be used as initial support for the reconstruction of cells and tissues. However, this approach may lead to the risk of activating an immune response in the host. Previously mentioned methods that result in good decellularization with complete cell removal can minimize this disadvantage (Wong et al. 2023).

In this chapter, there will be an approach to topics such as parameters to be considered for the development of an ideal scaffold, the advantage of using natural products in the preparation of ECMs in comparison with synthetic products, main natural raw materials and techniques for constructing a scaffold, and examples of applicability in different models. The information is provided for the purpose of promoting the perception that for building a scaffold, many variables must be considered. Moreover, there are problems to be mitigated. Our hope, esteemed reader, is that you will perceive that strategies and materials can be combined in different ways, thereby opening your mind to the countless applicability of a scaffold, and guiding questions for new approaches to research.

2.2 Parameters and Strategies for Constructing Scaffolds

Scaffolds can be divided into two categories: those made of biomaterials and the synthetic types. When considering the construction of scaffolds, either based on natural products, or not, different parameters must be considered (Fig. 2.1). Among the main parameters, there are the type of material to be used as the basis of the scaffold, its origin, its chemical and mechanical stability, whether decellularization is necessary before recellularization, cost-benefit ratio, whether the materials are biodegradable, whether they induce processes of rejection, and finally, whether they will achieve the purpose proposed during its construction (Da et al. 2021).

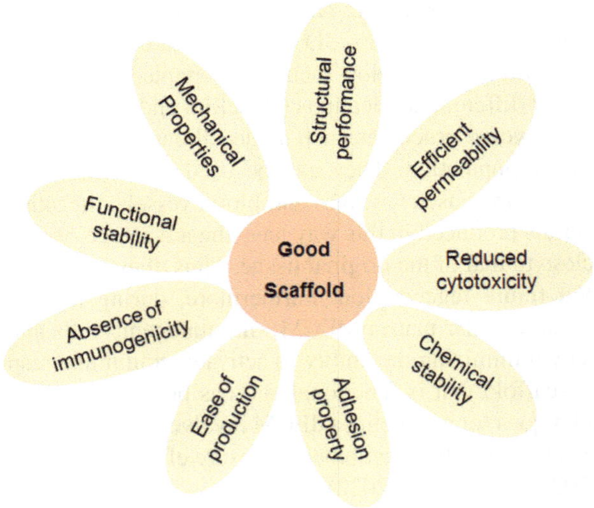

Fig. 2.1 Parameters considered important for the development of a good scaffold

In tissue engineering, the type of material and the technique that will be applied in the construction (of the scaffold) will be taken into account, considering the type of tissue that will be replaced. An example of this is the search for natural materials for application to hard tissues or those that contain cartilage. Hard tissues are denominated this way because of their ability to mineralize extracellular matrix (ECM). Therefore, a scaffold constructed for this purpose must be capable of withstanding and/or absorbing impact and pressure. Techniques such as rapid prototyping, including an additive layer manufacturing approach, can be achieved by 3D printing. "In vivo," some parameters to be considered for ECMs include their capacity for remodeling and absorption by the organism, and they may or may not have the ability to produce metabolites, exhibit functional dynamics, and show absence of inducing an immunological response (Abdollahiyan et al. 2021).

When using biomaterials such as Xantan gum (XG) and mastic gum (MG), thermal, chemical, traction, and compression parameters are used to evaluate stability and develop a better strategy for building scaffolds with interconnected pore structures. MG and XG are natural polysaccharides commonly used in the pharmaceutical, chemical, and food industries. Locust bean gum (LBG) is a cell wall polysaccharide from carob trees (*Ceratonia siliqua*) found in the Mediterranean region. They are highly biodegradable and biocompatible, eco-friendly in comparison with synthetic matrices. Analysis of the physicochemical parameters of these gums allows us to verify that the connected pore structures promote nutrient transport, infiltration, and cellular organization during repopulation. Another important parameter used in the analysis is the capacity for degradation of a given scaffold without causing large variations in pH, a crucial factor in the replacement of damaged tissue. Therefore, these gums are considered good candidates for use as scaffolds for soft and cartilaginous tissues (Bektas et al. 2021).

In studies that use porosity parameters, the integration obtained between these pores and heir mechanical stability is a long-standing concern when considering scaffold construction since intercommunications are necessary for both drug delivery and tissue engineering. The difference between pore sizes is significantly considered in 3D manufacturing. Different prototyping techniques can be used to provide the scaffold pores with different diameters. This difference is important and can allow cell organization and migration to influence the effectiveness of the tissue reconstruction process. Further studies are still needed regarding the relationship between pore size and the type of tissue to be reconstructed, irrespective of whether the scaffold is of natural or synthetic origin (Loh and Choong 2013).

Pore permeability in a fiber is also a concern and a parameter to be considered. In the case of scaffold production to replace bone tissue, these matrices must be capable of mimicking the permeability that occurs in bone trabeculae and the complexity of cancellous bone. Fluid dynamics is a parameter that can be evaluated by computational methods and can lead to the construction of scaffolds with different geometries. Different shapes such as cubic, Schoen gyroid, and octahedron influence the tortuosity and directly affect the permeability and shear stress of the ECM. By changing the diameter of the pores and the radius of curvature of a scaffold, for example, it is possible to see a significant improvement in permeability and porosity. Tortuosity of the fiber bed then becomes an extremely important parameter when one wishes to promote satisfactory nutrition in a cellular growth matrix, thus ensuring the correct maintenance of the structure and function of the organ (Prakoso et al. 2023).

In addition to porosity, biocompatibility, chemical stability, and structural performance are also parameters of significant importance in the development of scaffolds. In this context, these parameters can be analyzed together as they are closely related. The electrospinning technique used to construct gelatin nanofibers (GNFs) is a simple and economical method. Therefore, the cost-benefit ratio also becomes an important condition for analysis. It is possible to obtain gelatin by means of the partial degradation of collagen fibers. When the electrospinning technique is used, several processes can be evaluated, such as viscosity, surface tension, electrical conductivity, flow rate, temperature, and humidity. This technique has the advantage of preserving important factors in a good scaffold, such as biodegradability, low toxicity, and extensive biocompatibility. However, as it is a product made from collagen, it has little water resistance, low mechanical stability, and is highly sensitive to humidity. This behavior relative to water can significantly alter its shape and consequently its function, which, to some extent, limits its applicability. Making modifications to its structure, such as cross-linking, improves solubility and increases the possibility of its use in bone and skin regeneration (El-Seedi et al. 2023).

The same technique as that mentioned above, for example, has been used to develop scaffolds that resemble blood vessels with respect to property and function. The aim here is to replace surgeries such as heart bypass surgeries. In this case, mechanical properties become an important parameter to be evaluated. A good vascular graft must have high mechanical resistance as it will temporarily replace the three types of tunics of a blood vessel. Moreover, it must have compliance (matching to native vessels) and suture retention strength and provide a force of response close to that of the physiological levels of the vessel. Natural polymers such as collagen, gelatin, and fibrin have advantages such as flexibility, biocompatibility, low cost, and low antigenicity. However, they have low stability, high rate of degradation, and tendency to calcification. Furthermore, in the case of some polymers such as alginate, it is necessary to combine them with other biopolymers to minimize these disadvantages. For example, by combining collagen with elastin and imitating their conformational architecture, promising results have been obtained in functional simulation analyses. When studying mechanical forces that influence vascular function, expansion, contraction, and flexion are evaluated as criteria for classifying a good scaffold. The important challenge is to balance the mechanical parameters of resistance and pressure to mimic the ideal compliance present in blood vessels. Once again, shear stresses are considered, in addition to the need to mimic different conditions of stress present in the cardiovascular system (Ozdemir et al. 2022).

Combinations of techniques allow for a more efficient synthesis of matrices based on natural products (bioactive structures). The aim of these combinations is to improve the structure and function of a given scaffold. The implementation of C–H bond functionalization of heterocycles allows new tactics to be applied in building core architectures. This functionalization could, for example, allow the construction of more elaborate scaffolds with more complex structures more quickly and economically, as well as making them more stable, increasing their functionality and biological applicability (Zhang and Szostak 2022).

To maximize the functionality of a scaffold or increase its range of application and stability without promoting an immunogenic reaction, efforts have been made to improve properties by combining materials. Part of the material comes from natural products (elastin and gelatin) combined with synthetic products (PCL: polycaprolactone, PLGA: poly(lactic-co-glycolic acid) or PGA). However, the concept of pseudo-natural products has gained strength. In this concept, the material is synthetically constructed based on the chemical structure of the natural product, using bioinformatic tools and minimizing exploratory activities in nature. This type of product is being considered the evolution of the natural structure. By means of using libraries, it is possible to perform a combinatorial analysis, often allowing the prior study of different relevant parameters even before performing physical synthesis of the scaffold. Multidisciplinary work is increasingly necessary in this context. Based on the use of chemoinformatics, it has become possible to create scaffolds with the most varied design and applicability, minimizing the negative parameters that limit many techniques (Grigalunas et al. 2022).

2.3 Natural Products Used in the Production of Scaffolds

Biopolymers are natural structures that can be divided into three classes (Fig. 2.2). They all have the characteristic in common of being composed of monomeric chains of repeating units of covalently linked functional chemical groups. This characteristic makes biopolymers excellent candidates for the construction of scaffolds since they provide the gold standard characteristics of biocompatibility, antigenicity, and structure. Moreover, they do not produce toxic substances during degradation (Reddy et al. 2021).

One of the special characteristics of natural polymers is their ability to functionally bind to living tissues, thereby promoting interactivity and function. Polymer derived from lactic acid (PLA) is a thermoplastic aliphatic and biodegradable polyester. These polymers are synthesized from lactic acid monomers arising from the fermentation of renewable agricultural sources, which makes them considerably eco-friendly. This is a biocompatible, bioactive, and bioabsorbable type of polymer. When combined with inorganic molecules, PLA becomes a more stable and versatile matrix for pharmaceutical and biomedical applications (Gritsch et al. 2019).

Another natural product used in the construction of scaffolds is silk fibroin (SF). Silks are proteins found in the glands of different taxonomic families such as spiders and silkworms. However, the silk produced by spiders stiffens after contact with air, which reduces the chances of large-scale production. The yield of fibers obtained from silkworms is about ten times higher. Basically, the molecule consists of a primary protein structure (silk I) composed of a light chain (L26KDa) and a heavy chain (H390KDa) linked together by disulfide bonds forming an HL complex. Glycoproteins are linked to this complex, forming a conjugated molecule. The secondary structure (silk II) obtained by means of a regenerated fibroin solution composed of a crystalline and amorphous structure. This can be obtained by treating the former silk I with methanol and potassium chloride. Silk II is used in tissue engineering to produce scaffolds because it has demonstrated excellent mechanical properties such as high tensile strength, resilience, and greater elasticity than many synthetic fibers such as Kevlar and carbon fibers. They are biocompatible,

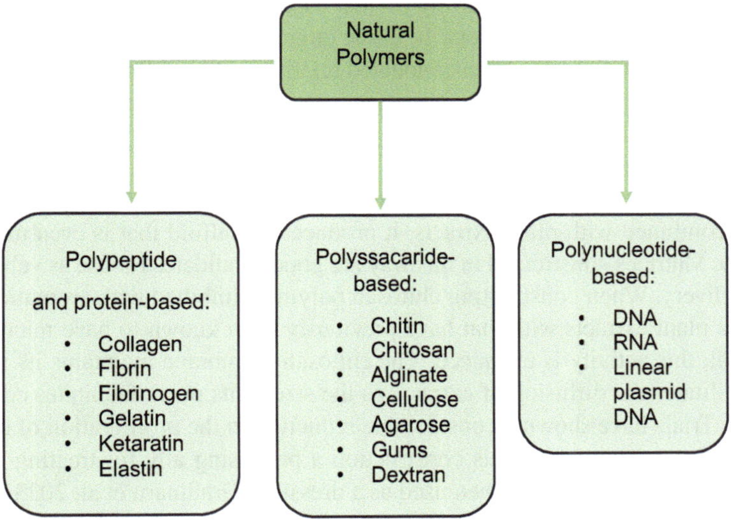

Fig. 2.2 Classes of natural polymers that can be used in the construction of scaffolds

biodegradable, and extremely malleable to work with. Since they have these characteristics, it is possible to obtain varied structures such as film, hydrogel, sponge, fiber, and 3D structures. This variety allows different applicability to be obtained, such as regeneration of cartilaginous tissue, tendon ligaments, and skin tissues, and has also shown promising results in the oriented migration of neurons for nerve regeneration, with future applicability in spinal cord injuries (Sun et al. 2021; Yap et al. 2022; Castro-Domínguez et al. 2023).

Electroactive biomaterials can provide direct electrical stimulation to cells, tissues, and organs. This characteristic has made these biomaterials strong candidates for use as scaffolds. Silver nanoparticles (AgNP) are known to be good electroactive conductors. The synthesis of this AgNP by means of formal routes includes the use of a reducing agent that would affect its use in biological systems due to toxicity. Sweet potato extract used as a green alternative for AgNP construction has been shown to be an easy approach without the need to use a reducing agent. When sweet potatoes are cooked, the process of formation of AgNPs is more efficient. When the newly synthesized particles were added to PCL, this provided a good scaffold with higher compressive strength and more hydrophilic properties. These characteristics allow for adequate use of the abovementioned scaffold as a support for cancellous bone since it promoted cell proliferation (Poongodi et al. 2021; Wibowo et al. 2021).

Polysaccharides are also formed by monomeric structures with a repetitive pattern that is generally well known. Considering "green chemistry," gums such as tragacanth gum (TG), easily extracted from the stems of *Astragalus* species, is a biodegradable, biocompatible, nontoxic polymer. Moreover, it has significant stability against the activity of microbial degradation. This also makes gums good candidates of the production of ECMs. Their physical-chemical properties can change according to the season in which they are collected. Chemical modifications can easily be made to their structures for the purpose of producing linear or 3D scaffolds, which increases the versatility of their use. Construction models based on GT include hydrogels and nanofibers. In tissue engineering GT is used in regenerative medicine, and studies have been conducted relative to their differentiation into stem cells and cellular activity that requires a support matrix. Another application of GT is drug delivery by means of controlled release of medications to reduce microbial proliferation (Taghavizadeh Yazdi et al. 2021).

Chitosan also fits into the classes of polysaccharides. When this type of biopolymer is combined with plant extracts, it produces a scaffold that is even more eco-friendly. Matrices constructed in this way are good candidates for use as vehicles for drug delivery. When constructing chitosan polymers imbibed with aromatic or herbaceous plant extracts with that have previously been known to have microbicidal potential, this activity is enhanced. The chitosan membrane maintains its integrity and facilitates the diffusion of extracts as the size of its pores facilitates controlled release. Trials have shown a considerable reduction in the proliferation of different microbial species, making this construction a promising ally for treating infected burn wounds, for example, when used as a dressing (Gradinaru et al. 2023).

The process of wound cicatrization can be optimized by using scaffolds. When they are constructed in a composite manner with the use of natural and synthetic products, their efficiency for this purpose improves. The challenge of using this type of construction in skin tissue regeneration is related to providing an adequate platform for the cellular migration of connective tissue cells since this tissue is responsible for promoting skin repair. Moreover, it allows the delivery of various factors associated with cell survival, proliferation, and differentiation. Different structures, such as porous, fibrous, hydrogel and microsphere types, can be proposed. As always, each type has advantages and disadvantages. This occurs because the skin is an inhospitable environment, making it difficult to reproduce interaction between scaffolds and skin tissue. The types that tend to present the best results are those with a porosity and 3D structure that are a combination of natural products such as collagen, cellulose, and chitosan with synthetic PEG, PCL, or PLGA (Chaudhari et al. 2016; Peng et al. 2022).

Another interesting biomaterial for use in tissue engineering is the membranous extracellular matrix. Its properties allow it to be used to treat chronic wounds such as diabetic ulcers, venous ulcers, and pressure ulcers. These matrices can be derived directly from tissues such as the pericardium or peritoneum of human or animal origin. Moreover, the acellular dermal matrix and amniotic membrane can be adapted for clinical applicability in cell regeneration of diseased organs. A process of decellularization and antigenic assays must be performed prior to application. To improve the mechanical properties and functionality of these scaffolds, the matrices can be combined by physical and/or chemical crosslinking with other biopolymers such as chitosan or one of its derivatives. In this way, the hemostatic and analgesic effect can be improved while preserving the antimicrobial property. The new construction generally acquires solubility and sensitivity to pH. Physical treatment with ultraviolet (UV) irradiation can sterilize the scaffold without the need to use an additional substance for this purpose. UV irradiation has shown an improvement in the stability of the scaffold, reducing its biodegradation. However, the capacity for mimicking the functionality of the organ in addition to having good mechanical properties is still the major challenge in the use of biomaterials (Da et al. 2021; Varinelli et al. 2023).

2.4 Examples of Scaffold Fabrication Techniques Using Natural Products and Their Applicability

Different techniques can be used to manufacture scaffolds using natural products. These are classified into two categories: conventional and rapid prototyping procedures. Some examples of techniques are listed in Table 2.1. The choice of technique is related to the objective one wishes to achieve and the type of tissue for which the scaffold was developed. Nowadays, in addition to analyzing the parameters and correct choice of biopolymer to be used in the synthesis, there is also concern about use of most eco-friendly process. This must promote the least stress to the host that will benefit from this technology.

Unified synthesis involves a synthetic approach that allows the construction of tropane-related scaffolds. Tropane is an alkaloid with different biological functions

Table 2.1 Some examples of fabrication techniques for manufacturing scaffolds and their applications

Technique	Process	Application	References
Unified synthesis	Molecular modifications: Hydrogenation, amination, diastereoselectivity, and cycloaddition	Screening libraries	Al-Tel et al. (2020), Lowe et al. (2020)
Microfluidics	Fluid processing through channels forming geometric scaffolds with cross junctions	Drug delivery, increase in biocompatibility	Rosellini and Cascone (2023)
Spinning	Passage of natural polymer solution into a syringe collected by coagulation and phase separation systems. There are variations of this technique	Nerve regeneration, cell repopulation in and migration into heart or other valves	Boehm et al. (2023)
Freeze drying	Sublimation of water from a mixture of biopolymers at controlled temperature and pressure	Structural basis for cell repopulation	Wu et al. (2021)
3D bioprinting, extrusion	Printing using different three-dimensional extracellular matrix programs	Support for cell repopulation and tissue regeneration of heart valves	Maxson et al. (2019), Bie et al. (2023), Munoz-Perez et al. (2023)
Chemoinformatics and chemoenzymatic methods	Combination of fragments derived from natural products using enzymes and bioinformatic analysis at the molecular level	Chemical screening libraries	Karageorgis et al. (2020), Chakrabarty et al. (2021)
3D extrusion	Biopolymer passed through a matrix designed to make it acquire the matrix shape	Cell repopulation and bone regeneration	Bie et al. (2023)
Leaching	Production of thin-walled 3D scaffold by adding a salt to the solvent solution, followed by washing to remove this salt	Increase in porosity for bone regeneration and tissue engineering	Gomez-Cerezo et al. (2021)
Gas foam	Natural polymer solution molded into solid discs exposed to gas foaming agents at high pressure, followed by reduction in pressure for thermodynamic stability	Production of polymeric matrices for use in cell transplantation	Fanovich et al. (2023)
Stereolithography	Synthesis of layered scaffolds using UV for polymer solidification	Favoring cell adhesion due to uniform porosity	Field et al. (2021)

(continued)

Table 2.1 (continued)

Technique	Process	Application	References
Decellularization	Combination of physical, chemical, and enzymatic methods for the production of cell-free scaffold	Recellularization of host cells, tissue, and organ regeneration	Barbulescu et al. (2022), Xu et al. (2024)

and is present in different medications such as atropine and HIV-1 entry inhibitors. Tropane fused with different natural products involves the chemical fusion of chemical groups to its nucleus, by means of hydrogenation, amination, and diastereoselectivity to form the 3D configuration of the scaffold. The structure can be confirmed by X-ray crystallography and purified by mass-directed HPLC (high-performance liquid chromatography). These scaffolds can be used to screen other functional chemical groups present in different microorganism libraries by means of their capacity for interaction. When using this approach, it is, for example, possible to guide new discoveries of more specific medications in the treatment of diseases caused by *Plasmodium falciparum*. The 3D diversity achieved with stereochemical techniques is often used as a platform for guided synthesis, which allows scaffold construction with a degree of complexity (Al-Tel et al. 2020; Lowe et al. 2020).

2.4.1 Microfluidic Technique

The microfluidic technique involves the manipulation of microfluids. Devices with integrated microchannels promote separation and transport of fluids through channels. The result leads to the production of geometric and fluid scaffolds in a T-junction, cross-junction, or Y-configuration at an angle of 180°. The fluid phases form a hydrodynamic flow that increases the viscous shear force. The geometry can be customized for different applications. An example is the encapsulation of cell types or natural molecules to improve the biocompatibility of the scaffold or its use for drug delivery (Rosellini and Cascone 2023).

2.4.2 Spinning Technique

The spinning technique involves the passage of a vicious natural polymer solution in a syringe followed by collection for use in a coagulation and phase separation system. The filaments may have diameters of various sizes. Modifications to this technique will be possible. Dry spinning involves transforming a polymer solution into a solid thread. Electrospinning involves adding a source of high-voltage energy to the syringe, thereby promoting electrostatic repulsion, leading to the formation of a conical structure with fibers of different diameters. In melt-spinning, the substance that forms the fiber is melted and pumped into an air chamber and solidifies into continuous filaments. The variations allow the formation of a scaffold with different porosities, which can favor cell migration, bone or cartilage tissue reconstruction, and nerve regeneration (Boehm et al. 2023).

2.4.3 Freeze-Drying

The freeze-drying process can be used to produce scaffolds from a mixture of biopolymers. Freeze-drying is also known as ice shaping. The polymer chose is dissolved in a solvent. Subsequently it is loaded into a mold and placed in the freezer for solidification or freezing. The cooling process can occur using chemicals such as methanol and dry ice or liquid nitrogen. Then, the sublimation process takes place to remove the water. The entire process occurs at controlled temperature and pressure. This process produces scaffolds with high porosity, of which the structure can be confirmed by analysis using scanning electron microscopy. Freeze-drying can be combined with electrospinning to optimize the ECM structure. Scaffolds constructed from pectin- and alginate-based products can provide a functional structure for cell adhesion, repopulation, and differentiation in a perfusion bioreactor (Wu et al. 2021; Cao et al. 2022).

2.4.4 3D Printers

Three-dimensional printers capable of printing scaffolds constructed with different biopolymers can be used to construct three-dimensional matrices. The printer can be programmed according to the parameters of the tissue desired to be printed. The parameters such as temperature, continuity, number of layers, crosslinking, displacement, and orientation of the fibers to be printed can be adjusted. The scaffold produced can be used as temporary support for recellularization and tissue remodeling of heart valves and epithelial and muscle tissue, after prior analysis of the mechanical and functional properties of the ECM (Maxson et al. 2019).

2.4.5 Chemoinformatics Method

Tools that involve chemoinformatics allow the synthesis of various scaffolds based on natural products. In this strategy, the combination of fragments derived from natural products is often used to synthesize new bioactive compounds in a molecular approach. The resulting construction is called a pseudonatural product because bioinformatic techniques are used, but they preserve the biological relevance of natural products. Patterns of similarity and connectivity allow the chemical interaction of molecules that do not normally interact in nature. Generally, the aim is to establish cyclic and spiral connections that have the greatest probability of providing a stable scaffold. The resulting matrix can be used to screen protein libraries in the search for new targets for designing medications (Karageorgis et al. 2020).

2.4.6 Chemoenzymatic Method

Chemoenzymatic methods can also be used to produce biocompatible and bioactive scaffolds that are considerably like natural products. Enzymes are used to construct complex molecules that are combined by using bioinformatic tools. The process can be optimized to produce structures that enable targeted delivery of medications and antimicrobial agents. The strategy is based on selective oxidation assays that take advantage of the chemical specificity of oxidation enzymes (Chakrabarty et al. 2021).

2.4.7 3D Extrusion Process

The extrusion process can be used to manufacture 3D scaffolds with complex structures. In this process, structures are produced continuously. The biopolymer is forced through a matrix and acquires the predetermined shape of the matrix designed for it. This can be of the full, transversal, or hollow type. This is an interesting method for constructing scaffolds that provide a basis for cell repopulation and bone regeneration, provided that the parameters are adjusted to the purpose for which reconstruction is desired (Bie et al. 2023; Munoz-Perez et al. 2023).

2.4.8 Solvent Casting and Particle Leaching

Leaching of particles allows the construction of 3D matrices with thin walls or membranes. These membranes are prepared by adding particles of a salt to the polymer solvent solution. The solvent is evaporated, and the membrane is washed with water to leach the salt. In this way, a scaffold with multiscale porosity containing micro-, meso-, and macrospores is obtained at a low cost. This technique, for example, allows the incorporation of hydroxyapatite into biodegradable polymer mixtures, improving their mechanical properties and their applicability in tissue engineering for bone regeneration. To cite another example, PLA can be chemically synthesized through direct polycondensation of lactic acid and ring-opening polymerization (ROP). The condensation process is conducted under low pressure to reduce the amount of water required. For large-scale production and obtaining higher yields, ROP is performed as it is more convenient to control the molecular weight. Processing techniques also involve particle leaching, phase separation, and electrospinning that produces porous and biodegradable scaffolds (Gritsch et al. 2019; Gomez-Cerezo et al. 2021).

2.4.9 Gas Foaming

Gas foaming is a technique that is incorporated to avoid the use of cytotoxic solvents at high temperatures. However, the resulting material has closed pores, thus limiting its applicability. The polymer used for constructing the scaffold is mixed with a salt and molded to form solid discs. Subsequently, the discs are exposed to gas-foaming agents (nitrogen or carbon dioxide) at high pressure up to the point of saturation. After this, the gas is then reduced to ambient pressure to allow thermodynamic stability, resulting in polymeric matrices that can be used in cell transplantation and mimic the hierarchical architecture of complex tissues (Fanovich et al. 2023).

2.4.10 Stereolithography

Stereolithography is a rapid prototyping technique that is commercially available. A light-curable polymer is generally used as a raw material. UV light is used to irradiate the material so that it solidifies. This process occurs in layers and is always covered by the next layer. The resulting material is structured in a 3D geometry and allows application as a basis for cell adhesion because it favors the uniformity and interconnectivity of the pores (Field et al. 2021).

2.4.11 Decellularization

Decellularization is one of the best-known techniques for producing scaffolds. This consists of a combination of chemical, physical, and enzymatic methods. Chemical methods may involve mechanical agitation, immersion, perfusion, and freeze-thaw cycles. Moreover, chemical methods involve the use of nonionic detergents such as Triton X-100, treatments with acids and bases of hypo- or hypertonic solutions. Enzymatic methods involve treatment with trypsin and nucleases. The purpose of physical and chemical treatments is to obtain complete cell lysis. Enzymatic treatments perform lysis of protein and genetic material (DNA and RNA). Figure 2.3 shows a sequence of events for performing chicken heart decellularization, combining chemical methods with Triton X-100 and physical methods with perfusion and immersion under agitation. After completion of the process, the scaffolds are washed to remove cellular debris, and the resulting matrix is tested for antigenicity and is then sterilized. To optimize mechanical strength, the decellularized scaffold can receive cross-linking treatment to preserve the 3D structure of the matrix. This matrix can then be used for cell repopulation, reconstruction of tissues, heart valves, and organs with host cells (Barbulescu et al. 2022; Xu et al. 2024).

Throughout this chapter, it is possible to see that the production of scaffolds can be carried out using different technological approaches and types of biopolymers. Therefore, when choosing a technique, biopolymer and scaffold construction approach, the intended applicability must be considered. Some processes and criteria were shown throughout the chapter. Figure 2.4 summarizes some examples of applicability for a scaffold.

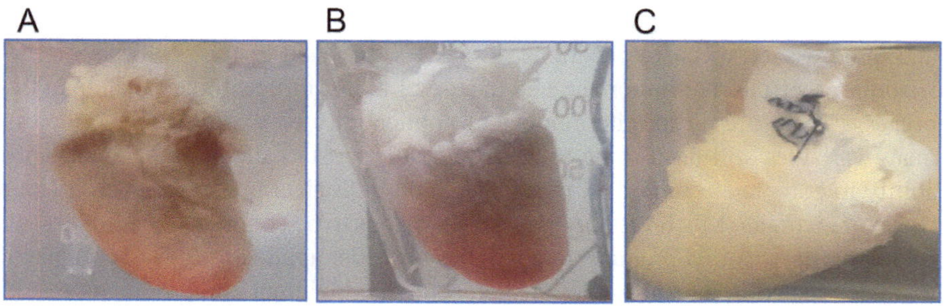

Fig. 2.3 Sequence of chicken heart decellularization events using chemical and physical methods. (**a–c**) Gradual loss of cells. (Source: own, Bioengineering Laboratory, Department of Mechanical Engineering, UFMG)

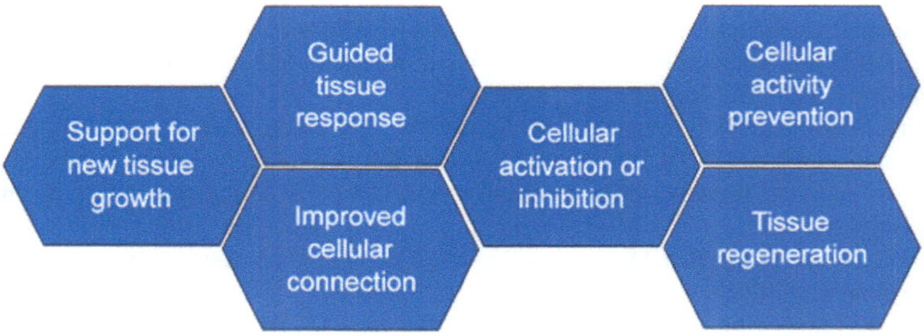

Fig. 2.4 Possible scaffold applications in tissue engineering

2.5 Conclusions and Perspectives

Throughout this chapter it was possible to come into contact with different approaches to the synthesis of scaffolds from natural products, compare functionality with artificial products, and analyze integration between the two in different applicability. It could be observed that to build a scaffold, different parameters must be considered. These parameters not only guarantee the functionality of a scaffold but also the safety of the process.

Natural polymers are more appreciated for use because most of them promote greater biocompatibility, biodegradability, and functionality and have better ability to mimic tissues and organs when compared with synthetic polymers. Much research still needs to be conducted, however, to resolve issues related to the mechanical stability of scaffolds based on natural polymers.

The techniques available for producing scaffolds are often not new and have previously been used for other purposes. But perhaps the greatest insight was the one obtained from a scientific thought that many techniques we know can be adapted to applications that have not yet been thought of.

Tissue engineering has, therefore, been shown to be an increasingly multidisciplinary area. The need for teamwork with professionals who have different expertise has become increasingly necessary. The quest for natural products needs to be conducted in a way that guarantees the integrity of ecosystems without aggression and unnecessary species collections. Processes that lead to the synthesis of eco-friendlier scaffolds are important, and the use of bioinformatic tools can help in this process. There are an infinite number of applicability of scaffolds. Nevertheless, many variables still need to be resolved, and the most challenging of them is the ability to completely mimic all mechanical and biological processes involved in a tissue or an organ. Much further research still needs to be conducted to build a scaffold of entirely natural products that will have broader applications. Research needs to be conducted in a responsible manner, safely and without compromising results since this line is directly linked to the future of biotechnology.

References

Abdollahiyan P, Oroojalian F, Hejazi M, de la Guardia M, Mokhtarzadeh A (2021) Nanotechnology, and scaffold implantation for the effective repair of injured organs: an overview on hard tissue engineering. J Control Release 333:391–417

Al-Tel TH, Srinivasulu V, Ramanathan M, Soares NC, Sebastian A, Bolognesi ML, Abu-Yousef IA, Majdalawieh A (2020) Stereocontrolled transformations of cyclohexadienone derivatives to access stereochemically rich and natural product-inspired architectures. Org Biomol Chem 18(42):8526–8571

Bakhtiari H, Nouri A, Khakbiz M, Tolouei-Rad M (2023) Fatigue behaviour of load-bearing polymeric bone scaffolds: a review. Acta Biomater 172:16–37

Barbulescu GI, Bojin FM, Ordodi VL, Goje ID, Barbulescu AS, Paunescu V (2022) Decellularized extracellular matrix scaffolds for cardiovascular tissue engineering: current techniques and challenges. Int J Mol Sci 23(21):13040

Bektas EI, Gurel Pekozer G, Kok FN, Torun Kose G (2021) Evaluation of natural gum-based cryogels for soft tissue engineering. Carbohydr Polym 271:118407

Bie H, Chen H, Shan L, Tan CY, Al-Furjan MSH, Ramesh S, Gong Y, Liu YF, Zhou RG, Yang W, Wang H (2023) 3D printing and performance study of porous artificial bone based on HA-ZrO(2)-PVA composites. Materials (Basel) 16(3):1107

Boehm CA, Donay C, Lubig A, Ruetten S, Sesa M, Fernández-Colino A, Reese S, Jockenhoevel S (2023) Bio-inspired fiber reinforcement for aortic valves: scaffold production process and characterization. Bioengineering 10(9):1064

Brown M, Li J, Moraes C, Tabrizian M, Li-Jessen NYK (2022) Decellularized extracellular matrix: new promising and challenging biomaterials for regenerative medicine. Biomaterials 289:121786

Cao L, Zhao H, Qian M, Shao C, Zhang Y, Yang J (2022) Construction of polysaccharide scaffold-based perfusion bioreactor supporting liver cell aggregates for drug screening. J Biomater Sci Polym Ed 33(17):2249–2269

Castro-Domínguez C, Lozano-Picazo P, Álvarez-López A, Garrote-Junco J, Panetsos F, Guinea GV, Elices M, Rojo FJ, González-Nieto D, Colchero L (2023) Axonal guidance using biofunctionalized straining flow spinning regenerated silk fibroin fibers as scaffold. Biomimetics 8(1):65

Chakrabarty S, Romero EO, Pyser JB, Yazarians JA, Narayan ARH (2021) Chemoenzymatic total synthesis of natural products. Acc Chem Res 54(6):1374–1384

Chaudhari AA, Vig K, Baganizi DR, Sahu R, Dixit S, Dennis V, Singh SR, Pillai SR (2016) Future prospects for scaffolding methods and biomaterials in skin tissue engineering: a review. Int J Mol Sci 17(12):1974

Costa A, Naranjo JD, Londono R, Badylak SF (2017) Biologic scaffolds. Cold Spring Harb Perspect Med 7(9):a025676

Da L-C, Huang Y-Z, Xie H-Q, Zheng B-H, Huang Y-C, Du S-R (2021) Membranous extracellular matrix-based scaffolds for skin wound healing. Pharmaceutics 13(11):1796
El-Seedi HR, Said NS, Yosri N, Hawash HB, El-Sherif DM, Abouzid M, Abdel-Daim MM, Yaseen M, Omar H, Shou Q, Attia NF, Zou X, Guo Z, Khalifa SAM (2023) Gelatin nanofibers: recent insights in synthesis, bio-medical applications and limitations. Heliyon 9(5):e16228
Fanovich MA, Di Maio E, Salerno A (2023) Current trend and new opportunities for multifunctional bio-scaffold fabrication via high-pressure foaming. J Funct Biomater 14(9):480
Field J, Haycock JW, Boissonade FM, Claeyssens F (2021) A tuneable, photocurable, poly(caprolactone)-based resin for tissue engineering-synthesis, characterisation and use in stereolithography. Molecules 26(5):1199
Gomez-Cerezo MN, Pena J, Ivanovski S, Arcos D, Vallet-Regi M, Vaquette C (2021) Multiscale porosity in mesoporous bioglass 3D-printed scaffolds for bone regeneration. Mater Sci Eng C Mater Biol Appl 120:111706
Gradinaru LM, Barbalata-Mandru M, Enache AA, Rimbu CM, Badea GI, Aflori M (2023) Chitosan membranes containing plant extracts: preparation, characterization and antimicrobial properties. Int J Mol Sci 24(10):8673
Grigalunas M, Brakmann S, Waldmann H (2022) Chemical evolution of natural product structure. J Am Chem Soc 144(8):3314–3329
Gritsch L, Conoscenti G, La Carrubba V, Nooeaid P, Boccaccini AR (2019) Polylactide-based materials science strategies to improve tissue-material interface without the use of growth factors or other biological molecules. Mater Sci Eng C Mater Biol Appl 94:1083–1101
Karageorgis G, Foley DJ, Laraia L, Waldmann H (2020) Principle and design of pseudo-natural products. Nat Chem 12(3):227–235
Krishani M, Shin WY, Suhaimi H, Sambudi NS (2023) Development of scaffolds from bio-based natural materials for tissue regeneration applications: a review. Gels 9(2):100
Loh QL, Choong C (2013) Three-dimensional scaffolds for tissue engineering applications: role of porosity and pore size. Tissue Eng Part B Rev 19(6):485–502
Lowe RA, Taylor D, Chibale K, Nelson A, Marsden SP (2020) Synthesis and evaluation of the performance of a small molecule library based on diverse tropane-related scaffolds. Bioorg Med Chem 28(9):115442
Maxson EL, Young MD, Noble C, Go JL, Heidari B, Khorramirouz R, Morse DW, Lerman A (2019) In vivo remodeling of a 3D-bioprinted tissue engineered heart valve scaffold. Bioprinting 16:e00059
Munoz-Perez E, Perez-Valle A, Igartua M, Santos-Vizcaino E, Hernandez RM (2023) High resolution and fidelity 3D printing of Laponite and alginate ink hydrogels for tunable biomedical applications. Biomater Adv 149:213414
Ozdemir S, Yalcin-Enis I, Yalcinkaya B, Yalcinkaya F (2022) An investigation of the constructional design components affecting the mechanical response and cellular activity of electrospun vascular grafts. Membranes 12(10):929
Peng Y, Lu M, Zhou Z, Wang C, Liu E, Zhang Y, Liu T, Zuo J (2022) Natural biopolymer scaffold for meniscus tissue engineering. Front Bioeng Biotechnol 10:1003484
Poongodi R, Chen Y-L, Yang T-H, Huang Y-H, Yang KD, Lin H-C, Cheng J-K (2021) Bio-scaffolds as cell or exosome carriers for nerve injury repair. Int J Mol Sci 22(24):13347
Prakoso AT, Basri H, Adanta D, Yani I, Ammarullah MI, Akbar I, Ghazali FA, Syahrom A, Kamarul T (2023) The effect of tortuosity on permeability of porous scaffold. Biomedicines 11(2):427
Reddy MSB, Ponnamma D, Choudhary R, Sadasivuni KK (2021) A comparative review of natural and synthetic biopolymer composite scaffolds. Polymers (Basel) 13(7):1105
Rosellini E, Cascone MG (2023) Microfluidic fabrication of natural polymer-based scaffolds for tissue engineering applications: a review. Biomimetics (Basel) 8(1):74
Sun W, Gregory DA, Tomeh MA, Zhao X (2021) Silk fibroin as a functional biomaterial for tissue engineering. Int J Mol Sci 22(3):1499
Taghavizadeh Yazdi ME, Nazarnezhad S, Mousavi SH, Sadegh Amiri M, Darroudi M, Baino F, Kargozar S (2021) Gum tragacanth (GT): a versatile biocompatible material beyond borders. Molecules 26(6):1510

Varinelli L, Guaglio M, Brich S, Zanutto S, Belfiore A, Zanardi F, Iannelli F, Oldani A, Costa E, Chighizola M, Lorenc E, Minardi SP, Fortuzzi S, Filugelli M, Garzone G, Pisati F, Vecchi M, Pruneri G, Kusamura S, Baratti D, Cattaneo L, Parazzoli D, Podesta A, Milione M, Deraco M, Pierotti MA, Gariboldi M (2023) Decellularized extracellular matrix as scaffold for cancer organoid cultures of colorectal peritoneal metastases. J Mol Cell Biol 14(11):mjac064

Wibowo A, Tajalla GUN, Marsudi MA, Cooper G, Asri L, Liu F, Ardy H, Bartolo P (2021) Green synthesis of silver nanoparticles using extract of Cilembu sweet potatoes (Ipomoea batatas L var. Rancing) as potential filler for 3D printed electroactive and anti-infection scaffolds. Molecules 26(7):2042

Wong SK, Yee MMF, Chin K-Y, Ima-Nirwana S (2023) A review of the application of natural and synthetic scaffolds in bone regeneration. J Funct Biomater 14(5):286

Wu YX, Ma H, Wang JL, Qu W (2021) Production of chitosan scaffolds by lyophilization or electrospinning: which is better for peripheral nerve regeneration? Neural Regen Res 16(6):1093–1098

Xu P, Kankala RK, Wang S, Chen A (2024) Decellularized extracellular matrix-based composite scaffolds for tissue engineering and regenerative medicine. Regen Biomater 11:rbad107

Yap JX, Leo CP, Mohd Yasin NH, Show PL, Chu DT, Singh V, Derek CJC (2022) Recent advances of natural biopolymeric culture scaffold: synthesis and modification. Bioengineered 13(2):2226–2247

Zhang Y, Szostak M (2022) Synthesis of natural products by C-H functionalization of heterocycless. Chemistry 28(22):e202104278

Zhang X, Chen X, Hong H, Hu R, Liu J, Liu C (2022) Decellularized extracellular matrix scaffolds: recent trends and emerging strategies in tissue engineering. Bioact Mater 10:15–31

Applications of Natural Product-Based Scaffold in Tissue Engineering

3

Gurpreet Kaur

Abstract

Tissue engineering (TE) is gaining great attention by creating bio-artificial tissues or biological replacements that improve, maintain, and repair damaged tissue or organs. Scaffolds are crucial for TE as they serve as a transient platform that offers direction and structural support during the development of new tissues. It promotes multiplication and adhesion of cells and the regeneration of the extracellular matrix, bones, muscles, and nerves. It also acts as a template for the rebuilding of tissue defects. Scaffolds have been developed using a variety of materials. Natural product-based scaffold materials have garnered the most interest in the engineering of tissues due to their exceptional mechanical qualities, bio-absorbability, and ecofriendly. Natural product-based scaffold is harmless, biocompatible, and do not emit any harmful substances as they break down. There is not yet a single material that satisfies every need of the scaffold. Therefore, it is recommended to combine two or more components to get the desired characteristics of scaffold. This review describes the various applications of numerous biopolymers with natural origins like hyaluronic acid, collagen, silk fibroin, chitosan, and gelatin, in regenerative medicine.

Keywords

Biocompatibility · Collagen · Icariin · Regenerative medicine · Scaffold

G. Kaur (✉)
Department of Zoology, Akal University, Talwandi Sabo, Punjab, India

M. Kumar et al. (eds.), *Natural Product Inspired Scaffolds*,
https://doi.org/10.1007/978-981-97-3111-4_3

3.1 Introduction

Tissue engineering holds inordinate hope for addressing the rising necessity for ground-breaking remedies to cure the disease conditions. It is an interdisciplinary field where scaffolds, growth factors, and cells are combined to form new tissue. It is a promising alternative to organ transplantation, which is plagued by an ongoing donor scarcity (Lauritano et al. 2020; Serafin et al. 2021). Patients with damaged and wounded tissues, as well as organ failure, are becoming more common as the world population and life expectancy increase (Giwa et al. 2017). To meet these demands, novel sources have been searched to create scaffolds which can help in the rejuvenation of tissues and organs. A three-dimensional (3D) porous biomaterial known as scaffolds offers an environment that is conducive to the repair and regeneration of tissues and organs by cells. To preserve and restore tissue function, extracellular matrix-like cells and growth factors are implanted into the scaffold, which serves as a template. Moreover, scaffolds are able to transport bioactive compounds as a physical barrier against the invasion of native tissues that can cause hinderance in tissue regeneration and repair (Akter 2016; Chocholata et al. 2019). A scaffold should have biocompatibility, high porosity, biodegradability, and mechanical qualities (Biswal et al. 2023). These should provide cells with both structural support and behavioral cues, in addition to serving as a supporting framework for the cells (Rashid et al. 2023). A solid matrix called the extracellular matrix (ECM) surrounds most tissue cells, except for blood cells. The extracellular matrix serves as a stabilizing force and are responsible for the mechanical characteristics of tissues. Thus, the scaffold chosen for created tissue needs to resemble the ECM of that particular tissue (Chan and Leong 2008). Tissue engineering thrives on the optimal cell selection, focused cell isolation and expansion, and biomaterial selection for scaffold construction. However only single polymer cannot provide every necessary scaffold quality, combining it with other polymers can provide the desired property (Dehghan et al. 2021). Synthetic polymers like epoxies (Kumar et al. 2020, 2021a, b, c; Kumar and Saini 2023) can be used for scaffolds, but mostly natural polymers like chitosan, alginate, collagen, gelatin, cellulose, and hyaluronic acid have been employed to form scaffolds for the regeneration and repair of tissues (He et al. 2023). This review aims to explore the applications of different natural product-based scaffold in tissue engineering with an emphasis on skeletal muscle, bone, central and peripheral nerve rejuvenation, cardiac muscle, and cornea substitutes.

3.2 Characteristics of Scaffolds

Scaffold's microarchitecture includes pore diameter and interconnectivity among the pores. To begin with, pore volume needs to be sufficient for the movement of cells and attachment onto scaffolds. It is recommended that 75–100-μm pore size is adequate for in vitro models, and 200–500-μm diameter of pores is enough for in vivo models to facilitate infiltration and vascularization of optimal tissue (Roseti et al. 2017; Perić Kačarević et al. 2020). In addition, a network of interconnected porous structures is needed to offer a greater scaffold surface area for cell adhesion (Verma and Verma 2020). Grafted scaffolds into live organisms should eventually decompose chemically or enzymatically, as they serve merely as a transient platform for growing cells or tissues. Biodegradability is the rate at which the scaffold materials decompose (Chocholata et al. 2019). The pace at which new bone is formed or tissue regeneration occurs should ideally be proportionate to the scaffold's biodegradation rate. The biomaterial scaffolds will be replaced by new tissues through a process known as "creeping substitution" once they have been effectively generated and integrated with the host bone (Roberts and Rosenbaum 2012). In addition, the scaffold needs to be extremely biocompatible in order to promote cell attachment and growth. To avoid major inflammatory reactions that might impair recovery or result in the body rejecting the treatment, there should be very little long-term immunological response. Even in the event of inflammatory responses, they ought to heal in a maximum of 2 weeks (Nikolova and Chavali 2019). Moreover, scaffolds should have the capacity to communicate with the surrounding cellular constituents of engineered tissues; it is also termed as bioactivity of scaffolds. It helps in the prevention of scarring by promoting appropriate cell differentiation, neoformation, and integration in the host in contrast to conventional passive biomaterials (Roseti et al. 2017). In addition, the scaffold materials must have inherent mechanical qualities that are comparable to those of natural tissues to reduce the likelihood of stress shielding and eventual refracture. Elastic modulus, tensile strength, fracture toughness, and fatigue are a few instances of mechanical qualities (Roseti et al. 2017; Gurumurthy and Janorkar 2021). The ideal parameter for scaffold is given in Fig. 3.1.

3.3 Natural Product-Based Scaffold

Natural polymers are also known as biopolymers and these are derived from natural sources. They are divided into two categories: biomaterials polysaccharides in nature like silk, chitosan, alginate, and gelatin and biomaterials based on proteins like collagen and fibrin. Biopolymers promote cell adhesion and differentiation, are extremely biocompatible, and are devoid of toxins. However, they are extremely susceptible to high temperatures and have little mechanical strength (Rahmani Del Bakhshayesh et al. 2018). They have traits in common with soft tissues: they are bioactive and meet the requirements for biodegradability and biocompatibility. Additionally, they are renowned for their environmental safety and adaptability to various uses (Krishani et al. 2023). Natural biopolymers such as silk fibrin, collagen, gelatin, and chitosan have garnered significant interest and are commonly used in scaffold production. Moreover, the breakdown products of naturally occurring macromolecules can be transported through the metabolic route and do not cause cytotoxicity. Natural polymer-based scaffolds have appropriate cell-scaffold responses and possess good cell affinity, resembling an extracellular matrix (Jalali Monfared et al. 2019; Asadpour et al. 2020).

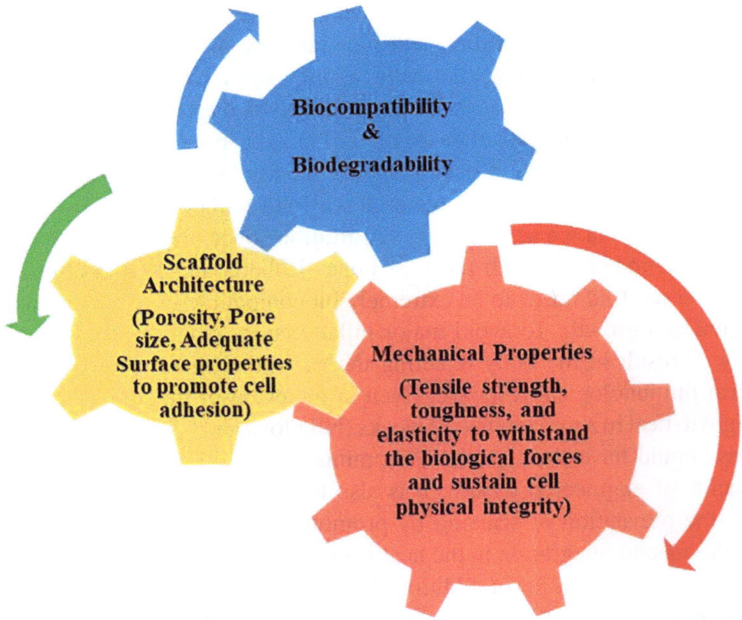

Fig. 3.1 Requirements for an ideal scaffold

3.4 Applications

Natural product-based scaffolds have various biomedical applications which are shown in Fig. 3.2.

3.4.1 Collagen-Based Scaffold

Collagen is highly biodegradable and biocompatible protein present in the animal tissues. As a result, it is used in medicine as a scaffold, a medication carrier, and a wound dressing (Zarei et al. 2021). Various studies suggested that collagen-based scaffolds has a variety of biological uses like musculoskeletal, cardiac, neuronal, cartilage, and bone tissue engineering (Grabska-Zielińska et al. 2021; Szychlinska et al. 2020; Jiang et al. 2020; Bayrak and Yilgor Huri 2018).

3.4.1.1 Repair of the Cornea

Corneal disease, which mostly results from chemical or thermal burns, excessive contact lens use, and mechanical damage, causes millions of blind people worldwide. A significant lack of superior allogeneic corneal tissues limits practical use of keratoplasty. Many studies concentrated on developing natural biomaterial-based corneal tissue replacements as a solution to the scarcity of corneal donors (Zhang et al. 2021; Liu et al. 2023). However, adequate mechanical qualities, great

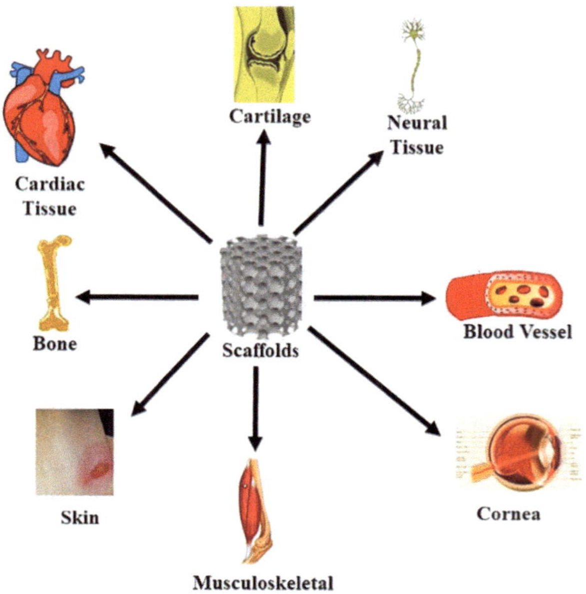

Fig. 3.2 Applications of natural product-based scaffolds in tissue engineering

transparency, and acceptable biocompatibility are needed for an artificial cornea to perform its normal physiological functions. Among the natural biopolymers, collagen is the essential protein in the animal connective tissues. It is used as a scaffold in medicine due to its biodegradability and biocompatibility (Zarei et al. 2021). Zhang et al. (2016) monitored physical and biological characteristics of a scaffold made of collagen I from rat tail. Optical coherence tomography imaging verified that the scaffold preserved the form of the cornea with a reflectivity profile similar to a typical human cornea and a central thickness of 502 μm. The tensile strength of cross-linked scaffold was observed to be 2.21 MPa in contrast to the noncross-linked scaffold where it was found to be 1.23 MPa. The migration of epithelial cells and keratocytes had been detected in the scaffold and were arranged in a lamellar fashion. This study depicted the potential of collagen scaffold as a substitute corneal tissue for transplantation into humans. In another investigation, cell-free recombinant human collagen scaffold stimulated the renewal of the nerves of the cornea, which remained firm for 4 years and did not reveal any refutation (Fagerholm et al. 2014). Moreover, Zargar et al. (2023) made three-layer nanofiber scaffold with silk fibroin and collagen and was also loaded with aloe vera and epithelial growth factor (EGF) in order to promote corneal regeneration by electrospinning and electrospraying techniques. This nanofiber scaffolds revealed right thickness, shape, and acceptable biological and engineering features, and it is a promising option as a substitute for damaged corneal epithelial layers. In another investigation, a scaffold was constructed by blending the poly(ε-caprolactone) (PCL) and collagen type I. This scaffold displayed lamellar architecture similar to the natural cornea and had capacity to stop differentiation of keratocytes and avoid fibrosis. In addition, it also reinvigorated propagation of limbal stromal stem cells. RNA sequencing demonstrated that several extracellular matrix-linked genes were found to be boosted in the cells. The PCL/collagen scaffold's superior biocompatibility and lack of immunogenicity indicate its promising potential in corneal bioengineering (Xu et al. 2022).

3.4.1.2 Cardiac Tissue Engineering

Collagen is a highly valued protein and exhibited protective potential in preventing myocardial infarction. Joanne et al. (2016) administered epicardially nanofibrous collagen scaffold seeded with cardiomyocytes in a dilated cardiomyopathy animal model. This scaffold showed stabilization of dilated cardiomyopathy in in vitro and in in vivo model. In another investigation, rat bone marrow-derived mesenchymal stem cells (MSCs) were implanted onto the PCT/collagen scaffold for the formation of MSC patch. It showed adaptable mechanical characteristics and biocompatibility, restored the damage due to myocardial heart infraction, and promoted angiogenesis (Wang et al. 2023a, b, c, d). Moreover, the combination of 3D collagen scaffold and zebularine therapy promoted differentiation in mesenchyme cells of the heart, upgrade cell engraftment at the infarcted site, lessen infarct size, and caused improvement in the function of the heart (Khan et al. 2022a, b). In addition to it, unique bimodal electrospun collagen/PLGA scaffold induced significant decline in cardiac remodeling due to its robust long-term stem cell engraftment properties, and consequently it restored the functioning of the heart (Wee et al. 2022).

3.4.1.3 Neuronal Tissue Engineering

Spinal cord injury (SCI) is still a significant morbidity in people, and its restoration is a problem. Even though neural stem cells demonstrated significant promise in the cure of spinal cord injuries, their use has been limited since in the wounded area they largely converted into astrocytes rather than neurons, indicating a strong demand for efficient methods to control neuronal differentiation. Baicalein is a clinically approved drug with several pharmacological effects, and it is functionalized with collagen scaffolds for the monitoring its effect on spinal cord injury. The baicalein collagen scaffolds exhibited physical features similar to the widely used collagen. Research conducted on primary neural stem cells in vitro and on SCI rats in vivo have demonstrated that these baicalein collagen scaffolds can promote neurogenesis and axon extension while also lowering glial scar development and astrocyte production. Additionally, implantation of baicalein collagen scaffolds caused improvement in the motor functional recovery in rats with spinal cord injury. Thus, the baicalein collagen scaffolds induced neuronal differentiation to aid in the healing of SCI and other damage to the central nervous system (Qian et al. 2023). In another investigation, a scaffold was made from collagen, stromal cell-derived factor, nanoparticle-encapsulated PLGA, and paclitaxel. When this scaffold was implanted into the injury site, they created a conduit channel for neural regeneration. It assisted movement and differentiation of neural cells into neurons, promoting motor function recovery and reducing the formation of glial scars following spinal cord injury (Li et al. 2023a, b, c). Moreover, administration of collagen scaffold functionalized with mesenchymal stem cells derived from umbilical cord showed marked improvement in the cognitive function, daily life activities, and intracerebral hemorrhage-associated encephalomalacia (Li et al. 2023a, b, c).

3.4.1.4 Bone and Musculoskeletal Tissue Engineering

The mammalian body depends on the musculoskeletal system for its physical support and mobility. Musculoskeletal problems are very inflated to society and have a significant impact on mobility and quality of life. Large musculoskeletal abnormalities may be treated alternatively with considerable potential in this emerging field of musculoskeletal tissue engineering. Musculoskeletal tissue engineering makes extensive use of both natural and synthetic polymers because of their high biocompatibility and biodegradability (Ye et al. 2022). An electrically conductive 3D collagen scaffold guided myoblast development and organization. The incorporation of polypyrrole nanoparticles further increased myotube development and maturation (Basurto et al. 2021). Apart from this, addition of β-tricalcium phosphate or calcium silicate in collagen composite scaffolds exhibited more linked porous structures and increased compressive modulus. This scaffold demonstrated no cytotoxicity against osteoblast cells and showed acceptable biocompatibility (Animut et al. 2023). Likewise, highly porous scaffolds based on chitosan and collagen coated with 30% of magnesium, selenium, strontium, and zinc-substituted calcium phosphate showed excellent viability and proliferation of human bone marrow-derived mesenchymal stromal cells and effective osteogenic differentiation (Ressler et al. 2023). Moreover, a scaffold made from naringin, nano-hydroxyapatite, and collagen (N/HA/C) revealed desirable biocompatibility and stimulated osteogenic differentiation of bone marrow mesenchymal stem cells (BMSCs). Additionally higher expression of genes related to osteogenic differentiation, such as osteocalcin, osteopontin, and BMP-2, as well as high levels of alkaline phosphatase, had been noticed. The N/HA/C scaffold dramatically aided in the restoration of bone tissues and healing of skull abnormalities when given to rats with defects in their skulls; this suggested that N/HA/C scaffold have enormous potential for use in the restoration of bone tissue (Zuo et al. 2022).

3.4.2 Hyaluronic Acid

Hyaluronic acid is an anionic glycosaminoglycan present in synovial fluid and extracellular matrix (ECM) (Kobayashi et al. 2020). It is negatively charged, extremely hydrophilic, and viscoelastic due to the presence of carboxyl groups (Abatangelo et al. 2020). This molecule is one of the main intracellular constituents of connective tissue, and it is crucial for proliferation, movement, and differentiation of cells (Yasin et al. 2022). In addition to its numerous physiological uses, hyaluronic acid's backbone comprises functional groups that are commonly employed in the construction of scaffolds with structural and space-filling qualities. Hyaluronic acid also regulates the amount of water in tissues and matrices (Collins and Birkinshaw 2013; Hemshekhar et al. 2016).

3.4.2.1 Restoration of the Cornea

It is still difficult to make a scaffold that have a structure similar to the natural cornea which can solve the significant lack of donors for the treatment of corneal blindness. Transparent, flexible, biocompatible, and having enough mechanical strength to suture or glue onto the eye are the characteristics of the perfect biomaterial for corneal tissue engineering (Kong and Mi 2016). Hyaluronan hydrogel scaffold caused the proliferation of corneal epithelial stem cells, and regenerated corneal epithelium keeps features that are comparable to those of the original corneal epithelium (Chen et al. 2017). In another study, hyaluronic acid modified gelatin-methacrylate 3D-printed scaffolds stimulated corneal stromal cells with signals which gave them guidance for the spatial and directional organization and caused the remodulation of ECM (Wang et al. 2022). Moreover, mammalian adipose-derived stem cells on a scaffold made of hyaluronic acid also has the potential to be utilized as a reservoir of keratocytes for the restoration of extracellular matrix-like material where the corneal stroma has been injured (Espandar et al. 2012). However, hyaluronic acid forms lymph vessels in the limbus, which can be a barrier to its utilization for repair of corneal tissue (Sun et al. 2019).

3.4.2.2 Cardiac and Cartilage Tissue Engineering

Hyaluronic acid possesses a large number of therapeutic applications which are exploited to cure various physiological complications related to cardiac and cartilage tissue. Electrospun microfiber scaffold of poly(glycerol sebacate) (PGS) and PCL with a hybrid hydrogel consisting of methacrylated gelatin and hyaluronic acid enhanced the mechanical characteristics for heart valve engineering. Incorporation of hydrogel in the scaffold enhanced the three-dimensional dispersion of mitral valvular interstitial cells. The integrated scaffold showed more appropriate three-dimensional structure for the construction of heart valve tissue in comparison to hydrogel or electrospun scaffolds used alone (Eslami et al. 2014). Moreover, it has been seen that addition of hyaluronic acid increased the initial fiber diameter and cell-scaffold interactions in fibrin-based scaffolds. It caused more scaffold compaction and rigidity and improved cell alignment and less fibrin fiber bundling by the cells (Lei et al. 2021). Apart from this, scaffolds based on the 3D-printed polycaprolactone loaded with hyaluronic acid assisted in the regeneration of bone in two patients (Özdemir et al. 2023). In another investigation, a 3D scaffold was formulated by silk fibroin, hyaluronic acid, gelatin, chondroitin sulfate, and aloe vera. This composite showed proliferation of mesenchymal stem cells in human bone marrow which depicted its usage in cartilage tissue engineering (Wuttisiriboon et al. 2023). Likewise, Correia et al. (2011) demonstrated that the chitosan/hyaluronic acid composite scaffolds boosted cell attachment and were observed to be noncytotoxic. Hyaluronic acid increased the formation of cartilage extracellular matrix and a better pore network structure. Thus chitosan/hyaluronic acid scaffolds may be beneficial in the restoration of cartilage.

3.4.3 Gelatin

Gelatin is a naturally occurring polymer, resembles its progenitor, collagen. It has been extensively researched as a hydrogels, scaffolds, and drug delivery systems in tissue engineering (Bello et al. 2020). Gelatin has a lot of potential applications as a scaffolding material among natural polymeric materials. The gelatin-based scaffolds have outstanding biocompatibility, biodegradability, and porous structure. Blends of gelatin with natural or synthetic polymers consistently result in very effective matrices with enhanced biomechanical and bio-affinity properties for the scaffolds (Hoque et al. 2015).

3.4.3.1 Bone Tissue Regeneration

It is still extremely difficult to repair large bone defects, which makes it clinically necessary to develop innovative bone biomaterials with angiogenic and osteogenic characteristics to promote the regeneration of vascularized bone. Natural or synthetic polymers have better biocompatibility and biodegradability and share structural similarities with extracellular matrix components. Gelatin scaffolds loaded with bioglass was able to enhance angiogenesis, viability, and differentiation of osteoblasts (Aldhaher et al. 2023). A new scaffold was made with decellularized bone particles isolated from rabbit femurs (dPTs, approximately 100 μm in diameter) and combined in varying amounts with gelatin. This 3D-printed gelatin/dPTs composite scaffold showed more porosity and provided an appropriate microenvironment for preosteoblasts. These scaffolds boosted cell adhesion, proliferation, and migration in preosteoblasts with negligible toxicity, indicating the scaffolds' possible use in bone tissue (Kara et al. 2022). In another study, cerium oxide nanoparticles (CNPs) revealed remarkable antioxidant potential; thus, these were added in gelatin and chitosan to form nanocomposite scaffolds by freeze-drying. In comparison to chitosan-gelatin scaffold, these nanocomposites showed superior outcomes in terms of physicochemical, mechanical, and biological characteristics. In vitro and in vivo experiments on preosteoblasts revealed that the scaffolds made of chitosan/gelatin/CNPs were cyto-friendly. This scaffold also revealed high inhibition zones in the microbicidal testing against both gram-positive and gram-negative bacteria. The chitosan/gelatin/CNP scaffold displayed biocompatibility, 79% porosity, and 1.83 calcium/phosphorus ratio. Thus the results reinforced the superiority of these scaffolds in fixing the bone deformities (Bhushan et al. 2023). Moreover, leonurine hydrochloride-coated gelatin, sodium alginate, and nano-attapulgite composite hydrogel scaffolds were also reported by Tan et al. (2023) which can promote bone growth. Leonurine hydrochloride (LH) has the potential to increase the deposition of collagen and vascular development; it can aid in the healing of full-thickness cutaneous wounds. In vitro studies demonstrated that LH-loaded scaffolds, as opposed to pure polymer scaffolds, enhanced the generation of osteoblasts and also upregulated the expression of osteogenic genes. In another investigation, Kalidas and Sumathi (2023) reported a scaffold made from gelatin, polyvinyl alcohol, and silk fiber. The scaffold showed maximum tensile strength of 34 MPa and also exhibited considerable porosity. It also showed microbicidal activity against *E. coli*, *C. albicans*, and *S. aureus*. This scaffold showed apatite production on the surface of composites. Various examples of scaffold type used in bone regeneration are given in Table 3.1.

3.4.3.2 Cardiac Tissue Engineering

It is still difficult to engineer the heart tissue to resemble the hierarchical structure of the cardiac tissue. One of the most promising approaches to precisely design complex tissue structures is 3D printing. This work uses a mixture of alginate and gelatin to create scaffolds with a unique angular shape that mimics heart architecture using 3D printing. These angular 3D-printed structures shown promising qualities for cardiac tissue engineering, including high mechanical strength, suitable swelling, and degrading properties for both endothelial and cardiac cells (Ketabat et al. 2023). In addition to it, electroconductive scaffolds was made by mixing

Table 3.1 Different scaffolds used in the regeneration of the bone tissue

Scaffold type	Type of cell	Main findings	References
Hydroxyapatite-collagen	In vitro: Human dental pulp cells	Biocompatibility Enhanced differentiation and angiogenesis in osteoblasts in SD rats	Xie et al. (2023)
Concentric mineralized silk fibroin	Human monocyte (THP-1 cell line)	More cell adhesion Upregulated expression of markers of osteoblasts	Bosio et al. (2023)
Poly-3-hydroxybutyrate-graphene oxide chitosan	Human osteosarcoma MG63 cells	Increased cell adhesion More ALP activity	Motiee et al. (2023)
Polycaprolactone-carbonated hydroxyapatite-marine atelocollagen	Osteoblasts	Enhanced differentiation in osteoblasts	Kim et al. (2022)
Gelatin-chitosan loaded with glutaraldehyde	Bone marrow mesenchymal stromal cells	Higher levels of alkaline phosphatase and osteocalcin	Banafati Zadeh and Zamanian (2022)
Gelatin-polycaprolactone-nanohydroxyapatite	Human osteoblast cell line	Efficient cell adhesion	Gautam et al. (2021)
PCL-gelatin-hydroxyapatite-vitamin D3	Human adipose-derived mesenchymal stem cells	Biocompatible Boosted osteogenic differentiation	Sattary et al. (2019)
Alginate-hydroxyapatite aerogel scaffold	Murine fibroblasts Human bone marrow mesenchymal stem cells	Nontoxic Enhanced the bone regeneration process	Iglesias-Mejuto and García-González (2021)
Silk fibroin-GYY4137 scaffold	Human bone marrow mesenchymal stromal cells	Increment in proliferation and adhesion in cells	Gambari et al. (2019)

poly(3,4-ethylenedioxythiophene)-poly(styrenesulfonate) with a biomimetic polymer network of gelatin and tested for cardiac tissue regeneration. The scaffolds were found to be nontoxic to H9C2 cardio-myoblasts and promote cell adhesion, proliferation, and spreading on scaffolds, particularly in the early stages of the process, which is critical for a good outcome following implantation (Furlani et al. 2023).

Heydari and Esmaeili (2022) reported the scaffold made from polyurethane/chitosan/gelatin/heparin, and it was screened for its usage in cardiovascular tissue engineering. It was observed that this scaffold was observed to be nontoxic and exhibited 91% cell survival after 7 days. Similarly electrospun scaffolds like furfuryl-gelatin alone, f-gelatin with PCL, and coaxial scaffolds with PCL (core) and f-gelatin (sheath) unveiled exceptional biocompatibility and enhanced the multiplication and adhesion in the mammalian cardiomyocytes (Nagiah et al. 2022).

3.4.4 Chitosan (CS)

Chitosan is a polysaccharide that is extensively found in the exoskeletons of insects and the shells of crustaceans. It is obtained after the deacetylation of chitin, and it possess nontoxicity, biocompatibility, biodegradation, and antibacterial activity. Chitosan-based biomaterials also possess several tissue engineering applications in different medical fields (Xiang et al. 2023).

3.4.4.1 Wound Healing

Chitosan (CS) is a well-known naturally occurring polymer which is appropriate for wound healing due to its biodegradable, biocompatible, bioadhesive, and antibacterial properties (Biswal et al. 2023). Proline-conjugated chitosan scaffold (P-CS) was prepared for healing of wounds. The scratch wound test on the L929 cell line revealed the 53.35–72.96% wound closure for P-CS in contrast to the native CS scaffold where healing capacity of wounds were observed to be 39%. Moreover, cell viability test indicated that the scaffold is not toxic to L929 cells (Linju and Rekha 2023). In another investigation, polyvinyl alcohol and chitosan nanofibrous scaffold (PVA/CS) coated with flaxseed extract displayed more water absorption, porosity, and wettability. This scaffold showed noteworthy antioxidant activity and can hinder the development of gram-positive and gram-negative bacteria in contrast to the PVA/CS scaffold without flaxseed extract. It has been seen that the flaxseed-coated scaffold exhibited fast healing of injuries and 100% repair of the injured area within 2 days (Doostan et al. 2023). Moreover, rhubarb chitosan/silk fibroin (RCS/SF) scaffold has a 3D porous structure which makes it pro-angiogenic as well as microbicidal. It also possess properties like good swelling power and biocompatibility. Within 2 weeks, RCS/SF scaffolds facilitated improvement in the collagen deposition, faster healing, augmented neovascularization, and reepithelialization in contrast to the control group. This scaffold is an ideal candidate for the cure of diabetic wounds and developing innovative therapeutics for diabetic ulcers due to its pro-healing qualities and effective release of carbonized rhubarb, which has quick hemostatic and excellent sterilizing effects (Wang et al. 2023a, b, c, d).

3.4.4.2 Nervous Tissue Regeneration

The quest for novel therapeutic strategies to combat central and peripheral nervous systems injuries has long been fueled by the dearth of effective treatments as well as the drawbacks of currently available therapies (Ojeda-Hernández et al. 2020). Given this, the mechanical strength, porosity, biodegradability, and biocompatibility of chitosan make it a desirable material. Additionally, nerve damage has been repaired using it alone or in conjunction with other biomaterials, adhesion molecules, cells, or growth factors. Furthermore, increased neuronal cell adhesion to the membrane of the nervous system is one of the possible methods by which chitosan speeds up the healing process. Additionally, it was shown that chitosan fibers might show improved Schwann cell adhesion and migration, reassembling axons to the nervous system's Bungner bands (Pahlevanzadeh et al. 2020). Chitosan scaffolds may provide as an ideal alternative to traditional biomaterials to stimulate regeneration in the peripheral and central nervous systems. Pooshidani et al. (2021) prepared a conductive scaffold from polycaprolactone and chitosan. This scaffold displayed desirable interconnected porosity and hydrophilicity up to 75–80% after the addition of 40% polyethylene oxide and gold nanoparticles. In another study, a chitosan scaffold composed of graphene oxide was shown to aid in the functional healing of a damaged spinal cord in rats. An 18–87-μm porous scaffold made of chitosan and graphene oxide was implanted into a murine spinal cord. This composite caused the angiogenesis, migration of neurons, and regeneration of neural tissue in order to restore injured spinal cord nerve tissue. Additionally, improvement in somatosensory potential and locomotor function in rats may be considerably noticed after the incorporation of chitosan-graphene oxide scaffold (Yang et al. 2021). Apart from this, chitosan scaffold with esterase-activable release was monitored for the revival of the peripheral nervous system. This scaffold boosted the growth of Schwann cells and markedly enhanced target muscle weight and sciatic nerve function recovery in a murine model (Wang et al. 2023a, b, c, d).

3.4.5 Alginate

Alginate is a biodegradable and biocompatible polymer derived from brown algae. This polysaccharide is identical with glycosaminoglycan of ECM which make its use as a scaffold more promising in the repair and regeneration of tissues (Axpe and Oyen 2016). Alginate is also frequently used in the development of bioinks for cell-laden bioprinting skeletal muscle constructions (Popov et al. 2017; Sonaye et al. 2022).

3.4.5.1 Wound Dressing

Alginate-hybrid materials (AHM) has a huge potential as a wound dressing material, and they may effectively inhibit the entry of harmful germs and assist wound exudations to drain rapidly, which is beneficial to healing. Other multifunctional polymers are coupled with alginate and its byproducts to create alginate hybrid materials with enhanced features like hydrophilicity, mechanical qualities, and cell adhesion (Varaprasad et al. 2020). Ahmad et al. (2023) reported a nonwoven hemp alginate hydrogel composite for wound dressing. Alginate hydrogel is one of the superlative materials for wound healing, whereas hemp is an environmentally friendly substance with strong anti-inflammatory properties. It has been noted that alginate content determined the resulting composite's performance. Moreover, Bîrcă et al. (2023) also revealed the antibacterial potential of various preparations of alginate-based hydrogels loaded with hyaluronic acid and silver nanoparticles and also monitored the efficiency of these composites in the cure of wounds. It has been seen that all the formulations of alginate exhibited desired pore diameters ranging from 66 to 99 μm and swelling index. In addition to it, addition of hyaluronic acid caused enhancement in the hydrophilic quality of the hydrogels and induced dispersion in the silver nanoparticles by the hydrogel pores. It was observed that alginate hydrogel in combination of silver nanoparticles and hyaluronic acid revealed microbicidal activity against *S. aureus* and *E. coli* in contrast to the alginate hydrogel made with only silver nanoparticles. Likewise, simvastatin-loaded alginate bilayer membranes exhibited more time for the drug release as compared to the monolayer membranes, and it also displayed no cytotoxicity against human keratinocytes (Monteiro et al. 2023). Apart from this, Puccetti et al. (2023) reported the alginate AgCl nanoparticles composite films for healing of wounds with negligible cytotoxicity. These composite films displayed antibacterial activity against *Staphylococcus epidermidis*, *Pseudomonas aeruginosa*, and *Staphylococcus aureus*. Therefore, the use of composite films of Ag/AgCl alginate nanoparticles in wound dressing can prevent the infections without any toxicity to tissue cells. Moreover, antiseptic wound dressing patches laden with cephalexin monohydrate were formulated from various concentrations of sodium alginate and carboxymethyl cellulose. The patches were observed to possess robust tensile strength (3.83–13.83 MPa), elasticity (74–98%), and biodegradability. These patches showed superiority in the prevention of invasion of microbes in the wound dressings due to less moisture uptake and 47–80% drug release profile. Moreover, the patches exhibited non-cytotoxicity against HaCaT cells and also lead to 66.7–74.3% migration of cells to the wounds for healing; thus, it depicted the biocompatibility of alginate patches for the cure of infected injuries (Feketshane et al. 2023). In another investigation, hydrogels made from sodium alginate, microcrystalline cellulose, and carrageenan loaded with the active components isolated from *Amaranthus spinosus* and *Rubia cordifolia* revealed good biocompatibility and moisture retention for wound dressings. Additionally, these hydrogel wound dressings exhibited excellent antimicrobial, procoagulant, and antioxidant activities (Li et al. 2023a, b, c). Wang et al. (2023a, b, c, d) reported a hydrogel made from sodium alginate and platelet-rich plasma for the healing of wounds, and this hydrogel showed potential in the cell proliferation and vascular regeneration. The hydrogel assisted in the closure of wounds when it applied to the injured skin in rats. Similarly, alginate hydrogel-electrospun silk fibroin fibers enhanced multiplication and dispersion in the cells and caused collagen synthesis, which aided in the healing of wounds (Ghalei et al. 2018). Likewise, Abou-Okeil et al. (2018) prepared a wound dressing from hyaluronic acid/sodium alginate film crosslinked with cations (Ca^{2+}, Zn^{2+}, and Cu^{2+}), and it exhibited strong antibacterial activity. Calcium alginate-graphene oxidate AHM showed strong protective efficacy against methicillin-resistant *Staphylococcus epidermidis* and *Staphylococcus aureus* and displayed low toxicity toward mammalian keratinocyte HaCaT cells (Martí et al. 2019).

3.4.5.2 Bone Tissue Engineering

In addition to chitin and chitosan, alginate is a popular polysaccharide which is used for the restoration of bone tissues due to its affordability, cytocompatibility, biodegradability, quick gelation, porosity, and encapsulation efficiency (Khan et al. 2022a, b). Eskandani et al. (2023) prepared nanofibrous scaffolds from alginate, polyaniline, and polyvinyl alcohol. It was observed that these scaffolds showed appropriate cytocompatibility and assist in the proliferation and adhesion of the cells. Moreover, the rate of hemolysis also indicated the hemocompatibilities of scaffolds. These scaffolds also displayed adequate protein adsorption and unveiled promising potential of developed scaffolds for the repair of bone tissues. In addition to it, nanocomposite scaffolds are made of hydroxyapatite, sodium alginate, and silica with varied concentration of graphene oxide. These scaffolds revealed strong cell proliferation viability and cell adhesion in murine pre-osteoblast cells (Khan et al. 2022a, b). Likewise osteogenesis was observed to be higher in silicon dioxide-alginate scaffolds in contrast to the alginate scaffolds made from silicon dioxide and platelet-rich plasma (Gholijani et al. 2022). In another study, calcium-cross-linked alginate dialdehyde-gelatin scaffolds monitored for their potential for the regeneration of the bone. It has been seen that scaffolds coated with mussel-derived polydopamine interact impulsively with osteocalcin and osteomodulin and lead to the development of a surface protein layer. Moreover, it also caused increment in the levels of alkaline phosphatase and osteoimage intensity (Ghorbani et al. 2022).

3.4.6 Silk Fibroin

Silk fibroin is a very resilient fiber that possesses comparable tensile strength to glass fiber and other synthetic organic fibers. Silk fibroin is less immunogenic, pro-inflammatory, and more biocompatible in contrast to collagen (Phillips et al. 2004). New prospects for the application of different types of silk in tissue regeneration have been generated by recent breakthroughs in our knowledge of the structure and processing of silk. This fiber will be used in the applications requiring strong mechanical qualities and delayed biodegradation, like skeletal muscle and bone tissues (Li and Sun 2022).

3.4.6.1 Bone Tissue Engineering

Silk fibroin (SF) has been utilized for bone tissue engineering due to its mechanical properties and biocompatibility. Silk fibroin, hydroxyapatite, and sodium alginate

(SF-HA-SA) scaffolds increased the propagation of cells and repaired the trabecular deformities in the bone and cartilage (Xu et al. 2023). Moreover, silk fibroin/alginate polylactic acid scaffold revealed upregulation in the expression of alkaline phosphatase and also caused increment in the deposition of calcium phosphate, which ultimately caused more osteogenic differentiation (Rajput et al. 2023). In another investigation, silk fibroin titanium oxide and hydroxyapatite nanoparticles were mixed to form scaffolds that are identical with mammalian bone structure. These scaffolds had a superior compressive modulus and more swelling power. They also demonstrated more adhesion and viability in the cells, as well as higher alkaline phosphatase activity and calcium content. The improved osteo-conductive ability of scaffolds makes them a viable option for maxillofacial and mouth surgery and bone augmentation (Watcharajittanont et al. 2023). Scaffold made from fibroin-based bone constructions can promote advanced bone tissue formation in as little as 5 weeks (Meinel et al. 2005). In nude mice, silk fibroin scaffolds also support the femoral defect repair process mediated by mammalian mesenchymal stem cells. In addition, the loading of bone morphogenic protein into a silk fibroin polyethylene oxide nano-fibrous composite scaffold containing human mesenchymal stem cells caused increment in the regeneration of a bone-like tissue (Li et al. 2006; Zhang et al. 2018). Animal studies have demonstrated that osteogenesis in bones improved when hydroxyapatite nanoparticles were added to the silk matrix (Bharadwaz and Jayasuriya 2020). Additionally, incorporation of graphene in SF scaffold enhanced the osteogenesis in the cells, and this scaffold can be used for the regeneration of the bone tissue (Ding et al. 2021). In another investigation, a 3D scaffold of polycaprolactone nanofibers was prepared and then biomineralized and coated with silk fibroin. It showed much quicker rate of repair in the bone tissue (Xiao et al. 2021).

3.4.6.2 Cartilage Tissue Regeneration

Cartilage is a nonvascular connective tissue that serves a vital function in preserving the body's ability to move by shielding the subchondral bones from excessive joint stress. Scaffolds based on silk fibroin loaded with gelatin/chondroitin sulfate/hyaluronate examined in human chondrocytes. It caused increment in the levels of aggrecans, glycosaminoglycans, and collagen X in the chondrocytes (Sawatjui et al. 2018). In another investigation, the collagen/silk fibroin scaffolds stimulated the chondrogenic development in adipose-derived stem cells under chondrogenic culture conditions. Collagen/SF scaffolds were successful in promoting osteochondral regeneration in New Zealand rabbits (Yu et al. 2023). Moreover, an antioxidant scaffold was made by mixing kartogenin-loaded liposomes with natural silk fibroin. It showed good biocompatibility in terms of more chondrocyte migration, adhesion, and proliferation. This scaffold stimulated cartilage matrix production by the induction of the nuclear factor erythroid-related factor 2/heme oxygenase-1 antioxidant signaling (Pan et al. 2023).

3.4.7 Other Emerging Natural Product-Based Scaffold

The hunt for novel natural product-based scaffolds is still a preferred goal in tissue engineering. Icariin, a major bioactive flavonoid found in the *Epimedium* plant, has the ability to promote cell proliferation, chondrocyte differentiation, and the differentiation of embryonic stem cells into cardiomyocytes. Moreover it has been noticed that icariin fusion into hydrogel scaffolds caused improvement in the collagen matrix and proteoglycan secretion in bone and cartilage tissue engineering (Seyedi et al. 2023). In addition to it, cellulose-based materials also displayed excellent biocompatibility, mechanical properties, high optical transparency, water absorption, and water retention properties which make it suitable for tissue engineering (Abdelhamid and Mathew 2022). Moreover, cellulose scaffolds were found to be effective material for the development and differentiation of 3D nerve cells (Hickey and Pelling 2019). Likewise, 3D nanocomposite scaffolds for cartilage regeneration were prepared by blending cellulose nanofibers with a chitosan/gelatin mixture. This scaffold exhibited excellent cell-cell interactions, high porosity, and moisture stability. As a result, it may be applied to bone tissue engineering to create extracellular matrix, which mimics natural cartilage by retaining moisture in its network of holes (Atiqah and Ansari 2019). Haider et al. (2023) reported the lignin nanoparticles/polycaprolactone nanofiber scaffold which showed nontoxicity to pre-osteoblastic cell line and displayed excellent cell proliferation and adhesive properties. This scaffold also caused improvement in the differentiation of osteoblasts which make it a useful biomaterial for bone tissue engineering.

3.5 Conclusion and Future Perspectives

Natural biomaterials have exceptional characteristics like excellent porosity, cell adhesion, biodegradability, affordability, and nontoxicity. Therefore, there is a rapid increase in research on creating scaffolds based on natural products for tissue regeneration applications. However, natural product-based scaffolds lacks mechanical strength and possess low immunogenicity, thus they are mixed with other polymers to address the primary shortcomings of both types of polymers and enhance their efficiency in the repair of tissues. Furthermore, addition of nanoparticles might enhance the scaffold's biological qualities when ingested by the body. Thus application of polymeric scaffolds in tissue engineering is constantly evolving in order to maximize their qualities and ultimately accelerate the healing process. Numerous biopolymers with natural origins, including hyaluronic acid, collagen, chitosan, and gelatin, are being thoroughly investigated for their potential as a functional substitute in regenerative medicine. More clinical studies of nanocomposite biopolymers should be prioritized in the future development of biopolymeric scaffolds and their composites in biomedical domains. Furthermore, additional research is needed to find the biopolymers for the synthesis of novel scaffolds that are apt for tissue engineering applications.

References

Abatangelo G, Vindigni V, Avruscio G, Pandis L, Brun P (2020) Hyaluronic acid: redefining its role. Cells 9(7):1743

Abdelhamid HN, Mathew AP (2022) Cellulose-based nanomaterials advance biomedicine: a review. Int J Mol Sci 23(10):5405

Abou-Okeil A, Fahmy HM, El-Bisi MK, Ahmed-Farid OA (2018) Hyaluronic acid/Na-alginate films as topical bioactive wound dressings. Eur Polym J 109:101–109

Ahmad F, Mushtaq B, Ahmad S, Rasheed A, Nawab Y (2023) A novel composite of hemp fiber and alginate hydrogel for wound dressings. J Polym Environ 31(6):2294–2305

Akter F (2016) Tissue engineering made easy. Academic Press, London

Aldhaher A, Shahabipour F, Shaito A, Al-Assaf S, Elnour AA, Teimourtash S, Elfadil AA (2023) 3D hydrogel/bioactive glass scaffolds in bone tissue engineering: status and future opportunities. Heliyon 9:e17050

Animut TY, Ningsih HS, Shih HH, Wu MH, Shih SJ (2023) Effect of calcium silicate and β-tricalcium phosphate reinforcement on the mechanical–biological properties of freeze-dried collagen composite scaffolds for bone tissue engineering applications. Ceramics 6(1):548–560

Asadpour S, Kargozar S, Moradi L, Ai A, Nosrati H, Ai J (2020) Natural biomacromolecule based composite scaffolds from silk fibroin, gelatin and chitosan toward tissue engineering applications. Int J Biol Macromol 154:1285–1294

Atiqah A, Ansari MN (2019) Nanostructure–polymer composites for soft-tissue engineering. In: Nanostructured polymer composites for biomedical applications. Elsevier, pp 105–115

Axpe E, Oyen ML (2016) Applications of alginate-based bioinks in 3D bioprinting. Int J Mol Sci 17(12):1976

Banafati Zadeh F, Zamanian A (2022) Glutaraldehyde: introducing optimum condition for cross-linking the chitosan/gelatin scaffolds for bone tissue engineering. Int J Eng 35(10):1967–1980

Basurto IM, Mora MT, Gardner GM, Christ GJ, Caliari SR (2021) Aligned and electrically conductive 3D collagen scaffolds for skeletal muscle tissue engineering. Biomater Sci 9(11):4040–4053

Bayrak E, Yilgor Huri P (2018) Engineering musculoskeletal tissue interfaces. Front Mater 5:24

Bello AB, Kim D, Kim D, Park H, Lee S (2020) Engineering and functionalization of gelatin biomaterials: from cell culture to medical applications. Tissue Eng Part B Rev 26:164–180

Bharadwaz A, Jayasuriya AC (2020) Recent trends in the application of widely used natural and synthetic polymer nanocomposites in bone tissue regeneration. Mater Sci Eng C 110:110698

Bhushan S, Singh S, Maiti TK, Das A, Barui A, Chaudhari LR, Joshi MG, Dutt D (2023) Cerium oxide nanoparticles disseminated chitosan gelatin scaffold for bone tissue engineering applications. Int J Biol Macromol 236:123813

Bîrcă AC, Gherasim O, Niculescu AG, Grumezescu AM, Neacşu IA, Chircov C, Vasile BŞ, Oprea OC, Andronescu E, Stan MS, Curuţiu C (2023) A microfluidic approach for synthesis of silver nanoparticles as a potential antimicrobial agent in alginate–hyaluronic acid-based wound dressings. Int J Mol Sci 24(14):11466

Biswal A, Purohit SS, Swain SK (2023) Chitosan based composite scaffolds in skin wound repair: a review. J Drug Deliv Sci Technol 9:104549

Bosio VE, Rybner C, Kaplan DL (2023) Concentric-mineralized hybrid silk-based scaffolds for bone tissue engineering in vitro models. J Mater Chem B 11(33):7998–8006

Chan BP, Leong KW (2008) Scaffolding in tissue engineering: general approaches and tissue-specific considerations. Eur Spine J 17:467–479

Chen D, Qu Y, Hua X, Zhang L, Liu Z, Pflugfelder SC, Li DQ (2017) A hyaluronan hydrogel scaffold-based xeno-free culture system for ex vivo expansion of human corneal epithelial stem cells. Eye 31(6):962–971

Chocholata P, Kulda V, Babuska V (2019) Fabrication of scaffolds for bone-tissue regeneration. Materials 12(4):568

Collins MN, Birkinshaw C (2013) Hyaluronic acid based scaffolds for tissue engineering—a review. Carbohydr Polym 92(2):1262–1279

Correia CR, Moreira-Teixeira LS, Moroni L, Reis RL, van Blitterswijk CA, Karperien M, Mano JF (2011) Chitosan scaffolds containing hyaluronic acid for cartilage tissue engineering. Tissue Eng Part C Methods 17(7):717–730

Dehghan M, Mehrizi MK, Nikukar H (2021) Modeling and optimizing a polycaprolactone/gelatin/polydimethylsiloxane nanofiber scaffold for tissue engineering: using response surface methodology. J Text Inst 112(3):482–493

Ding X, Huang Y, Li X, Liu S, Tian F, Niu X, Chu Z, Chen D, Liu H, Fan Y (2021) Three-dimensional silk fibroin scaffolds incorporated with graphene for bone regeneration. J Biomed Mater Res A 109(4):515–523

Doostan M, Doostan M, Mohammadi P, Khoshnevisan K, Maleki H (2023) Wound healing promotion by flaxseed extract-loaded polyvinyl alcohol/chitosan nanofibrous scaffolds. Int J Biol Macromol 228:506–516

Eskandani M, Derakhshankhah H, Jahanban-Esfahlan R, Jaymand M (2023) Biomimetic alginate-based electroconductive nanofibrous scaffolds for bone tissue engineering application. Int J Biol Macromol 249:125991

Eslami M, Vrana NE, Zorlutuna P, Sant S, Jung S, Masoumi N, Khavari-Nejad RA, Javadi G, Khademhosseini A (2014) Fiber-reinforced hydrogel scaffolds for heart valve tissue engineering. J Biomater Appl 29(3):399–410

Espandar L, Bunnell B, Wang GY, Gregory P, McBride C, Moshirfar M (2012) Adipose-derived stem cells on hyaluronic acid–derived scaffold: a new horizon in bioengineered cornea. Arch Ophthalmol 130(2):202–208

Fagerholm P, Lagali NS, Ong JA, Merrett K, Jackson WB, Polarek JW, Suuronen EJ, Liu Y, Brunette I, Griffith M (2014) Stable corneal regeneration four years after implantation of a cell-free recombinant human collagen scaffold. Biomaterials 35(8):2420–2427

Feketshane Z, Adeyemi SA, Ubanako P, Ndinteh DT, Ray SS, Choonara YE, Aderibigbe BA (2023) Dissolvable sodium alginate-based antibacterial wound dressing patches: design, characterization, and *in vitro* biological studies. Int J Biol Macromol 232:123460

Furlani F, Campodoni E, Sangiorgi N, Montesi M, Sanson A, Sandri M, Panseri S (2023) Electroconductive scaffolds based on gelatin and PEDOT: PSS for cardiac regeneration. Int J Biol Macromol 224:266–280

Gambari L, Amore E, Raggio R, Bonani W, Barone M, Lisignoli G, Grigolo B, Motta A, Grassi F (2019) Hydrogen sulfide-releasing silk fibroin scaffold for bone tissue engineering. Mater Sci Eng C 102:471–482

Gautam S, Sharma C, Purohit SD, Singh H, Dinda AK, Potdar PD, Chou CF, Mishra NC (2021) Gelatin-polycaprolactone-nanohydroxyapatite electrospun nanocomposite scaffold for bone tissue engineering. Mater Sci Eng C 119:111588

Ghalei S, Nourmohammadi J, Solouk A, Mirzadeh H (2018) Enhanced cellular response elicited by addition of amniotic fluid to alginate hydrogel-electrospun silk fibroin fibers for potential wound dressing application. Colloids Surf B: Biointerfaces 172:82–89

Gholijani A, Tavanafar S, Zareifard N, Vojdani Z, Namavar MR, Emami A, Talaei-Khozani T (2022) *In situ* casting of platelet rich plasma/SiO_2/alginate for bone tissue engineering application in rabbit mandible defect model. J Dent 23(2 Suppl):349

Ghorbani F, Kim M, Monavari M, Ghalandari B, Boccaccini AR (2022) Mussel-inspired polydopamine decorated alginate dialdehyde-gelatin 3D printed scaffolds for bone tissue engineering application. Front Bioeng Biotechnol 10:940070

Giwa S, Lewis JK, Alvarez L, Langer R, Roth AE, Church GM, Markmann JF, Sachs DH, Chandraker A, Wertheim JA, Rothblatt M (2017) The promise of organ and tissue preservation to transform medicine. Nat Biotechnol 35(6):530–542

Grabska-Zielińska S, Sionkowska A, Carvalho Â, Monteiro FJ (2021) Biomaterials with potential use in bone tissue regeneration—collagen/chitosan/silk fibroin scaffolds cross-linked by EDC/NHS. Materials 14(5):1105

Gurumurthy B, Janorkar AV (2021) Improvements in mechanical properties of collagen-based scaffolds for tissue engineering. Curr Opin Biomed Eng 17:100253
Haider MK, Kharaghani D, Sun L, Ullah S, Sarwar MN, Ullah A, Khatri M, Yoshiko Y, Gopiraman M, Kim IS (2023) Synthesized bioactive lignin nanoparticles/polycaprolactone nanofibers: a novel nanobiocomposite for bone tissue engineering. Biomater Adv 144:213203
He L, Yin J, Gao X (2023) Additive manufacturing of bioactive glass and its polymer composites as bone tissue engineering scaffolds: a review. Bioengineering, 10(6):672
Hemshekhar M, Thushara RM, Chandranayaka S, Sherman LS, Kemparaju K, Girish KS (2016) Emerging roles of hyaluronic acid bioscaffolds in tissue engineering and regenerative medicine. Int J Biol Macromol 86:917–928
Heydari S, Esmaeili A (2022) Synthesize of polyurethane/chitosan/Vicia ervilia protein/gelatin/heparin-coated Astragalus gossypinus scaffold for cardiovascular tissue engineering. Process Biochem 122:205–213
Hickey RJ, Pelling AE (2019) Cellulose biomaterials for tissue engineering. Front Bioeng Biotechnol 22(7):45
Hoque ME, Nuge T, Yeow TK, Nordin N, Prasad RG (2015) Gelatin based scaffolds for tissue engineering—a review. Polym Res J 9(1):15
Iglesias-Mejuto A, García-González CA (2021) 3D-printed alginate-hydroxyapatite aerogel scaffolds for bone tissue engineering. Mater Sci Eng C 131:112525
Jalali Monfared M, Nasirinezhad F, Ebrahimi-Barough S, Hasanzade G, Saberi H, Tavangar SM, Asadpour S, Aryan L, Barabadi Z, Ai J (2019) Transplantation of miR-219 overexpressed human endometrial stem cells encapsulated in fibrin hydrogel in spinal cord injury. J Cell Physiol 234(10):18887–18896
Jiang JP, Liu XY, Zhao F, Zhu X, Li XY, Niu XG, Yao ZT, Dai C, Xu HY, Ma K, Chen XY (2020) Three-dimensional bioprinting collagen/silk fibroin scaffold combined with neural stem cells promotes nerve regeneration after spinal cord injury. Neural Regen Res 15(5):959
Joanne P, Kitsara M, Boitard SE, Naemetalla H, Vanneaux V, Pernot M, Larghero J, Forest P, Chen Y, Menasché P, Agbulut O (2016) Nanofibrous clinical-grade collagen scaffolds seeded with human cardiomyocytes induces cardiac remodeling in dilated cardiomyopathy. Biomaterials 80:157–168
Kalidas S, Sumathi S (2023) Mechanical, biocompatibility and antibacterial studies of gelatin/polyvinyl alcohol/silk fibre polymeric scaffold for bone tissue engineering. Heliyon 9:e16886
Kara A, Distler T, Polley C, Schneidereit D, Seitz H, Friedrich O, Tihminlioglu F, Boccaccini AR (2022) 3D printed gelatin/decellularized bone composite scaffolds for bone tissue engineering: fabrication, characterization and cytocompatibility study. Mater Today Bio 15:100309
Ketabat F, Maris T, Duan X, Yazdanpanah Z, Kelly ME, Badea I, Chen X (2023) Optimization of 3D printing and in vitro characterization of alginate/gelatin lattice and angular scaffolds for potential cardiac tissue engineering. Front Bioeng Biotechnol 11:1161804
Khan MU, Abd Razak SI, Rehman S, Hasan A, Qureshi S, Stojanović GM (2022a) Bioactive scaffold (sodium alginate)-g-(nHAp@SiO_2@GO) for bone tissue engineering. Int J Biol Macromol 222:462–472
Khan I, Haneef K, Malick TS, Naeem N, Ahmad W, Salim A, Mohsin S (2022b) Combination of mesenchymal stem cells and three-dimensional collagen scaffold preserves ventricular remodeling in rat myocardial infarction model. World J Stem Cells 14(8):633
Kim SC, Heo SY, Oh GW, Yi M, Jung WK (2022) A 3D-printed polycaprolactone/marine collagen scaffold reinforced with carbonated hydroxyapatite from fish bones for bone regeneration. Mar Drugs 20(6):344
Kobayashi T, Chanmee T, Itano N (2020) Hyaluronan: metabolism and function. Biomol Ther 10(11):1525
Kong B, Mi S (2016) Electrospun scaffolds for corneal tissue engineering: a review. Materials 9(8):614
Krishani M, Shin WY, Suhaimi H, Sambudi NS (2023) Development of scaffolds from bio-based natural materials for tissue regeneration applications: a review. Gels 9(2):100

Kumar M, Saini JS (2023) Computational analysis of bidirectional carbon fibre/epoxy nanocomposites using different failure criteria. Proc Inst Mech Eng E J Process Mech Eng 237(4):1299

Kumar M, Saini JS, Bhunia H (2020) Performance of mechanical joints prepared from carbon-fiber-reinforced polymer nanocomposites under accelerated environmental aging. J Mater Eng Perform 29:7511

Kumar M, Saini JS, Bhunia H (2021a) Investigations on MWCNT embedded carbon/epoxy composite joints subjected to hygrothermal aging under bolt preloads. Fibers Polym 22(7):1975

Kumar M, Saini JS, Bhunia H, Chowdhury SR (2021b) Behavior of mechanical joints prepared from EB cured CFRP nanocomposites subjected to hygrothermal aging under bolt preloads. Appl Compos Mater 28:296

Kumar M, Saini JS, Bhunia H, Chowdhury SR (2021c) Aging of bolted joints prepared from electron-beam-cured multiwalled carbon nanotube-based nanocomposites with variable torques. Polym Compos 42(8):4104

Lauritano D, Limongelli L, Moreo G, Favia G, Carinci F (2020) Nanomaterials for periodontal tissue engineering: chitosan-based scaffolds. A systematic review. Nanomaterials 10(4):605

Lei Y, Bortolin L, Benesch-Lee F, Oguntolu T, Dong Z, Bondah N, Billiar K (2021) Hyaluronic acid regulates heart valve interstitial cell contraction in fibrin-based scaffolds. Acta Biomater 136:124–136

Li G, Sun S (2022) Silk fibroin-based biomaterials for tissue engineering applications. Molecules 27(9):2757

Li C, Vepari C, Jin HJ, Kim HJ, Kaplan DL (2006) Electrospun silk-BMP-2 scaffolds for bone tissue engineering. Biomaterials 27(16):3115–3124

Li XY, Deng WS, Wang ZQ, Li ZC, Chen SL, Song Z, Zhang Q, Liang J, Chen XY (2023a) Injectable collagen scaffold with human umbilical cord-derived mesenchymal stem cells promotes functional recovery in patients with spontaneous intracerebral hemorrhage: phase I clinical trial. Neural Regen Res 18(9):1999

Li Z, Man J, Wang X, Zhang Y (2023b) Biodegradable sodium alginate/carrageenan/cellulose composite hydrogel wound dressings containing herbal extracts for promoting blood coagulation and wound healing. Macromol Biosci 23(1):2200235

Li Z, Xu P, Shang L, Ma B, Zhang H, Fu L, Ou Y, Mao Y (2023c) 3D collagen porous scaffold carrying PLGA-PTX/SDF-1α recruits and promotes neural stem cell differentiation for spinal cord injury repair. J Biomater Sci Polym Ed 11:1–24

Linju MC, Rekha MR (2023) Proline conjugated chitosan as wound healing material: in vitro studies on the influence of the scaffold on collagen production and wound healing. Int J Biol Macromol 242:124688

Liu Y, Liu H, Zhang C, Kong Y, Chen C, Gao W, Xi X, Deng L (2023) Preparation and investigation of a novel antibacterial collagen-based material loaded with gentamicin following surface modification with citric acid for corneal tissue engineering. Int J Biol Macromol 253:126791

Martí M, Frígols B, Salesa B, Serrano-Aroca Á (2019) Calcium alginate/graphene oxide films: reinforced composites able to prevent Staphylococcus aureus and methicillin-resistant Staphylococcus epidermidis infections with no cytotoxicity for human keratinocyte HaCaT cells. Eur Polym J 110:14–21

Meinel L, Fajardo R, Hofmann S, Langer R, Chen J, Snyder B, Vunjak-Novakovic G, Kaplan D (2005) Silk implants for the healing of critical size bone defects. Bone 37(5):688–698

Monteiro RT, Da Silva TF, Moreira Filho RF, Vasconcelos NF, Nogueira KB, Petrilli R, Andrade FK, Vieira RS (2023) Simvastatin-loaded alginate bilayer membranes for wound dressing applications. J Drug Deliv Sci Technol 22:104701

Motiee ES, Karbasi S, Bidram E, Sheikholeslam M (2023) Investigation of physical, mechanical and biological properties of polyhydroxybutyrate-chitosan/graphene oxide nanocomposite scaffolds for bone tissue engineering applications. Int J Biol Macromol 247:125593

Nagiah N, El Khoury R, Othman MH, Akimoto J, Ito Y, Roberson DA, Joddar B (2022) Development and characterization of furfuryl-gelatin electrospun scaffolds for cardiac tissue engineering. ACS Omega 7(16):13894–13905

Nikolova MP, Chavali MS (2019) Recent advances in biomaterials for 3D scaffolds: a review. Bioact Mater 4:271–292

Ojeda-Hernández DD, Canales-Aguirre AA, Matias-Guiu J, Gomez-Pinedo U, Mateos-Díaz JC (2020) Potential of chitosan and its derivatives for biomedical applications in the central nervous system. Front Bioeng Biotechnol 8:389

Özdemir E, Familiari F, Huri PY, Firat A, Huri G (2023) Use of 3D-printed polycaprolactone+ hyaluronic acid-based scaffold in orthopedic practice: report of two cases. J 3D Print Med 7(1):3DP001

Pahlevanzadeh F, Emadi R, Valiani A, Kharaziha M, Poursamar SA, Bakhsheshi-Rad HR, Ismail AF, RamaKrishna S, Berto F (2020) Three-dimensional printing constructs based on the chitosan for tissue regeneration: state of the art, developing directions and prospect trends. Materials 13(11):2663

Pan Z, Hou M, Zhang Y, Liu Y, Tian X, Hu X, Ge X, Zhao Z, Liu T, Xu Y, Yang H (2023) Incorporation of kartogenin and silk fibroin scaffolds promotes rat articular cartilage regeneration through enhancement of antioxidant functions. Regen Biomater 10:074

Perić Kačarević Ž, Rider P, Alkildani S, Retnasingh S, Pejakić M, Schnettler R, Gosau M, Smeets R, Jung O, Barbeck M (2020) An introduction to bone tissue engineering. Int J Artif Organs 43(2):69–86

Phillips DM, Drummy LF, Conrady DG, Fox DM, Naik RR, Stone MO, Trulove PC, De Long HC, Mantz RA (2004) Dissolution and regeneration of *Bombyx mori* silk fibroin using ionic liquids. J Am Chem Soc 126(44):14350–14351

Pooshidani Y, Zoghi N, Rajabi M, Haghbin Nazarpak M, Hassannejad Z (2021) Fabrication and evaluation of porous and conductive nanofibrous scaffolds for nerve tissue engineering. J Mater Sci Mater Med 32:1–2

Popov A, Malferrari S, Kalaskar DM (2017) 3D bioprinting for musculoskeletal applications. J 3D Print Med 1(3):191–211

Puccetti M, Donnadio A, Ricci M, Latterini L, Quaglia G, Pietrella D, Di Michele A, Ambrogi V (2023) Alginate Ag/AgCl nanoparticles composite films for wound dressings with antibiofilm and antimicrobial activities. J Funct Biomater 14(2):84

Qian L, Yang K, Liu X, Zhang L, Zhao H, Qiu LZ, Chu Y, Hao W, Zhuang Y, Chen Y, Dai J (2023) Baicalein-functionalized collagen scaffolds direct neuronal differentiation toward enhancing spinal cord injury repair. Biomater Sci 11(2):678–689

Rahmani Del Bakhshayesh A, Mostafavi E, Alizadeh E, Asadi N, Akbarzadeh A, Davaran S (2018) Fabrication of three-dimensional scaffolds based on nano-biomimetic collagen hybrid constructs for skin tissue engineering. ACS Omega 3(8):8605–8611

Rajput M, Nilawar S, Chatterjee K (2023) Embedding silk fibroin-alginate hydrogel in a 3D-printed porous poly (lactic acid) bone tissue scaffold augments stem cell function. Regen Eng Transl Med 26:1–3

Rashid AB, Showva NN, Hoque ME (2023) Gelatin-based scaffolds—an intuitive support structure for regenerative therapy. Curr Opin Biomed Eng 2:100452

Ressler A, Ohlsbom R, Žužić A, Gebraad A, Frankberg EJ, Pakarinen TK, Ivanković H, Miettinen S, Ivanković M (2023) Chitosan/collagen/Mg, Se, Sr, Zn-substituted calcium phosphate scaffolds for bone tissue engineering applications: a growth factor free approach. Eur Polym J 194:112129

Roberts TT, Rosenbaum AJ (2012) Bone grafts, bone substitutes and orthobiologics: the bridge between basic science and clinical advancements in fracture healing. Organogenesis 8(4):114–124

Roseti L, Parisi V, Petretta M, Cavallo C, Desando G, Bartolotti I, Grigolo B (2017) Scaffolds for bone tissue engineering: state of the art and new perspectives. Mater Sci Eng 78:1246–1262

Sattary M, Rafienia M, Kazemi M, Salehi H, Mahmoudzadeh M (2019) Promoting effect of nano hydroxyapatite and vitamin D3 on the osteogenic differentiation of human adipose-derived stem cells in polycaprolactone/gelatin scaffold for bone tissue engineering. Mater Sci Eng C 97:141–155

Sawatjui N, Limpaiboon T, Schrobback K, Klein T (2018) Biomimetic scaffolds and dynamic compression enhance the properties of chondrocyte-and MSC-based tissue-engineered cartilage. J Tissue Eng Regen Med 12(5):1220–1229
Serafin A, Murphy C, Rubio MC, Collins MN (2021) Printable alginate/gelatin hydrogel reinforced with carbon nanofibers as electrically conductive scaffolds for tissue engineering. Mater Sci Eng C 122:111927
Seyedi Z, Amiri MS, Mohammadzadeh V, Hashemzadeh A, Haddad-Mashadrizeh A, Mashreghi M, Qayoomian M, Hashemzadeh MR, Simal-Gandara J, Taghavizadeh Yazdi ME (2023) Icariin: a promising natural product in biomedicine and tissue engineering. J Funct Biomater 14(1):44
Sonaye SY, Ertugral EG, Kothapalli CR, Sikder P (2022) Extrusion 3D (bio)printing of alginate-gelatin-based composite scaffolds for skeletal muscle tissue engineering. Materials 15(22):7945
Sun M, Puri S, Mutoji KN, Coulson-Thomas YM, Hascall VC, Jackson DG, Gesteira TF, Coulson-Thomas VJ (2019) Hyaluronan derived from the limbus is a key regulator of corneal lymphangiogenesis. Invest Ophthalmol Vis Sci 60(4):1050–1062
Szychlinska MA, Calabrese G, Ravalli S, Dolcimascolo A, Castrogiovanni P, Fabbi C, Puglisi C, Lauretta G, Di Rosa M, Castorina A, Parenti R (2020) Evaluation of a cell-free collagen type I-based scaffold for articular cartilage regeneration in an orthotopic rat model. Materials 13(10):2369
Tan Y, Fan S, Wu X, Liu M, Dai T, Liu C, Ni S, Wang J, Yuan X, Zhao H, Weng Y (2023) Fabrication of a three-dimensional printed gelatin/sodium alginate/nano-attapulgite composite polymer scaffold loaded with leonurine hydrochloride and its effects on osteogenesis and vascularization. Int J Biol Macromol 249:126028
Varaprasad K, Jayaramudu T, Kanikireddy V, Toro C, Sadiku ER (2020) Alginate-based composite materials for wound dressing application: a mini review. Carbohydr Polym 236:116025
Verma P, Verma V (2020) Concepts of tissue engineering. In: Animal biotechnology. Academic Press, Cambridge, MA, pp 295–307
Wang R, Deng S, Wu Y, Wei H, Jing G, Zhang B, Liu F, Tian H, Chen X, Tian W (2022) Remodelling 3D printed GelMA-HA corneal scaffolds by cornea stromal cells. Colloid Interface Sci Commun 49:100632
Wang Y, Fan Z, Li Q, Lu J, Wang X, Zhang J, Wu Z (2023a) Construction of a myocardial patch with mesenchymal stem cells and poly (CL-co-TOSUO)/collagen scaffolds for myocardial infarction repair by coaxial electrospinning. J Mater Chem B 11(22):4980–4990
Wang T, Yi W, Zhang Y, Wu H, Fan H, Zhao J, Wang S (2023b) Sodium alginate hydrogel containing platelet-rich plasma for wound healing. Colloids Surf B: Biointerfaces 222:113096
Wang S, Zhang Y, Shi Y, He Q, Tan Q, Peng Z, Liu Y, Li D, Li X, Ke D, Wang J (2023c) Rhubarb charcoal-crosslinked chitosan/silk fibroin sponge scaffold with efficient hemostasis, inflammation, and angiogenesis for promoting diabetic wound healing. Int J Biol Macromol 253:126796
Wang Y, Zhu L, Wei L, Zhou Y, Yang Y, Zhang L (2023d) A bio-orthogonally functionalized chitosan scaffold with esterase-activatable release for nerve regeneration. Int J Biol Macromol 229:146–157
Watcharajittanont N, Tabrizian M, Ekarattanawong S, Meesane J (2023) Bone-mimicking scaffold based on silk fibroin incorporated with hydroxyapatite and titanium oxide as enhanced osteoconductive material for bone tissue formation: fabrication, characterization, properties, and in vitro testing. Biomed Mater 18(6):065007
Wee JH, Yoo KD, Sim SB, Kim HJ, Kim HJ, Park KN, Kim GH, Moon MH, You SJ, Ha MY, Yang DH (2022) Stem cell laden nano and micro collagen/PLGA bimodal fibrous patches for myocardial regeneration. Biomater Res 26(1):1–7
Wuttisiriboon K, Tippayawat P, Daduang J, Limpaiboon T (2023) Three-dimensional silk fibroin-gelatin/chondroitin sulfate/hyaluronic acid–aloe vera scaffold supports in vitro chondrogenesis of bone marrow mesenchymal stem cells and reduces inflammatory effect. J Biomed Mater Res B Appl Biomater 111:1557

Xiang W, Cao H, Tao H, Jin L, Luo Y, Tao F, Jiang T (2023) Applications of chitosan-based biomaterials: from preparation to spinal cord injury neuroprosthetic treatment. Int J Biol Macromol 26:123447

Xiao L, Wu M, Yan F, Xie Y, Liu Z, Huang H, Yang Z, Yao S, Cai L (2021) A radial 3D polycaprolactone nanofiber scaffold modified by biomineralization and silk fibroin coating promote bone regeneration *in vivo*. Int J Biol Macromol 172:19–29

Xie H, Ruan S, Zhao M, Long J, Ma X, Guo J, Lin X (2023) Preparation and characterization of 3D hydroxyapatite/collagen scaffolds and its application in bone regeneration with bone morphogenetic protein-2. RSC Adv 13(33):23010–23020

Xu W, Kong B, Xie H, Zhang J, Liu W, Liu S, Zhang Y, Yang F, Xiao J, Mi S, Xiong L (2022) PCL scaffold combined with rat tail collagen type I to reduce keratocyte differentiation and prevent corneal stroma fibrosis after injury. Exp Eye Res 217:108936

Xu Z, Li K, Zhou K, Li S, Chen H, Zeng J, Hu R (2023) 3D printing silk fibroin/hydroxyapatite/sodium alginate composite scaffolds for bone tissue engineering. Fibers Polym 24(1):275–283

Yang B, Wang PB, Mu N, Ma K, Wang S, Yang CY, Huang ZB, Lai Y, Feng H, Yin GF (2021) Graphene oxide-composited chitosan scaffold contributes to functional recovery of injured spinal cord in rats. Neural Regen Res 16:1829–1835

Yasin A, Ren Y, Li J, Sheng Y, Cao C, Zhang K (2022) Advances in hyaluronic acid for biomedical applications. Front Bioeng Biotechnol 10:910290

Ye B, Wu B, Su Y, Sun T, Guo X (2022) Recent advances in the application of natural and synthetic polymer-based scaffolds in musculoskeletal regeneration. Polymers 14(21):4566

Yu S, Shu X, Chen L, Wang C, Wang X, Jing J, Yan G, Zhang Y, Wu C (2023) Construction of ultrasonically treated collagen/silk fibroin composite scaffolds to induce cartilage regeneration. Sci Rep 13(1):20168

Zarei M, Samimi A, Khorram M, Abdi MM, Golestaneh SI (2021) Fabrication and characterization of conductive polypyrrole/chitosan/collagen electrospun nanofiber scaffold for tissue engineering application. Int J Biol Macromol 168:175–186

Zargar S, Asefnejad A, Azami M, Farokhi M (2023) Development of multi-layer nanofiber scaffolds for corneal epithelial layer using silk fibroin, collagen, aloe vera, and epithelial growth factor. Iran J Chem Chem Eng. https://doi.org/10.30492/IJCCE.2023.2012457.6239

Zhang J, Sisley AM, Anderson AJ, Taberner AJ, McGhee CN, Patel DV (2016) Characterization of a novel collagen scaffold for corneal tissue engineering. Tissue Eng Part C Methods 22(2):165–172

Zhang Y, Sun T, Jiang C (2018) Biomacromolecules as carriers in drug delivery and tissue engineering. Acta Pharm Sin B 8(1):34–50

Zhang Y, Li C, Zhu Q, Liang R, Xie C, Zhang S, Hong Y, Ouyang H (2021) A long-term retaining molecular coating for corneal regeneration. Bioact Mater 6(12):4447–4454

Zuo Y, Li Q, Xiong Q, Li J, Tang C, Zhang Y, Wang D (2022) Naringin release from a nano-hydroxyapatite/collagen scaffold promotes osteogenesis and bone tissue reconstruction. Polymers 14(16):3260

Natural Gum (Xanthan, Gellan, Arabic, Guar, Ghatti Gum, Etc.)-Based Bio-Scaffold and Their Application in Tissue Engineering

4

Swati Gajbhiye, Sanjay Dhoble, and Desmond Tobin

Abstract

Recent progress in tissue engineering (TE) has highlighted the importance of natural gums, including xanthan, gellan, Arabic, guar and ghatti gum, as promising bio-scaffold materials. This chapter offers a comprehensive exploration of fundamental concepts in TE, emphasising the critical role of natural gum bio-scaffolds in guiding cell growth and tissue regeneration. Each natural gum is examined for its molecular composition, physical properties and interactions with cells, with case examples showcasing successful applications in bone, cartilage, skin and vascular tissue regeneration. Challenges, such as controlled scaffold degradation and immune response, are addressed, and future research directions for optimising natural gum-based bio-scaffolds are discussed. Integrating natural gums into tissue engineering presents an innovative approach for enhanced biocompatibility and tailored mechanical properties, contributing to the advancement of regenerative medicine.

Keywords

Bio-scaffold · Xanthan · Gellan · Arabic · Guar · Ghatti gum

The original version of the chapter has been revised, the affiliation of the author Desmond Tobin has been updated. A correction to this chapter can be found at https://doi.org/10.1007/978-981-97-3111-4_12

S. Gajbhiye (✉) · S. Dhoble
Department of Physics, R.T.M. Nagpur University, Nagpur, India

D. Tobin
The Charles Institute of Dermatology, School of Medicine, The Conway Institute, University College Dublin, Belfield, Dublin 4, Ireland

M. Kumar et al. (eds.), *Natural Product Inspired Scaffolds*,
https://doi.org/10.1007/978-981-97-3111-4_4

4.1 Introduction

Professor Robert Nerem introduced the term 'tissue engineering' (TE) in 1988 where he delivered a talk on *Molecular and Cellular Biology* at symposia in UCLA. In his interpretation, he characterised TE as the convergence of both life sciences and engineering to explore the functional and structural aspects of natural and pathologic mammalian tissues. Moreover, TE involves the development of biological substitutes aimed at restoring, maintaining or enhancing the damaged tissue or lost due to different circumstances such as diseases, injuries or congenital disabilities. In essence, TE is dedicated to the improvement, rejuvenation and upkeep of compromised tissues induced by diverse factors (Eltom et al. 2019). TE focuses on restoring normal biological functions by replacing or regenerating specific organs, tissues and cells of the body (Khillar and Jaiswal 2022). The effectiveness of TE relies on three key components: the scaffold or cell support system, the source and type of cells and the cell signalling factors. Among these, the scaffold plays a pivotal role in TE success. This scaffold not only offers a framework for promoting cellular expansion but also facilitates the delivery of essential cellular nutrients and signalling molecules necessary for precise cellular differentiation toward specific tissues (Okamoto 2019). Further, TE aims at regenerating damaged tissues or reconstructing organs by creating a suitable three-dimensional (3D) microenvironment that fosters cell adhesion, proliferation and tissue formation. This environment is typically established using biomaterial structures designed to imitate the supportive architecture of the natural extracellular matrix (ECM). These biomaterial structures, which can take various forms like microspheres, nanofibrous films, micro-/nanoparticles, microbeads, foams or hydrogels (hydrogel is a three-dimensional network of hydrophilic polymers that provides a supportive environment for cell growth and tissue regeneration), rely on different polymers (natural, synthetic or their blends) depending on the targeted tissue application. Among these forms, hydrogels are favoured for providing a conducive 3D environment. Natural polymers possess unique advantages compared to synthetic counterparts for crafting hydrogels suitable for cell differentiation and regeneration of tissues. These advantages stem from their abundant availability, exceptional biocompatibility, straightforward processing and closely mimicking the natural extracellular matrix of tissues in a biological context (Kumar et al. 2018).

Over the past decade, natural gums (NGs) have attracted significant attention in the context of their versatility in developing polymeric systems across multiple sectors like food, pharmaceuticals, cosmetics and biomedical fields. Among these, TE stands out as a prominent area where NGs have been extensively explored. Their utilisation in TE research has introduced novel possibilities and avenues in this domain (Pal et al. 2023). Polysaccharides, found abundantly in nature, derive from diverse sources such as algae (e.g. alginate), microbes (e.g. dextran, xanthan gum), plants (e.g. pectin, guar gum) and animals (like chitosan). Among these, natural polysaccharides showcase remarkable properties that have led to their extensive utilisation in various biomedical applications. They exhibit nontoxic characteristics, possess high swelling capacity through simple chemical modifications, maintain stability across wide pH ranges and offer a diverse array of chemical structures. This versatility of natural polysaccharides allows them to offset specific constraints like low mechanical, thermal and chemical stability. The mechanical characteristics of scaffolds are crucial in TE. The amalgamation of various materials, including natural and synthetic resins, along with polymers, enhances strength, facilitates cell adhesion and ensures compatibility with surrounding tissues. Attaining high tensile strength necessitates a strong affinity between the copolymer and the scaffold, while robust interfacial bonding improves compressive strength and load transfer efficiency (Prasadh and Wong 2018). Moreover, their susceptibility to microbial and enzymatic degradation, in some contexts, can even be viewed as advantageous (Nayak and Pal 2012). These NG polysaccharides display various characteristics like biodegradability, non-toxicity and biocompatibility, rendering them promising biomaterials in numerous biomedical applications. Their advantages over semisynthetic and synthetic polymers are evident. However, natural polymers do present challenges, including potential risks of pathogen transfer and variations in porosity. Despite these challenges, polysaccharide polymers are utilised as scaffolds due to their porosity and extensive surface area, serving as substitutes for the extracellular matrix (ECM). These materials offer robust support for cellular growth, directing cell regeneration through their surface properties and pronounced geometric structure (Koyyada and Orsu 2021). Such scaffold fabrication for tissue engineering generally involves two primary approaches: the top-down and bottom-up methods. In the top-down approach, the initial step involves constructing scaffolds, followed by the introduction of cells into their microstructures. Techniques employed for crafting three-dimensional (3D) bio-scaffolds through the top-down approach encompass electro-spinning, freeze-drying, phase separation and self-assembly. On the other hand, the bottom-up approach concentrates on creating microscale tissue building blocks with precise microarchitecture, subsequently assembling these units to form larger tissue structures. Various methods for producing these building blocks include generating cell sheets, encapsulating cells in hydrogels, forming self-assembled cell aggregates and utilising bioprinting techniques (Indurkar et al. 2021). Thus, the potential for creating personalised living tissues and organs has significantly improved due to the progress in 3D bioprinting methods. Despite the presence of flexible regulations, the licensing of tissue-engineered medical products (TEMPs) remains limited. Regulatory authorities and stakeholders must understand the intricacies of these innovative products, facilitating a streamlined approval process for both 3D bioprinted tissues and TEMPs. Additionally, clinicians and researchers must grasp the scientific prerequisites, understand standard protocols, undergo essential training in clinical trials and adhere to good manufacturing practices (GMPs). These actions are necessary to meet the requirements set forth by regulatory bodies and global standards (Sekar et al. 2021).

This chapter delves into the multifaceted landscape of NG as bio-scaffold materials, aiming to elucidate their role, characteristics and applications in TE. An in-depth analysis of each NG compound forms the crux of this chapter, unravelling their molecular compositions, unique physical attributes and interactions with cellular environments. Drawing insights from recent studies and real-world applications, this chapter provides a panoramic view of the successful integration of natural gum-based bio-scaffolds in various TE paradigms, encompassing regeneration in the bone, cartilage, skin and vascular tissues. Challenges, considerations and future prospects surrounding the utilisation of natural gums in tissue engineering constitute another focal point of this discourse. In essence, the integration of NG into the realm of tissue engineering heralds a novel paradigm in bio-scaffold design, harnessing their inherent biocompatibility and tailored mechanical properties. The in-depth exploration of xanthan, gellan, Arabic, guar, ghatti gum and their kin (similar substances that are structurally or functionally similar) within this chapter promises a nuanced understanding of their potential contributions toward advancing the frontiers of regenerative medicine.

4.2 Properties of Natural Gums for Scaffold Fabrication

Tissue engineering arises as a promising approach for the rejuvenation of trabecular bone and cartilage, given their limited natural regenerative capabilities following substantial damage. Various biomaterials based on natural gums are currently employed in TE methodologies for the restoration of articular cartilage. These gums exhibit versatility in processing conditions and demonstrate a broad spectrum of applications (Anandan et al. 2019). 'Gum' denotes a group of polysaccharides found in nature that find broad applications in various industries due to their capacity to form gels or create viscous solutions. These natural gums are classified based on their origins, characteristics and chemical structures shown in Fig. 4.1. Gums are complex polysaccharides sourced from diverse origins, such as the endosperm of plant seeds (e.g. guar gum), extracts from seaweeds (e.g. agar), bacteria (e.g. xanthan gum), exudates from trees or shrubs (e.g. Arabic gum, karaya gum (KG), plant exudates (e.g. tragacanth and tragacanth) (these resins and gums, are used in scaffold fabrication as components or additives to modify their properties) and animal sources (e.g. chitosan derived from the deacetylation of chitin) (Mirhosseini and

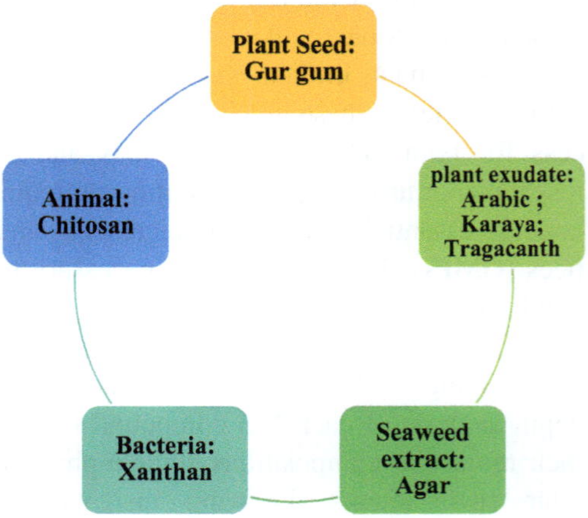

Fig. 4.1 Natural bio-scaffold materials for tissue engineering

Amid 2012). Moreover, naturally occurring polymers are currently highly significant, with polysaccharides holding a significant position in biomaterials research and various industrial applications because of their widespread availability, non-toxic nature and environmentally friendly characteristics. Among polysaccharides, natural gums, belonging to one specific category (hydrocolloids), are extensively utilised to increase solution viscosity which is effective even in small amounts. These lipophilic substances are mainly sourced from plants or microbes. Because of their biological origin, gum molecules exhibit notable variations in linear chain length, molecular weight and branching features. Upon hydrolysis, natural gums yield a mixture of xylose, mannose, arabinose, uronic acids, galactose and other components (Sharma et al. 2018). In tissue engineering, matrices are formulated to facilitate cellular activities, promoting both differentiation and proliferation to generate new tissues. These approaches enable the development of hybrid constructs suitable for implantation, aiming to stimulate tissue regeneration or replace organs facing failure. Various materials, particularly biodegradable polymers, have been proposed for scaffold fabrication. Therefore, these natural-based gum polymers present a notable advantage by closely resembling biological macromolecules, which the biological environment readily recognises and metabolically deals with. Because of their resemblance to the (ECM), natural polymers may also help alleviate the risks of immunological reactions, toxicity and chronic inflammation, commonly associated with synthetic polymers. For instance, studies have shown that natural polymers like chitosan and hyaluronic acid possess biocompatible properties that mimic those of the ECM, thereby reducing the likelihood of adverse reactions. Additionally, the biodegradability and non-toxic nature of natural polymers make them favorable candidates for biomedical applications, offering safer alternatives to synthetic counterparts (Mano et al. 2007; Satchanska et al. 2024).

In a recent study, cryogel scaffolds were created by combining three types of natural biomaterials: xanthan gum (XG), locust bean gum (LBG) and mastic gum (MG) to evaluate the cytocompatibility of these cryogels and their potential applications in polysaccharide-based tissue engineering and drug delivery. The swollen cryogels demonstrated mechanical properties akin to other cryogels which are polysaccharide-based, with average values of tensile and compressive modulus falling within the range of 3.5–11.6 kPa and 82–398 kPa, respectively. The release profile of the small molecule Kartogenin from varying concentrations and ratios of cryogels ranged from 32% to 66% over a 21-day incubation period. Considering their physical, mechanical and chemical attributes, LBG-XG (LX) and LBG-XG-MG (LXM) polysaccharide-based cryogels present themselves as promising candidates for applications in cartilage and other soft tissue engineering, as well as drug delivery (Bektas et al. 2021). Similarly ceramic honeycomb extrusion is a promising technique for producing robust porous ceramics, which provide a three-dimensional framework that supports cell attachment, proliferation, and differentiation, ultimately leading to the formation of new tissue but several challenges impede its full realisation especially in the development of an extrudable paste suitable for ceramic honeycomb extrusion. Overcoming issues such as the intricate design of a honeycomb die and the development of an extrudable paste are crucial as this involves formulating a paste that has the right rheological properties to pass through the extrusion die smoothly while maintaining its shape and structural integrity. Therefore, natural gums play a pivotal role in addressing these challenges, as they are commonly used to enhance the rheological properties of ceramic pastes, modifying the viscosity and flow behavior of ceramic pastes thereby facilitating the extrusion process and enhancing the final properties of the ceramic material. Thus, in one of the studies, guar gum was chosen as a binder for its water-soluble polysaccharide properties, forming a gel network ideal for suspending ceramic particles. Guar gum specifically, was selected due to its hydrogels exhibiting high shear-thinning properties along with strong stiffness properties, essential for effective extrusion. Rheological assessment confirmed the achievement of ceramic pastes with comparable qualities, utilising hydroxyapatite (HA) as the model ceramic. The result was a porous hydroxyapatite (HA) scaffold, exhibiting porosity comparable to that found in a commercial hydroxyapatite graft but possessing higher compressive strength. Importantly, this study highlights the versatile application of guar gum in various ceramic fabrication processes, leveraging its biological and water-soluble properties to improve both rheological and mechanical characteristics (Elbadawi et al. 2017).

Developing an ideal plant-based hydrogel that accurately mimics human tissue properties remains a significant challenge. In response, an innovative multicomponent hydrogel with natural, biocompatible and sustainable attributes was formulated. The formulated hydrogel, formed by combining tragacanth gum cellulose nanofibrils (TOCNF) and lignin nanoparticles, exhibits synergistic effects. Leveraging lignin's antioxidant properties, the rheological features of tragacanth gum and the analgesic properties of TOCNF, the hydrogel is deemed suitable for TE applications. The shear-thinning behaviour of hydrogel attributed to TOCNF, and the adjustable rheological characteristics influenced by tragacanth gum (TG) enable the scaffolds 3D printing with outstanding printability. These scaffolds are characterised with a high swelling ratio, demonstrated an enhanced degradation rate with increasing TG content in the hydrogels. Analysis of the mechanical properties indicates that these hydrogels create soft gels with stiffness levels situated between that of the skin and muscle tissues. This example underscores the successful improvement of rheological properties through the incorporation of natural gums in the hydrogel formulation, showcasing its potential for applications in tissue engineering (Polez et al. 2022).

4.2.1 Xanthan Gum-Based Bio-Scaffolds in Tissue Engineering

Xanthan gum (XG) has lately gained significant interest as a biomaterial in the creation of TE scaffolds or matrices. These scaffolds, based on xanthan gum, offer an optimal setting for cell adhesion and growth due to their tailored surface chemistry, nanotopology and stiffness. For example, due to its tailored surface chemistry, the scaffolds that are made of natural and conducting polymers along with the combination of EMF a more pronounced proliferation of fibroblast can be seen in the electro polymerization of poly pyrrole in xanthan hydrogel which is an electroactive scaffold used for cell proliferation. because of the higher hydrophobicity and large surface roughness these are better scaffolds for fibroblasts which offers a new strategy for cell proliferation (Aavani et al. 2021; Bueno et al. 2015). It stands out as a microbial polysaccharide extensively applied in bone tissue engineering. It presents as a white powder or light yellow under room temperature conditions. Notably, XG possesses commendable water solubility, resistance to acid-base conditions, thermal stability, distinctive rheological properties and biocompatibility. Leveraging these attributes, XG is efficiently integrated into composite materials through interactions with different substances, contributing to its broad use in diverse TE applications, with a particular emphasis on bone TE (Su et al. 2022). This XG is primarily derived from the *Xanthomonas campestris* a gram-negative bacterium, i.e. NRRL B-1459. These bacteria are characterised by their motility and strict aerobic nature, where oxygen serves as an essential nutrient for their growth and the production of XG. The manufacturing process of XG involves a considerable cost, primarily attributed to the use of glucose as the single source of carbon. The polyelectrolyte microstructure of XG is significantly influenced by the ionic strength of the solution. Adjusting the ionic strength can cause the folding down of XG side chains to the backbone, leading to a structured and stable form with an increased glass transition temperature. Additionally, XG features reactive functional groups, including carboxylic and hydroxyl groups, offering possibilities for modification and functionalisation. This improves properties such as its physicochemical nature, broadening its range of applications (Abu Elella et al. 2021). It is categorised as a naturally occurring polysaccharide. Its structural foundation comprises β-D-glucose units linked similarly to cellulose. Specifically, a trisaccharide unit, consisting of β-D-mannose-(1–4)-α-D-glucuronic acid-(1–2)-α-D-mannose, is connected to the $O_{(3)}$ position of every alternate glucose residue. Ketal bonds has a crucial function in connecting residues of pyruvic acid to approximately half of the terminal mannose units. Moreover, the acetate groups are linked to these terminal mannose residues. This distinctive molecular structure contributes to the unique properties and functionalities of XG (Nsengiyumva and Alexandridis 2022). Therefore, XG, produced by Xanthomonas bacteria, boasts excellent water solubility properties and remarkable biocompatibility. Its branched polymeric chains confer upon it the ability to achieve high viscosity at low concentrations, as well as a favourable shear-thinning property. Moreover, it exhibits notable thermal stability, making it a versatile and valuable ingredient in TE applications (Kumar et al. 2018). In an aqueous environment, XG undergoes two distinct conformational changes: at lower temperatures, it adopts an ordered and rigid double helical strand structure, while at higher temperatures, it transforms into a disordered and flexible coil structure. The midpoint transition temperature (Tm) for this conformational shift ranges from 40 to 50 °C, affected by the level of ions of the solution. When the temperature remains below the Tm threshold, the ordered double-helical strand structure facilitates the creation of a 3D network within the XG solution. Consequently, the solution exhibits characteristics akin to a weak gel, displaying a shear-thinning property under stress (Liu and Yao 2015). Injectable hydrogels composed of a blend of XG and methylcellulose exhibit promising characteristics. Upon injection through a syringe, this blend recovers high viscosity and forms a hydrogel at body temperature. This behaviour is attributed to xanthan gum's double helical strand structure, which imparts excellent shear-thinning properties. In rat studies, this hydrogel formulation demonstrates potential for long-term drug delivery applications, and in the near future, it can be used for humans as well (Liu and Yao 2015). In another study, a secondary chemical crosslinking network into XG through the conjugation of dopamine (Dopa) via an amidation reaction was introduced. Initially, the hydrogels exhibited excellent thixotropic properties during the early stage of secondary crosslinking. However, after the formation of the secondary network, the thixotropic behaviour disappeared, but the mechanical strength, swelling ratio, and degradation rate of the hydrogels significantly improved. Both in vitro and in vivo studies confirmed that the XG–Dopa/HRP/H2O2 hydrogel was injectable, biocompatible, and biodegradable. This study demonstrates that a shear-thinning hydrogel network is an effective strategy for maintaining injectability while enhancing stability and mechanical strength (Yao 2015).

Xanthan gum (XG) finds diverse applications in healthcare, serving as a crucial component in the management of dysphagia (difficulty swallowing) through the thickening of fluids. Additionally, it demonstrates utility as a mucoadhesive for the buccal mucosa, is employed in eyedrop formulations for dry eyes and functions as a carrier for delivering omega-3 polyunsaturated fatty acids (PUFA), among other applications. Notably, XG shares rheological and viscosity similarities with hyaluronic acid (HA), yet it exhibits greater stability than HA. Furthermore, XG plays a vital role in supporting the regeneration of cartilage by suppressing chondrocyte apoptosis, downregulating MMP-1 and -3 protein expressions and safeguarding cartilage integrity. FDA approval underscores its bio-adhesive properties, wound healing properties and biocompatible properties, all these paving way for its versatile usage in diverse medical and pharmaceutical realms. Despite these attributes, the inherent physical limitations of XG (such as relatively low mechanical strength, potential difficulties in processing or handling)necessitate innovative approaches for enhancing its mechanical strength. In this context, a novel physical cross-linking method has been developed, involving the blending of silk fibroin (SF) and XG biopolymers through ultrasonication, presenting an chemical cross-linking alternative. The resulting scaffolds offer stability, tunability and cytocompatibility, making them promising candidates for use in regenerative medicine and soft TE. Mesenchymal stem cells (MSCs), osteoblast-like cells, and dental pulp stem cells are often preferred for working with these gums due to their ability to differentiate into various cell types relevant to TE and their compatibility with biomaterial scaffolds (Madappura and Madduri 2023). This strategic combination aims to leverage the beneficial characteristics of both SF and XG, creating a platform conducive to connective tissue regeneration (Byram et al. 2020). Beyond these properties, the attributes that have captured the attention of researchers in utilising XG as an anionic component in the polyelectrolyte matrix are noteworthy. This include XG, rich in carboxylic groups, which facilitates interacts with calcium, serving as nucleation points for hydroxyapatite mineralisation and enhancing the osteo-inductive properties of implants. XG scaffolds have demonstrated effectiveness in stimulating osteoblast growth and inducing elevated alkaline phosphatase (ALP) activity (Bueno et al. 2014). Recent research explored tri-composite scaffolds consisting of nanohydroxyapatite (NHA) within xanthan gum-chitosan (XG-CHI) polyelectrolyte complexes (PEC). These scaffolds displayed positive mechanical characteristics and capacity for swelling, showing promise for bone tissue engineering (BTE). They also showcased high bioactivity forming a dense apatite layer during simulated body fluid (SBF) mineralization , biocompatibility with MG-63 cells (human osteosarcoma cell line) and antibacterial properties. The enhanced functionality, encompassing protein adsorption, facilitating ALP activity and increasing expression of extracellular matrix proteins osteopontin (OPN) and osteocalcin (OCN), confirmed their osteogenic potential. XG-CHI/NHA scaffolds displayed superior physical, mechanical and biological properties compared to the control CHI/NHA scaffold, highlighting their potential for effective applications in scaffold-assisted bone tissue regeneration (Zia et al. 2020). In the realm of biomimetic (i.e., imitation of biological processes or models) hydrogels, which are skilfully derived from natural polysaccharides, specialised biomaterials have emerged, showcasing substantial applications in the realm of TE. The concept of mimicking extracellular matrix (ECM) using such hydrogels has gained substantial recognition. In such investigation, the researchers had developed bio-scaffolds utilising a combination of dialdehyde alginate and XG, exploring their potential for TE. The resulting scaffolds exhibited a well-defined interconnected porous network structure, a highly favourable characteristic for the intended application (such as wound healing, cartilage repair, and bone regeneration). These scaffolds demonstrated commendable mechanical properties, protein adsorption efficiency, thermostability and degradability. The collective findings unequivocally endorse the efficacy of these scaffolds as proficient biomaterials suitable for various application of TE (Jena et al. 2022). Existing literature also highlights successful synergies between polypyrrole (PPy) and polysaccharides in the context of TE. Specifically, electropolymerisation of PPy within xanthan hydrogels (XCA) resulted in the development of electroactive XCAPPy scaffolds, featuring approximately (15 ± 3) wt.% PPy and a thickness of $(40 \pm 10)\mu m$. XCAPPy films exhibited enhanced strain with a unique PPy chain alignment compared to pristine XCA hydrogels. The increased hydrophobicity and surface roughness of XCAPPy scaffolds, in contrast to XCA alone, made them superior for supporting fibroblasts. The application of an electromagnetic field (EMF) at 0.4 T significantly boosted fibroblast proliferation on both XCA and XCAPPy scaffolds. Combining natural and conducting polymers with EMF shows potential for innovative strategies to promote cell proliferation in TE applications (Bueno et al. 2015).

In a recent study, researchers proposed a straightforward method for creating dense composite polysaccharide-based membranes designed for guided tissue and bone regeneration. The mucoadhesive properties of chitosan (C) and xanthan gum (XG) were harnessed to fabricate polyelectrolyte-based complex membranes. Hydroxyapatite (HA), a compound mimicking native bone tissues, was synthesised through aqueous precipitation, providing a material conducive to bone regeneration.

Successful integration of HA with the muco-adherent polysaccharides, chitosan and xanthan, yielded dense composite membranes. Importantly, the combination did not alter the crystallinity or phase behaviour of HA. The resulting membranes exhibited mechanical resistance and thermal stability under physiological conditions, showcasing their potential for effective application in guided tissue and bone regeneration. The study further delved into investigating the influence of varying concentrations of HA on the morphological, mechanical properties and physiochemical properties of these membranes. Additionally, the proliferation analysis of dental pulp stem cells (DPSC) seeded on different membrane surfaces was conducted, offering insights into the biomaterial's suitability for guided bone and tissue regeneration applications (Barbosa et al. 2023). Tissue-engineered scaffolds designed to function as substitutes for periosteum play a crucial role in boosting bone regeneration, particularly in severely injured tissues. One of the studies focuses on a strategic chemical modification of chitosan, a polysaccharide known for its osteoconductive properties, by the introduction of phosphate groups into its structure. The resulting phosphorylated polymer (Chp) served as a key component in the fabrication of chitosan-xanthan-based scaffolds tailored for periosteal tissue engineering. These scaffolds exhibited non-toxicity toward adipose tissue-derived stem cells, although there remains room for further enhancement of cell adhesion and proliferation on the scaffold surfaces. The observed mineralisation on the surface of all formulations signifies promising characteristics for their potential application in bone regeneration. The research successfully produced matrices that are extremely porous and mechanically reinforced, exhibiting physical and physicochemical characteristics similar to scaffolds made with unmodified chitosan complexed with XG. Importantly, formulations with (Chp) showed an enhanced ability to adsorb cytochrome C, indicating potential in vivo applications where they could gather native bone morphogenetic proteins (BMPs) and promote osteogenesis. This work contributes to the advancement of biomaterials with tailored properties, holding significant promise for effective bone tissue engineering applications (Bombaldi de Souza et al. 2020).

Xanthan gum stands out as one of the most captivating biopolymers in contemporary research, offering myriad applications owing to its remarkable structural diversity, rheological features, biodegradability, biocompatibility and non-toxicity. The exceptional properties of XG have spurred substantial research endeavours across diverse studies and applications. Furthermore, the burgeoning body of literature on this intriguing polymer indicates a trajectory of sustained growth in research activities. As the versatility of XG continues to unfold, it is anticipated that the volume of research exploring its manifold applications will see a continuous rise in the scientific landscape.

4.2.2 Gellan Gum-Based Bio-Scaffolds in Tissue Engineering

Gellan gum emerges as a noteworthy anionic polysaccharide with the capacity to fulfil various pivotal roles in engineered tissues, particularly following suitable modifications and blends (Stevens et al. 2016). The biopolymer Gellan gum was discovered in 1978 by CP Kelco (San Diego, USA) through an extensive screening operation aimed at identifying polysaccharides from soil and water bacteria exhibiting valuable rheological properties. Subsequently, Gellan gum has garnered widespread usage, primarily serving as a multifunctional agent for gelling, stabilising and suspending agent. It has received approval from both the US FDA and the EU EMA(E418) for these specified purposes (Ferris et al. 2013).

Gellan gum is a high-molecular-weight linear anionic exopolysaccharide derived through the microbial fermentation of the *Sphingomonas paucimobilis* microorganism. This natural biocompatible polysaccharide is industrially produced by the bacterium *Sphingomonas* (*Pseudomonas*) *elodea* (ATCC 31461), which resides on the algae *Elodea canadensis*. The production process entails inoculating a culture of *S. elodea* in a medium containing nitrogen, inorganic salts and glucose. Following the depletion of glucose, the gum is collected from the fermentation medium after precipitation with alcohol (Bacelar et al. 2016). It is composed of tetra-saccharide-repeating units, consisting of 1,3-β-D-glucose (Glc), 1,4-β-D-glucuronic acid (GlcA), 1,4-β-D-glucose (Glc) and 1,4-α-L-rhamnose (Rha), with one carboxyl side group. This structure of GG reveals repeating tetramers of L-rhamnose, D-glucuronic acid and two D-glucose subunits, as illustrated in Fig. 4.2. Gellan gum also incorporates acetate and glycated functionalities (Muthukumar et al. 2019).

As the gellan gum's repeating unit is a tetra-saccharide consisting of L-rhamnose, D-glucuronic acid and two D-glucose residues, which are similar to other polysaccharides, the lysozyme secreted by monocytes and neutrophils has the capability to degrade the backbone. The degradation kinetics are subject to control by adjusting the degree of cross-linking and the nature of the cross-linking bond. This manipulation can be attained through modifications in physical or chemical linkages. Similar to the adjustments influencing degradation profiles, changes in mechanical properties can also be introduced by varying the type and degree of cross-links. The design specifications for the mechanical properties of scaffolds made by using Gellan gum are contingent on the tissue type. For instance, softer substrates (below 2 kPa) are more suitable for liver tissue, while harder implants (exceeding 40 kPa) are necessary for application of bone-tissue engineering (Xu et al. 2018). Further,

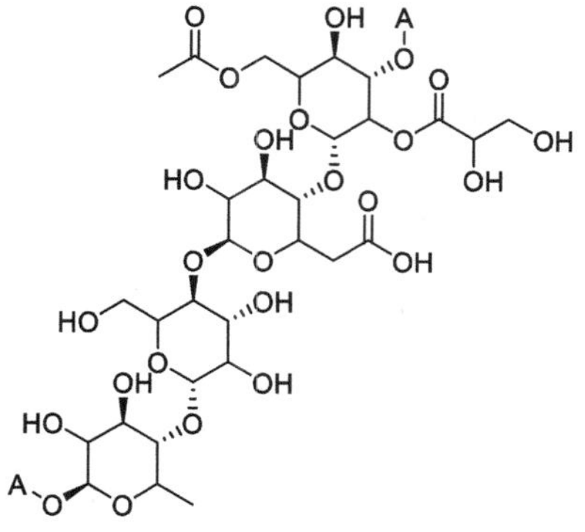

Fig. 4.2 The structure of gellan gum

Gellan gum is considered noncytotoxic and can undergo simple processing without the use of harsh reagents. This leads to the formation of transparent gels that resist acid stress and also heat. Similar to other polysaccharides like carrageenan or alginate, gellan gum's gelation process is ionotropic, necessitating the existence of cations for the creation of a stable hydrogel structure (Oliveira et al. 2010). Hence, gellan gum arises as a naturally existing, large linear polysaccharide that shares structural and biological similarities with the extracellular matrix. Its biocompatibility, biodegradability and injectability make it an apt matrix material for crafting diverse composite materials. In recent years, gellan gum-based hydrogels, integrated with various organic and inorganic elements, have attracted considerable interest in the realm of TE. These composite materials are recognised as multifunctional biomaterials, boasting remarkable mechanical properties, biocompatibility, low cytotoxicity and other advantageous features (Wu et al. 2024). It also holds promise in TE beyond basic cell viability in a 1% solution cross-linked with alpha-modified minimum essential medium (αMEM), as indicated in one of the studies. The tetra-saccharide molecule's structural variations, including high or low acyl forms in different ratios, provide a wide range of options for tailoring rheological characteristics. This versatility allows the creation of diverse culture matrices, from soft elastic gels to firm brittle gels. This, in turn, opens up opportunities to explore how cells respond to different 3D environments—a crucial aspect highlighted in various studies. Gellan gum's tuneable properties position it as a favourable contender for such investigations. The straightforward method of immobilising cells within 3D constructs, thanks to the cross-linking of gellan solutions with millimolar concentrations of cations, poses minimal risk to the immobilised cell population. This simplicity underscores gellan's potential as a viable material for tissue engineering scaffolds (Smith et al. 2007). In covalently modifying gellan gum by attaching short peptides, a study introduced a reliable method, enhancing gellan gum's cell attachment properties. This improvement is expected to broaden gellan gum's use in TE for cell scaffolding and as a bio-ink in 3D bioprinting (Ferris et al. 2015).

Recent studies highlight gellan gum as a versatile biomaterial for TE hydrogel formation. Researchers explored strategies like oxidation, purification, blending and reductive chain scission to overcome processing limitations. They assessed viscoelastic and mechanical properties, establishing correlations with polymer flow and hydrogel-forming capabilities. Native gellan gum and purified native gellan gum showed shear-thinning behaviour suitable for printing, with similarities in gelation. Oxidised gellan gum exhibited reduced viscosity, increased toughness and introduced aldehydes, while scissored gellan gum had a lower molecular weight. These modifications enhance gellan gum's processing capabilities without compromising hydrogel formation in TE (Gering et al. 2021).

Moreover, in general bioprinting stands out as a prominent bio fabrication technique employed for crafting 3D tissue-like materials. This innovative approach involves the spatially controlled assembly of cells, biomaterials and biomolecules to recreate the intricate architectures found in native tissues. Leveraging the advantages of 3D printing, bioprinting provides comprehensive control over fabrication, modelling and scaffold design particularly with the incorporation of 3D porous biomaterial designs. These porous structures enhance cell-cell contact and cell-matrix interactions while facilitating the diffusion of oxygen, nutrients and waste. The bio-fabrication of hydrogel-based constructs holds significance promise in developing 3D systems that can serve as valuable tools for drug screening and implantable devices (Aldana et al. 2021). Advanced fabrication technologies, particularly 3D printing, continue to progress, the creation of artificial tissue for personalised healing to become a reality. However, inks derived from polymers often fall short in terms of meeting expectations related to scaffold integrity, mechanical strength and the ability to stimulate the formation of tissue. Addressing these challenges involves both the development of new printable formulations and the adaptation of existing printing methods, which is a crucial focus of contemporary bio fabrication research. To expand the printability window, diverse strategies have been devised, making use of gellan gum. This has led to significant breakthroughs in the construction of 3D hydrogel scaffolds that closely mimic authentic tissues, allowing for the fabrication of more intricate systems (Cernencu and Ioniță 2023). However, the brittle gel characteristics of low-acyl gellan gum could limit its application in bioprinting. A recent study improved 3D printing suitability by introducing high-acyl gellan gum to low-acyl gellan gum). These blends showed excellent shape fidelity, confirming 3D printability for creating clinically relevant constructs. With low swelling, long-term stability in α-MEM and favourable wet-state mechanical properties, the blends are advantageous for bioprinting. In low-acyl gellan gum (3%) and high-acyl gellan gum (3%) bioprinted blends, mesenchymal stem cells maintained over 80% viability postprinting, offering potential for enhancing tissue engineering (Akkineni et al. 2022).

Thus, gellan gum emerges as an extensively researched natural polysaccharide, playing a versatile role as an excipient in diverse dosage forms. Its notable attributes, such as non-toxicity, biodegradability, high water holding capacity, swift gelation in the presence of cations and mucoadhesive potential, establish it as a valuable component in the field of TE. The recent exploration of gellan gum in the realm of TE, particularly in cartilage reconstruction, signifies its potential for addressing the limited self-repair capabilities of cartilage following degeneration or mechanical damage. Thus, Gellan gum is well-suited for cartilage regeneration due to its non-toxicity, biodegradability, high water holding capacity, rapid gelation in the presence of cations, and mucoadhesive properties. These attributes create an ideal environment for cartilage cell growth and scaffold formation, making gellan gum a promising biomaterial for addressing cartilage damage and promoting regeneration. As we delve into the innovative applications of gellan gum in TE, especially in the context of cartilage regeneration, this recognises the imperative role of novel biomaterials in replacing damaged tissue and fostering regeneration.

4.2.3 Arabic Gum-Based Bio-Scaffolds in Tissue Engineering

Gum Arabic (GA), or also known as acacia gum, is derived from the exudate of *Acacia senegal* and *Acacia seyal* trees within the Leguminosae family. This complex, branched heteropolysaccharide is typically neutral or slightly acidic, composed primarily of 1,3-linked β-D-galactopyranosyl units and features additional constituents such as L-arabinose, L-rhamnose and D-glucuronic acid. The side chains consist of two to five 1,3-linked β-D-galactopyranosyl units, connecting to the main chain through 1,6-linkages. Some sources suggest gum Arabic is a mixture of polysaccharides and glycoproteins, presenting light-orange or pale-white pieces that are soluble in water (Patel and Goyal 2015). It is a complex heteropolyelectrolyte with a 1,3-linked β-galactopyranose backbone and 1,6-linked galactopyranose side chains ending in glucuronic acid or 4-O-methylglucuronic acid residues. About 2% protein is covalently linked to carbohydrates through serine and residues of hydroxyproline, resulting in arabinogalactan-protein complexes. Upon hydrolysis, constituents like D-galactopyranose, L-arabinose, L-rhamnose and D-glucuronic acid are obtained. Moreover, based on its chemical composition and biochemical properties, gum Arabic is comprised of two polysaccharides. One is a high-molecular-weight glycoprotein, primarily composed of 90% carbohydrate, while the other is a low-molecular-weight polysaccharide fraction consisting of sugars. The protein component is covalently bonded to carbohydrates, constituting approximately 2.5% of the crude protein content. Therefore, GA is widely used in various industries for its favourable low viscosity and surface activity (Li et al. 2021; Fadlelmoula 2018). Gum Arabic holds the 'Generally Recognised as Safe (GRAS)' status according to the US Food and Drug Administration (US FDA). Its recognised safety allows for diverse applications in routine use. To serve as scaffolding in tissue engineering, the polymeric organic matrix needs to integrate inorganic reinforcements. This incorporation is essential to emulate the nanocomposite structure found in bones, enhancing both biological and mechanical properties (Mirza et al. 2020).

In a unique approach to scaffold development using a modified gum Arabic cross-linked gelatin scaffold, a novel non-cytotoxic and cytocompatible material was created without external cross-linking agents. Here, gelatin underwent successful cross-linking with oxidised gum Arabic, achieved through periodate oxidation, resulting in the formation of the scaffold. Assessment of cytotoxicity confirmed the non-cytotoxic and cytocompatible nature of the material. Additionally, cell adhesion studies revealed the nonadherent characteristics of the scaffold. This non - cell adherent and cytocompatible scaffold hold promise for applications that can be valuable in diverse aspects of TE, drug screening and cancer research. Additionally, its bio-compatibility and unique properties may find applications in creating supportive environments for growth of cells and function in TE (Sarika et al. 2014).

The latest research centres on creating composite scaffolds using collagen and oxidised gum Arabic. In this procedure, collagen underwent cross-linking with oxidised gum Arabic, resulting in the formation of a porous scaffold. Manipulating the oxidation degree of gum Arabic allowed for the control of polysaccharide quantities, impacting the swelling behaviour of the scaffold. Rigorous assessments were conducted to evaluate the scaffold's physical attributes, biocompatibility, stability and its ability to support growth of cells. Results showed customisable polysaccharide integration based on gum Arabic's oxidation state, impacting swelling characteristics. Crucially, the scaffolds exhibited non-toxicity toward mesenchymal stem cells and nucleus pulposus cells over 30 days, emphasising their potential in tailored tissue engineering applications (Rekulapally et al. 2021).

In the context of manufacturing, electrospinning proves to be a valuable method that leverages applied electrostatic force to counteract solution surface tension, producing fibres with diameters ranging from nano metres to micrometres. These fibres structurally resemble the (ECM) and are promising for TE applications in general. Various parameters, including solution properties, processing conditions and ambient factors, influence fibre morphology. Notably, solution viscosity has a crucial impact on the electrospinning process. Enhancing cell attachment and cytocompatibility in nanofibers for TE requires increased chitosan and gelatin content. Current methods involve corrosive or expensive solvents like TFA or HFIP. In a recent study, researchers developed a safer and cost-effective approach for chitosan-gelatin nanofiber preparation. Gum Arabic, serving as a viscosity-reducing agent, was mixed with an aqueous solution of acetic acid to create a solution with higher chitosan-gelatin (16 wt%) content. Electrospinning this solution resulted in the production of nanofibers rich in chitosan and gelatin, demonstrating favourable cytocompatibility for potential tissue engineering applications (Tsai et al. 2015).

The main goal of TE is to develop artificial tissues that can replace biological functions for healing wounds and TE. Studies have aimed to create a porous nanofiber scaffold through electrospinning to address deep skin damage. To mimic a scaffold resembling the natural ECM of the skin (such as collagen, elastin, and fibrin, to create scaffolds that can effectively support cell growth and tissue regeneration), a blend of GA, polycaprolactone (PCL) and corn protein (zein), was utilised with varied concentrations and ratios. Zein and GA served as the protein and polysaccharide components, respectively, while PCL contributed to scaffold elasticity, strength and degradation time. The results indicated that the manufactured PCL/zein/GA scaffolds exhibited a porous structure with a bimodal diameter distribution. These scaffolds demonstrated high hydrophilicity, favourable porosity (approximately 80%) and tensile strength ranging from 1.36 to 3 MPa, coupled with an elongation of 19.13–44.06%, making them suitable for skin tissue engineering (Pedram Rad et al. 2018). Another study using electrospun nanofibers, when utilised in cell culture, reported the ability to replicate the extracellular matrix, leading to improve in-vitro cell behaviour. In this context, the description outlines biocompatible-blended nanofibers composed of polyvinyl alcohol (PVA) and GA. These nanofibers aim to mimic the ECM for in-vitro cell cultures, showcasing the potential to deliver nanocomposites for various application in biomedical. Heat treatment is employed for nanofiber cross-linking, and the biocompatibility of the platform is assessed, highlighting its ability to create a cell culture-friendly environment. Furthermore, the with GA-AuNPs, which is a functionalised cross-linked nanofiber, showcase efficient adsorption onto the PVA/GA nanofibers' surface. This system successfully delivers nanocomposites to metastatic melanoma cells in-vitro, offering a promising in vitro tool for precision medicine with GA incorporation (Serio et al. 2021). Moreover, scaffolds incorporating plant extracts demonstrate enhanced capability for cell proliferation. Their novel approach involved the construction of bioinspired alginate/gum Arabic hydrogels for the delivery of protein such as mitsugumin, chronic human wound healing and targeting cell membrane treatment. This hydrogel exhibited biphasic-kinetic release behaviour. Facilitating swift protein mitsugumin delivery, it accelerates wound reepithelisation and ensures sustained protein release for chronic wound treatment, owing to its adaptable structure and distinctive physicochemical attributes; such a hydrogel holds promise as a valuable delivery vehicle for the healing of chronic wounds (Farshidfar et al. 2023).

In conclusion, GA has proven to be an important biomaterial in TE, offering a versatile solution for scaffold development. Recognised for its safety, GA has diverse applications, including cross-linking with gelatin to create non-cytotoxic scaffolds suitable for various purposes. Studies also demonstrate the adjustability of GA in composite scaffolds, impacting swelling behaviour and exhibiting nontoxic behaviour toward different cell types. Electrospinning techniques incorporating GA provide a safer and cost-effective method for nanofiber fabrication, enhancing cell attachment and cytocompatibility. Furthermore, GA contributes to the creation of scaffolds with porous nanofiber for skin TE and biocompatible nanofibers for in-vitro cell cultures. Its application extends to bioinspired hydrogels, showcasing enhanced cell proliferation capabilities and potential in delivering proteins for chronic wound healing. The biocompatibility, adjustability and cost-effectiveness of GA highlight its crucial role in advancing TE, offering potential breakthroughs in regenerative medicine and biomedical applications.

4.2.4 Guar Gum-Based Bio-Scaffolds in Tissue Engineering

Guar gum (GG) stands out as a compound that is a naturally derived polymeric form which is valued for its impressive biodegradability and biocompatibility in the environment. This nonionic, hydrophilic carbohydrate represents as an economical hydrocolloid polysaccharide and belongs to the latest generation of plant gums. GG, known for its ready availability, eco-friendliness and non-toxicity, proves to be an excellent choice of application in drug delivery (DD) for TE due to its biodegradable properties. Widely explored for, guar gum(GG)-based hydrogels, scaffolds, films and also including nanoparticles exhibit remarkable versatility. These non-toxic carriers exhibit significant potential for targeted drug delivery, meeting the demand for efficient and biocompatible systems in pharmaceuticals (Amjed et al. 2024). As GG stands out as an exceptional raw material with distinct advantages such as a high molecular weight, lengthy molecular chain and widespread availability, setting it apart from other biopolymers. These characteristics position GG as a favourable choice for the creation of innovative materials, functioning as an abundant supplier of natural polysaccharide macromolecules; GG is not only cost-effective but also exhibits exceptional water solubility. This polysaccharide possesses the inherent ability to significantly enhance the remediation capabilities of biopolymers. As a result, GG finds diverse applications in various fields (Mandal et al. 2023). The combination of abundant availability, low production costs, excellent functional properties, versatility, and long shelf life, contributes to guar gum's reputation as a cost-effective hydrocolloid polysaccharide for consumers and businesses (Baranwal 2022). GG is derived from the endosperm of *Cyamopsis tetragonoloba* seeds. It contains a galactomannan complex polysaccharide with D-galactose and D-mannose. When in contact with water, this polymer enhances viscosity and thickens the solution through hydrogen bonds. GG is valued for its versatile properties such as thickening, binding, emulsifying and gelling. Its rapid dissolution in cold water, film-forming ability, wide pH stability and biodegradability make it valuable across various industries (Thombare et al. 2016). GG Comprising a b(1→4) D-mannopyranose backbone linked with a(1→6) D-galactopyranose units in a 1:2 M ratio, GG boasts a distinctive molecular structure. Beyond its inherent properties, GG finds diverse commercial applications, attributed to its exceptional capacity to modify rheological properties (Tiwari et al. 2009).

Guar gum, derived from *Cyamopsis tetragonoloba* seeds, is easily modifiable for various applications. One study explored a scaffold design using GG, transformed into carboxymethyl guar gum (CMGG) with 10μm average pore diameters. The scaffold's impact on RAW 264.7 macrophage cells, NIH3T3 fibroblast cells and bone marrow-derived mesenchymal stem cells (BM-MSCs) was accessed. The cells were cultured on the scaffold under different conditions (untreated, treated with LPS for inflammation and treated with $CoCl_2$ for hypoxia). Over 48 h, MTT assays showed successful growth and proliferation, even under inflammatory and hypoxic conditions induced by LPS and $CoCl_2$ treatments. These initial results indicate the promise of the carboxymethyl guar gum scaffold for TE applications (Banerjee 2018). Guar gum is utilised in scaffold design for cell propagation due to its hydration and unique coil formations. Introducing chaotropic ions, like lithium, enhances guar gum homogenisation in polar solvents, allowing previously unattainable degrees of substitution. Blending carboxymethyl guar gum (CMGG) with a DS of 1.10 and polyvinyl alcohol (PVA) at a 2:1 ratio results in hydrophilic porous scaffolds for cell propagation. CMGG-PVA scaffolds exhibit 70–90% porosity and a tensile strength of 6.32 MPa, making them valuable in tissue engineering, demonstrating non-cytotoxicity and supporting cell propagation. This guar gum derivative scaffold represents a significant biomaterial advancement for biomedical applications because of its Hydration and Coil Formations, Enhanced Homogenization with Chaotropic Ions, Hydrophilic Porous Scaffolds, and Non-Cytotoxicity and Support for Cell Propagation (Kundu et al. 2018). Similarly in a study on collagen-poly(dialdehyde) guar gum hybrid scaffolds, poly(dialdehyde) guar gum fabricated to enhance collagen cross-linking and immobilise platelet-derived growth factor—BB. This covalent cross-linking improved thermal stability, swelling resistance and biodegradation. Microscopic examination revealed a 3D interconnected honeycomb porous structure. Immobilised growth factor exhibited sustained release for a duration extending up to the 13th day, demonstrating controlled delivery in-vitro. The hybrid scaffold significantly increased NIH 3T3 fibroblast cell density and proliferation, showcasing its potential for tissue engineering applications (Ragothaman et al. 2014). A study explores hydrogel application in soft tissue engineering, investigating viscoelastic properties' impact on scaffold fabrication. Characterised by G′, G″ and tan ∂ (G″/G′), the hydrogel's analysis focuses on freeze-dried tissue engineering scaffolds. Rheological and freeze-drying analyses were conducted on hydrogels with varying guar gum (GG) and gelatin (GE) concentrations. Higher tan ∂ values weakened scaffolds, while lower values produced harder scaffolds with deformities. Optimal scaffolds had a tan ∂ of 0.88, showing desirable porosity. Physical parameters like contact angle and surface roughness favoured soft TE. Human keratinocyte bioassays conducted on the optimised scaffold in-vitro demonstrated robust viability and proliferation over 14 days, highlighting its promise in soft TE (Indurkar et al. 2020). In skin tissue engineering, guar gum indole acetate (GGIA) ester, synthesised for the first time from guar gum, was cross-linked with hydrolysed chicken feather keratin. The resulting bio-based films, with a high degree of substitution (DS = 0.61), displayed intriguing surface chemistry, excellent biocompatibility and antimicrobial efficacy. With 63% porosity and a tensile strength at break of 6.4 MPa, GGIA-keratin film scaffolds prove promising for skin TE, offering a mix of structural integrity and antimicrobial properties (Das et al. 2021). Further in bone TE, a study explored a calcium-free phosphate approach to enhance apatite formation in guar gum-based hydrogels (P-GG). Synthesised through radical polymerisation, these hydrogels, containing acrylamide (AM) and bis[2-methacryloyloxy]ethyl phosphate (BMEP) as a cross-linker, displayed significant structural and mechanical characteristics influenced by GG content. The porous ladder-like fibrous topology facilitated superior cell growth and adhesion. The phosphate functionality promoted rapid biomineralisation in simulated body fluid (SBF) solutions. P-GG hydrogels demonstrated outstanding physical, chemical and biological characteristics, suggesting substantial promise in bone TE (Pal et al. 2023).

In conclusion, GG has emerged as a versatile and promising biomaterial for various applications in TE. Its natural origin, biocompatibility and biodegradability make it an attractive choice for scaffold design, delivery of drug systems and other application related to biomedical. The studies discussed above highlight the diverse modifications and applications of GG, ranging from its role in creating hydrogels, films and scaffolds to its use in promoting controlled delivery of drug. The unique properties of GG, such as its ability to form hydrogen bonds, alter rheological properties and enhance remediation capabilities, contribute to its effectiveness in TE. From the creation of carboxymethyl guar gum scaffolds for cell propagation to the synthesis of GG indole acetate ester for skin TE, each study demonstrates the adaptability of GG in addressing specific challenges in biomedical research. Furthermore, the exploration of phosphate cross-linked guar gum-based hydrogels in bone TE showcases the material's potential in enhancing biomineralisation and promoting cell growth. The findings suggest that guar gum-based materials could play a crucial role in advancing renewal medicine and TE applications. Overall, the comprehensive insights gained from these studies underscore the significance of GG in the ever-evolving field of biomaterials. As researchers continue to unlock its potential, GG stands poised as a valuable resource playing a key role in advancing novel solutions in TE and biomedical sciences.

4.2.5 Ghatti Gum-Based Bio-Scaffolds in Tissue Engineering

Gum ghatti, also known as Indian gum, is a complex anionic polysaccharide obtained from the *Anogeissus latifolia* tree. It appears as a reddish-grey powder or a dark-brown lump, nearly odourless, with limited solubility in cold water. Improved solubility occurs with breaking and heating at 90 °C. Gum ghatti's structure includes sugars like L-arabinose, D-mannose, D-xylose, D-galactose and D-glucuronic acid in a molar ratio of 48:29:10:5:10. The primary structure features a main chain of 1→6 linked β-D-galactopyranose units and alternating 4-O-substituted and 2-O-substituted α-D-mannopyranose units, with a side chain comprising single L-arabinofuranose units (Mittal et al. 2013). In the realm of TE, the exploration of biomaterials with heightened biocompatibility and biodegradability has led researchers to investigate extracts and gums of plant based. This polysaccharide gum, i.e. gum ghatti, has emerged as a standout candidate due to its sustainability, biodegradability, safety and cost-effectiveness. These plant-derived substances, acting as nontoxic surfactants, bring about enhanced biocompatibility and effective drug-delivery properties, rendering them valuable for diverse applications, including TE (Patra et al. 2021). A recent study delved into the impact of gum ghatti concentration on the porosity of hydroxyapatite (HAP), a vital component in TE. The GG-HAP composite, synthesised through a chemical precipitation method, demonstrated superior qualities, particularly in samples with a 10% Gum ghatti concentration after the sintering process in comparison to lower concentrations of Gum ghatti (1% and 5%). The Sin-10% Gum ghatti-HAP composite showcased a crystal size of 30 nm, 95% crystallinity and favourable pore characteristics. Biocompatibility tests confirmed inherent bioactivity, while a drug delivery profiles of naringenin exhibited a controlled release from Sin-10% Gum ghatti-HAP. These findings position the composite as a favourable choice for application in biomedical, particularly in the field of tissue engineering, where materials with such advantageous properties are sought after for supporting tissue regeneration and repair (Padmanabhan et al. 2021). Ghatti gum is known for its water-absorbing capacity and film-forming properties. Recent studies highlight its role in accelerating wound healing by reducing surface area and increasing tensile strength. In a novel approach, Gum ghatti and pectin were combined to enhance film strength, addressing potential integrity loss in wound exudates. Cross-linking Gum ghatti-pectin film with trivalent Al^{+3} ions resulted in a robust hydrogel for sustained release of ciprofloxacin hydrochloride. This biodegradable wound dressing, utilising natural polymers, showed promising results in in vivo wound healing tests (Sinha et al. 2022). (note: In wound healing, film formation refers to the process of applying a liquid or semi-liquid material onto the wound surface, which then dries or solidifies to form a thin, protective film. This film helps to cover the wound, protecting it from external contaminants while creating an optimal environment for healing. thus, combining Gum ghatti and pectin enhanced the film strength of the wound dressing, making it more robust and capable of withstanding the stresses encountered during application and wear. This improvement in film strength contributes to the effectiveness of the wound dressing in promoting wound healing.)

To advance 'green electrospinning,' it's crucial to use eco-friendly solvents and materials. Nanofibers (NFs) as polymeric scaffolds are preferred for TE because of its small size, promoting permeability and adhesion. Gum ghatti and polyvinyl alcohol (GG-PVA) were combined to create green nanofiber mats (NFM) for wound healing. The CL-1 cross-linked NFM were characterised using XRD, DSC, TGA and FTIR and displayed superior mechanical strength. AFM revealed well-established cross-links suitable for loading of drugs and cell proliferation. In vivo studies demonstrated accelerated wound healing within 5 days, and histological investigations confirmed the CL-1 sample's suitability as a nontoxic polymeric scaffold for TE. Thus, the combination of eco-friendly materials, gum ghatti and polyvinyl alcohol (GG-PVA), in cross-linked nanofiber mats, demonstrated accelerated wound healing in vitro mice model and holds promise as a nontoxic polymeric scaffold for tissue engineering (Dey et al. 2020). In another study on porous hydroxyapatite (HAP) scaffold fabrication, gum ghatti served as a natural binding agent using a template-assisted polymer sponge replication technique. The distinctive formulation utilised water and a plant extract of Gum ghatti as a solution in the polymer replication method to produce porous hydroxyapatite scaffolds. It was observed the formation of a mechanically stable, spongy bone architecture in the prepared HAP scaffold. The scaffold displayed a well-connected structure with micro- and macropores, varying in size from about 200–500µm. This characteristic makes the HAP scaffold a promising material for various applications in bone TE. Further to assess the suitability of the prepared scaffold, study conducted preliminary investigations, including cytotoxicity, biocompatibility and cell proliferation tests, using MDCK cell lines in vitro. The results indicated that the HAP scaffold displayed favourable properties; highlighting its potential, the material shows promise for use in bone TE applications. Notably, the bioactivity and biocompatibility studies further supported the viability of the proposed unique formulations. Thus, the developed porous scaffold materials hold promise for advancement in bone tissue engineering applications (Lett et al. 2020).

In conclusion, gum ghatti stands out as a versatile and sustainable polysaccharide gum sourced from the *Anogeissus latifolia* tree, offering unique properties for diverse applications, particularly in TE. Its complex structure contributes to biocompatibility and biodegradability. Gum ghatti has shown promise in hydroxyapatite composites, wound healing films, nanofiber mats and porous hydroxyapatite scaffolds. Notable findings include superior qualities in Gum ghatti-hydroxyapatite composites, Gum ghatti-pectin films with sustained drug release and Gum ghatti-polyvinyl alcohol nanofiber mats for accelerated wound healing. Gum ghatti role as a natural binding agent in porous hydroxyapatite scaffolds showcased a mechanically stable and interconnected structure, emphasising its potential in bone tissue engineering. Overall, these studies highlight gum ghatti's multifaceted capabilities in advancing various aspects of tissue engineering, ranging from composite materials to wound healing and porous scaffolds.

4.2.6 Other Natural Gum-Based Bio-Scaffolds in Tissue Engineering

As in the realm of TE, the advancement of matrices is crucial to support cell growth, encourage differentiation and promote proliferation, ultimately resulting in the generation of new tissues. These novel approaches allow for the development of hybrid constructs that can be implanted to initiate tissue regeneration or act as substitutes for compromised organs. Various materials, particularly biodegradable polymers that were discussed above in this chapter, have been proposed for scaffold fabrication. As natural-based polymers present a distinct advantage due to their similarity to biological macromolecules. This resemblance aligns with the biological environment, allowing for recognition and metabolic interaction, making them favourable candidates in the realm of TE (Mano et al. 2007). In recent years, the scientific literature has introduced an array of novel gum exudates from diverse sources, each subjected to thorough structural analysis and functional characterisation. This chapter further explores the classification of natural gum hydrocolloids in short, considering factors such as source, structural characteristics, functional properties and TE applications. Tissue engineering alternatives encompass a wide range, including bacterial polysaccharides like dextran, pullulan and cellulose; botanicals such as locust bean gum and pectin; tree gums like tragacanth, karaya and kondagogu; and seaweed polysaccharides including carrageenan, alginate and agar, all discussed further in this chapter (Ahmad et al. 2019).

Dextran, the initial microbial polysaccharide brought into commercial use, is synthesised by *Leuconostoc mesenteroides* and *Streptococcus mutans* bacteria. Its structural composition comprises linear α-1,6-glycosidic linkages with branching through α-1,3-glycosidic bonds connecting glucose monomers. This branching characteristic sets dextran apart from dextrin, which features branch α-1,4-glycosidic linkages (Ng et al. 2020). Dextran hydrogel scaffolds, resistant to protein adsorption and cell adhesion, are synthesised through radical polymerisation of methacrylate groups on dextran. The resulting cross-linked hydrogels form a sturdy network, creating macroporous, interconnected beaded wall morphology scaffolds by controlling liquid-liquid phase separation. The distinctive interconnected macroporous structure of dextran scaffolds proves advantageous in applications requiring such geometry, particularly in tissue engineering scenarios where tissue regeneration relies on cell penetration and nutrient diffusion (Lévesque et al. 2005). Similar to dextran, pullulan is a microbial derived polysaccharide mainly synthesized by yeast-like fungi such as *Aureobasidium pullulans*, *Cyttaria darwinii*, *Cyttaria harioti*, *Cryphonectria parasitica*, *Rhodotorula bacarum*, *Teloschistes flavicans* and *Tremella mesenterica*, can also synthesise pullulan. Pullulan's structure primarily comprises α-1,6-linked maltotriose units (G3), featuring repeating G3 units with two α-1,4-glycosidic bonds (Trinetta and Cutter 2016; Kırtel et al. 2017; Singh et al. 2015). Pullulan's unique physicochemical properties, stemming from α-(1→4) and α-(1→6) glycosidic linkages, pose challenges for adhesion of cells and spreading in bone TE. However, these issues can be addressed through techniques such as hydroxyapatite (HA) surface deposition and the incorporation of functional groups, cell-recognisable molecules like gelatin and arginine-glycine-aspartate ligands (Singh et al. 2016). In a recent study, researchers used bio-cross-linker-oxidised pullulan to stabilise collagen. Varying concentrations of the cross-linker were applied to collagen, producing scaffolds through freeze-drying. Higher cross-linker concentrations resulted in increased cross-linking, slower degradation and enhanced tensile strength. The 3% cross-linker scaffold exhibited highly porous and interconnected features. The swelling capacity increased with cross-linker concentration, and the scaffolds proved biocompatible with 3T3 fibroblast cell lines. This novel bio-cross-linker shows promise as a replacement for synthetic counterparts in biomaterial synthesis (Selvakumar and Lonchin 2020).

Cellulose stands out among biomaterials chosen for scaffold fabrication owing to its abundant and renewable nature. It is predominantly derived from plant cell walls, and through specific mechanical and chemical processes, nanoscaled forms of plant cellulose known as cellulose nanofibers (CNFs) and cellulose nanocrystals (CNCs) can be obtained. Cellulose can also be produced by certain microorganisms, leading to what is termed bacterial cellulose (BC). BC distinguishes itself by considering it the most biocompatible manifestation of cellulose, due to the absence of biogenic impurities like lignin and hemicellulose. Moreover, only mild chemical treatments are necessary to ensure the purity of BC such as washing with mild acids or alkalis to remove any remaining impurities and to enhance the purity of the bacterial cellulose. The nanofibrillar porous structure inherent in bacterial cellulose closely resembles the extracellular matrix, positioning BC as a highly recommended material for the construction of tissue engineering scaffolds. This unique structural similarity enhances its suitability for such applications, making bacterial cellulose an ideal choice in the realm of TE biomaterials (Oprea and Voicu 2020). Due to their remarkable swelling capacity, adhesion, elasticity and 3D network structure, hydrogels emerge as promising scaffolds that mimic the ECM. These hydrogels serve the purpose of encapsulating and delivering cells while promoting organised cell growth and morphogenesis. Within the realm of biomedical applications, cellulose and its derivatives have gained widespread use, attributed to their outstanding attributes such as biodegradability, biocompatibility, non-toxicity and functionality. What sets cellulose and its derivatives apart is their ability to significantly improve the mechanical characteristics of polymer matrices when compared to other materials. This augmentation in mechanical strength comes without compromising the biocompatibility and biodegradability of the overall material. Consequently, cellulose-based hydrogels stand out as a particularly advantageous choice for applications in cartilage TE, bone TE and also skin TE. Their incorporation can substantially improve the mechanical integrity of these tissues, highlighting the capacity of cellulose-based hydrogels to advance methods in TE (Su et al. 2022).

Locust bean gum, a prevalent natural polymer, is categorised as a non-starch polysaccharide with a galactose-to-mannose ratio of 1:4, classifying it as a galactomannan. The mannose units form a linear chain linked to galactopyranosyl residues in side chains, with variations depending on the plant origin. Commonly known as carob bean gum as well, it is extracted from the seeds of the leguminous plant *C. siliqua* Linn. of the Fabaceae family. The processing involves milling the brown pods or beans of the locust bean tree to obtain locust bean gum (Prajapati et al. 2013). The locust bean gum (LBG) has gained widespread use in hydrogel formulations across diverse industries such as food, pharmaceuticals, biomedicine, cosmetics and more. Its versatile physicochemical properties, including viscoelasticity, swellability and release-retarding capabilities, can be enhanced through grafting or synergies with compatible polymers. These modifications do not compromise LBG's inherent characteristics of being biocompatible, nontoxic, biodegradable and nonmutagenic. LBG has found application as a carrier in various drug delivery systems, and innovative techniques have led to novel versions of LBG through chemical modifications. Combined with other polymers and in grafted forms, LBG contributes to the fabrication of hydrogel-derived formulations with specific functions, responding to various stimuli necessary for delivering therapeutic actives tailored to specific applications (Prajapati et al. 2021).

Another natural gum is **pectin** is an excellent candidate, primarily due to its origin as a by-product from the industrial processing of fruits like oranges, apples and mango peels. Being a naturally occurring polysaccharide rich in carboxylic groups, pectin is a viable alternative to acrylic acid polymers. The latter are commonly used in wound dressing applications, but the utilisation of pectin helps circumvent environmental concerns associated with these synthetic materials. By repurposing waste materials such as pectin, we not only harness a valuable resource but also contribute to sustainable practices in wound care (Tortorella et al. 2021). The pursuit of sustainable biomaterials for 3D bioprinting in TE is a current focal point. In the study of green and nontoxic biomaterial introduces an innovative method involving the cross-linking of pectin sourced from citrus peels with (3-glycidyloxypropyl)trimethoxysilane (GPTMS) has been presented. The outcome is the development of freeze-dried porous pectin sponges, characterised by adjustable properties like water uptake, porosity and compressive modulus, all modulated through the careful control of GPTMS content. Notably, cell bio-compatibility remains unaffected, and the introduction of GPTMS enhances the pectin printability by elevating viscosity and yield stress. The versatility of this biomaterial is demonstrated through the successful bioprinting of intricate scaffolds featuring both micro- and macropores. This advancement highlights the considerable potential of GPTMS-cross-linked pectin as a biomaterial ink, offering a pathway for the fabrication of patient-specific scaffolds with promising implications for in vivo tissue regeneration (Lapomarda et al. 2020).

Sterculia gum(SG), recognised by its common names such as karaya or sterculia, is a biopolymer of medicinal significance. This gum is produced from trees of Sterculia species is known as gum karaya. the major sources of this gum is from Sterculia urens, Sterculia setigera and Sterculia villosa (Prasad et al. 2022). This gum is partly acetylated polysaccharides within its substituted rhamnogalacturonoglycan structure, SG undergoes partial hydrolysis to yield essential components, including D-glucuronic acid, D-galacturonic acid, D-galactose and L-rhamnose residues. Notably, SG has garnered distinction for its safety profile, characterised by being non-teratogenic, nontoxic, nonallergenic and non-mutagenic at moderate doses. Acknowledging these attributes, SG has earned the designation of being 'Generally Recognised as Safe (GRAS)' by the US Food and Drug Administration (US FDA). This recognition underscores its widespread application in various industries, including food, pharmaceuticals and beyond. The versatility and safety of SG make it a valuable resource, contributing to its extensive utilisation in diverse sectors (Nayak et al. 2019). In recent years, gum karaya has undergone synthetic modifications like grafting, cross-linking and more. These alterations enhance its chemical stability, thereby enhancing its physicochemical properties, and also its mechanical strength. The resulting modified gum karaya hydrogels have found d diverse applications in biomedical and pharmaceutical industries, showcasing the adaptability and expanded utility of this biopolymer (Raj et al. 2021).

A versatile polysaccharide, **gum tragacanth** (GT), is known for its biodegradability, biocompatibility and multifunctional properties as a thickener, stabiliser, emulsifier and cross-linker. Comprising tragacanth acid (60–70%) and tragacanthin, GT shows potential in drug delivery and wound healing. Recent studies use GT in electrospun nanofibrous scaffolds, often blended with materials like PVA and/or poly(ε-caprolactone) (PCL). This approach unveils remarkable attributes in collagenation and proliferation, particularly beneficial for wound healing applications. The innovative use of GT in these nanofibrous scaffolds showcases its promising role in advancing therapeutic strategies, emphasising its unique contributions to biomedical applications (Anita Lett et al. 2021).

Gum kondagogu (GK) is a natural biopolymer from *Cochlospermum gossypium* DC, which is a part of Bixaceae family. Characterised as rhamnogalacturonan gum, GK is chosen as a material for scaffold due to its water retention, biocompatibility, non-toxicity, biodegradability, porous structure, cost-effectiveness and abundant natural availability. These qualities make GK an attractive option for various applications, particularly in scaffold design (Rathore et al. 2022). Due to its cost-effectiveness, abundance and high biocompatibility, gum kondagogu (GK) has found diverse applications. It serves as a stabilising agent in synthesising nanoparticles like silver, gold and platinum; functions as an emulsifying agent and acts as a gelling agent in plant tissue culture. Additionally, GK is utilised as a hydrogel for colon drug delivery, a vehicle for ocular delivery and a drug release retardant in tablets. Its unique properties, including water absorption and mucoadhesive characteristics, make it an ideal material for developing wound dressings. Some research has explored composite sponges using collagen, gelatin and other naturally available gums, introducing a novel 3D sponge combining GK with gelatin for wound healing applications. A specifically developed composite sponge, containing GK, gelatin and ciprofloxacin (CP), demonstrated excellent swelling, porosity and biocompatibility. The sponge exhibited enhanced cell proliferation, positioning it as a potential candidate for wound dressing. Impregnated with ciprofloxacin, the sponge showcased a 95% drug release within 48 h. In vitro evaluations, including biocompatibility, cell adhesion and proliferation using NIH 3T3 fibroblast and Human keratinocytes cell lines, confirmed the sponge's high biocompatibility. Moreover, the ciprofloxacin-loaded sponge demonstrated significant antimicrobial activity against *S. aureus* and *E. coli*. In conclusion, this composite sponge holds promise as a valuable biomaterial for the application of dressing wounds (Rathore et al. 2016).

Alginate, a naturally occurring polysaccharide, is biocompatible, forms gels and is nontoxic and easy to process. Sodium alginate, oxidised alginate and other materials sourced from alginate are valuable applications in biomedical, including wound healing, drug delivery and TE. They can be processed into various 3D scaffolding materials like hydrogels, microspheres, microcapsules, sponges, foams and fibres, playing a crucial role in tissue healing and regeneration. The use of alginate in scaffolds supports cell growth, making it a pivotal component in various biomedical applications (Sahoo and Biswal 2021). Scaffolds based on alginate are extensively employed in bone TE (Valente et al. 2012). Additionally, alginate-based biomaterials present numerous benefits for cartilage treatment. Their potential lies in the ability to fine-tune mechanical properties, offering versatility in addressing specific needs. Moreover, the ease of the manufacturing process further enhances their attractiveness for applications in cartilage treatment (Farokhi et al. 2020). Alginate-based biomaterials offer significant potential for TE, easily integrating drugs and cells into the scaffolding matrix. The success of tissue constructs relies on the careful design of these scaffolds, considering their physical, chemical and biological properties. Importantly, these biomaterials are versatile in various tissues and organs, such as the cartilage, skin and bone, highlighting their potential for repair and regeneration applications (Sun and Tan 2013).

Alginate, agar and agarose stand out as the most abundant and extensively researched seaweed-based polysaccharides. Lately, there has been an increasing focus on another seaweed-based polysaccharide, carrageenan (CG), due to its versatile biological properties. The significance characteristics of CGs have prompted researchers to explore CG-based structures for diverse applications in the field of TE and regenerative medicine. In light of recent advancements in bio-fabrication techniques, particularly 3D bioprinting, scholars have turned their attention to the progress in this field. The emerging trends in the application of CGs in TE are notably focused on hydrogels, nanofibers and three-dimensional (3D) (bio)printed scaffolds (Jafari et al. 2022). Carrageenan, a natural polysaccharide, is ideal for controlled release strategies due to its exceptional gelling capabilities. Comprising alternating copolymers of 1,3-linked β-D-galactose and 1,4-linked α-D-galactose, carrageenan exhibits strong anionic properties with varying sulfation degrees. Biocompatibility studies confirm its safety at appropriate concentrations. κ-carrageenan is particularly suitable for constructing carrier systems in tissue engineering, offering efficient encapsulation and controlled release without the need for organic solvents. These findings underscore carrageenan's significant potential in bone TE applications (Santo et al. 2009).

4.3 Conclusions

In conclusion, this chapter delves into the exciting realm of tissue engineering with a specific importance on harnessing the potential of natural gums as bio-scaffold materials. The exploration encompasses a diverse array of natural gums derived from various sources, each exhibiting unique properties that make them invaluable in tissue regeneration efforts. Despite the challenges faced in integrating these natural gums into TE, their biocompatibility and tailored mechanical properties present a promising avenue for bio-scaffold design. The properties of natural gums for scaffold fabrication has been extensively examined, emphasising their versatility and applications in biomaterials for repairing bone and cartilage. This chapter provides discussions on specific natural gums like locust bean gum, xanthan gum, guar gum, mastic gum and other gums, shedding light on their potential contributions to regenerative medicine.

Detailed examinations of xanthan gum (XG), gellan gum , gum Arabic (GA), guar gum (GG) and gum ghatti reveal their distinct roles and applications in tissue engineering. XG stands out for creating a conducive microenvironment for cell attachment and proliferation, with manifold applications from cartilage regeneration to bone engineering. GG, Gellan gum a notable anionic polysaccharide, emerges as a versatile biomaterial, especially in cartilage reconstruction, offering promising prospects for tissue regeneration.

The chapter underscores the significant role of natural gums as promising candidates for natural materials as bio-scaffold in tissue engineering. Their diverse properties and applications across various tissues and organs, underscore their potential for advancing regenerative medicine and biomedical applications. The ongoing scientific exploration of these natural gums promises sustained growth and innovation in the field, opening new opportunities for the creation of efficient and sustainable solution for TE .

References

Aavani F, Biazar E, Heshmatipour Z, Arabameri N, Kamalvand M, Nazbar A (2021) Applications of bacteria and their derived biomaterials for repair and tissue regeneration. Regen Med 16(6):581–605

Abu Elella MH, Goda ES, Gab-Allah MA, Hong SE, Pandit B, Lee S et al (2021) Xanthan gum-derived materials for applications in environment and eco-friendly materials: a review. J Environ Chem Eng 9(1):104702

Ahmad S, Ahmad M, Manzoor K, Purwar R, Ikram S (2019) A review on latest innovations in natural gums based hydrogels: preparations & applications. Int J Biol Macromol 136:870–890

Akkineni AR, Elci BS, Lode A, Gelinsky M (2022) Addition of high acyl gellan gum to low acyl gellan gum enables the blends 3D bioprintable. Gels 8(4):199

Aldana AA, Valente F, Dilley R, Doyle B (2021) Development of 3D bioprinted GelMA-alginate hydrogels with tunable mechanical properties. Bioprinting 21:e00105

Amjed N, Zeshan M, Farooq A, Naz S (2024) Applications of guar gum polysaccharide for pharmaceutical drug delivery: a review. Int J Biol Macromol 257:128390

Anandan D, Madhumathi G, Nambiraj NA, Jaiswal AK (2019) Gum based 3D composite scaffolds for bone tissue engineering applications. Carbohydr Polym 214:62–70

Anita Lett J, Sagadevan S, Fatimah I, Hoque ME, Lokanathan Y, Léonard E et al (2021) Recent advances in natural polymer-based hydroxyapatite scaffolds: properties and applications. Eur Polym J 148:110360

Bacelar AH, Silva-Correia J, Oliveira JM, Reis RL (2016) Recent progress in gellan gum hydrogels provided by functionalization strategies. J Mater Chem B 4(37):6164–6174

Banerjee ER (2018) Carboxymethyl guar gum nanoscaffold as matrix for cell growth in vitro. J Lung Pulm Respir Res 5(1):00156

Baranwal J, Barse B, Fais A, Delogu GL, Kumar A (2022) Biopolymer: a sustainable material for food and medical applications. Polymers 14(5). https://doi.org/10.3390/polym14050983

Barbosa RM, da Rocha DN, Bombaldi de Souza RF, Santos JL, Ferreira JRM, Moraes ÂM (2023) Cell-friendly chitosan-xanthan gum membranes incorporating hydroxyapatite designed for periodontal tissue regeneration. Pharmaceutics 15(2):705

Bektas EI, Gurel Pekozer G, Kök FN, Torun Kose G (2021) Evaluation of natural gum-based cryogels for soft tissue engineering. Carbohydr Polym 271:118407

Bombaldi de Souza RF, Bombaldi de Souza FC, Thorpe A, Mantovani D, Popat KC, Moraes ÂM (2020) Phosphorylation of chitosan to improve osteoinduction of chitosan/xanthan-based scaffolds for periosteal tissue engineering. Int J Biol Macromol 143:619–632

Bueno V, Bentini R, Luiz C, Leandro B, Denise P (2014) Synthesis and characterization of xanthan–hydroxyapatite nanocomposites for cellular uptake. Mater Sci Eng C 37:195–203. https://doi.org/10.1016/j.msec.2014.01.002

Bueno VB, Takahashi SH, Catalani LH, de Torresi SIC, Petri DFS (2015) Biocompatible xanthan/polypyrrole scaffolds for tissue engineering. Mater Sci Eng C 52:121–128. ISSN 0928-4931. https://doi.org/10.1016/j.msec.2015.03.023

Byram PK, Sunka KC, Barik A, Kaushal M, Dhara S, Chakravorty N (2020) Biomimetic silk fibroin and xanthan gum blended hydrogels for connective tissue regeneration. Int J Biol Macromol 165:874–882

Cernencu AI, Ioniţă M (2023) The current state of the art in gellan-based printing inks in tissue engineering. Carbohydr Polym 309:120676

Das A, Das A, Basu A, Datta P, Gupta M, Mukherjee A (2021) Newer guar gum ester/chicken feather keratin interact films for tissue engineering. Int J Biol Macromol 180:339–354

Dey P, Bal T, Gupta RN (2020) Fabrication and in vitro evaluation of electrospun gum ghatti-polyvinyl alcohol polymeric blend green nanofibre mat (GG-PVA NFM) as a novel material for polymeric scaffolds in wound healing. Polym Test 91:106826

Elbadawi M, Mosalagae M, Reaney IM, Meredith J (2017) Guar gum: a novel binder for ceramic extrusion. Ceram Int 43(18):16727–16735

Eltom A, Zhong G, Muhammad A (2019) Scaffold techniques and designs in tissue engineering functions and purposes: a review. Adv Mater Sci Eng 2019:1–13

Fadlelmoula AA (2018) 22 - Dietary gum arabic as animal feed additive. In: Mariod AA (ed) , pp 261–267. https://doi.org/10.1016/B978-0-12-812002-6.00022-1. isbn:9780128120026Gum Arabic, Academic Press

Farokhi M, Jonidi Shariatzadeh F, Solouk A, Mirzadeh H (2020) Alginate based scaffolds for cartilage tissue engineering: a review. Int J Polym Mater Polym Biomater 69(4):230–247

Farshidfar N, Iravani S, Varma RS (2023) Alginate-based biomaterials in tissue engineering and regenerative medicine. Mar Drugs 21(3):189

Ferris CJ, Gilmore KJ, Wallace GG, in het Panhuis M (2013) Modified gellan gum hydrogels for tissue engineering applications. Soft Matter 9(14):3705

Ferris CJ, Stevens LR, Gilmore KJ, Mume E, Greguric I, Kirchmajer DM et al (2015) Peptide modification of purified gellan gum. J Mater Chem B 3(6):1106–1115

Gering C, Rasheed A, Koivisto JT, Párraga J, Tuukkanen S, Kellomäki M (2021) Chemical modification strategies for viscosity-dependent processing of gellan gum. Carbohydr Polym 269:118335

Indurkar A, Bangde P, Gore M, Agrawal AK, Jain R, Dandekar P (2020) Fabrication of guar gum-gelatin scaffold for soft tissue engineering. Carbohydr Polym Technol Appl 1:100006

Indurkar A, Pandit A, Jain R, Dandekar P (2021) Plant-based biomaterials in tissue engineering. Bioprinting 21:e00127

Jafari A, Farahani M, Sedighi M, Rabiee N, Savoji H (2022) Carrageenans for tissue engineering and regenerative medicine applications: a review. Carbohydr Polym 281:119045

Jena SR, Dalei G, Das S, Nayak J, Pradhan M, Samanta L (2022) Harnessing the potential of dialdehyde alginate-xanthan gum hydrogels as niche bioscaffolds for tissue engineering. Int J Biol Macromol 207:493–506

Khillar PS, Jaiswal AK (2022) Gums for tissue engineering applications. In: Polysaccharides of microbial origin. Springer, Cham, pp 997–1024

Kırtel O, Avşar G, Erkorkmaz BA, Öner ET (2017) Microbial polysaccharides as food ingredients. In: Microbial production of food ingredients and additives. Elsevier. [cited 2024 Feb 7]. pp 347–383. https://linkinghub.elsevier.com/retrieve/pii/B978012811520600012X

Koyyada A, Orsu P (2021) Natural gum polysaccharides as efficient tissue engineering and drug delivery biopolymers. J Drug Deliv Sci Technol 63:102431

Kumar A, Rao KM, Han SS (2018) Application of xanthan gum as polysaccharide in tissue engineering: a review. Carbohydr Polym 180:128–144. ISSN 0144-8617, 10.1016/j.carbpol.2017.10.009. https://www.sciencedirect.com/science/article/pii/S0144861717311554

Kundu S, Das A, Basu A, Ghosh D, Datta P, Mukherjee A (2018) Carboxymethyl guar gum synthesis in homogeneous phase and macroporous 3D scaffolds design for tissue engineering. Carbohydr Polym 191:71–78

Lapomarda A, De Acutis A, Chiesa I, Fortunato GM, Montemurro F, De Maria C et al (2020) Pectin-GPTMS-based biomaterial: toward a sustainable bioprinting of 3D scaffolds for tissue engineering application. Biomacromolecules 21(2):319–327

Lett JA, Sagadevan S, Shahnavaz Z, Latha MB, Alagarswamy K, Hossain MAM et al (2020) Exploration of gum ghatti-modified porous scaffolds for bone tissue engineering applications. New J Chem 44(6):2389–2401

Lévesque SG, Lim RM, Shoichet MS (2005) Macroporous interconnected dextran scaffolds of controlled porosity for tissue-engineering applications. Biomaterials 26(35):7436–7446

Li Z, Zhang L, Mao C, Song Z, Li X, Liu C (2021) Preparation and characterization of konjac glucomannan and gum arabic composite gel. Int J Biol Macromol 183:2121–2130

Liu J, Yao P (2015) Injectable thermoresponsive hydrogel composed of xanthan gum and methylcellulose double networks with shear-thinning property. Carbohydr Polym 132:490–498. ISSN 0144-8617. https://doi.org/10.1016/j.carbpol.2015.06.013

Madappura AP, Madduri S (2023) A comprehensive review of silkfibroin hydrogels for cell and drug delivery applications in tissue engineering and regenerative medicine. Comput Struct Biotechnol J 21:4868–4886. https://doi.org/10.1016/j.csbj.2023.10.012

Mandal S, Hwang S, Shi SQ (2023) Guar gum, a low-cost sustainable biopolymer, for wastewater treatment: a review. Int J Biol Macromol 226:368–382

Mano JF, Silva GA, Azevedo HS, Malafaya PB, Sousa RA, Silva SS et al (2007) Natural origin biodegradable systems in tissue engineering and regenerative medicine: present status and some moving trends. J R Soc Interface 4(17):999–1030

Mirhosseini H, Amid BT (2012) A review study on chemical composition and molecular structure of newly plant gum exudates and seed gums. Food Res Int 46(1):387–398

Mirza S, Jolly R, Zia I, Saad Umar M, Owais M, Shakir M (2020) Bioactive gum arabic/κ--carrageenan-incorporated nano-hydroxyapatite nanocomposites and their relative biological functionalities in bone tissue engineering. ACS Omega 5(20):11279–11290

Mittal H, Mishra SB, Mishra AK, Kaith BS, Jindal R (2013) Flocculation characteristics and biodegradation studies of gum ghatti based hydrogels. Int J Biol Macromol 58:37–46

Muthukumar T, Song JE, Khang G (2019) Biological role of gellan gum in improving scaffold drug delivery, cell adhesion properties for tissue engineering applications. Molecules 24(24):4514

Nayak A, Pal D (2012) Natural polysaccharides for drug delivery in tissue engineering applications. Everymans Sci 46:347–352

Nayak AK, Ahmed SA, Tabish M, Hasnain MS (2019) Natural polysaccharides in tissue engineering applications. In: Natural polysaccharides in drug delivery and biomedical applications. Elsevier, pp 531–548

Ng JY, Obuobi S, Chua ML, Zhang C, Hong S, Kumar Y et al (2020) Biomimicry of microbial polysaccharide hydrogels for tissue engineering and regenerative medicine—a review. Carbohydr Polym 241:116345

Nsengiyumva EM, Alexandridis P (2022) Xanthan gum in aqueous solutions: fundamentals and applications. Int J Biol Macromol 216:583–604

Okamoto M (2019) The role of scaffolds in tissue engineering. In: Handbook of tissue engineering scaffolds: volume one. Elsevier, Amsterdam, pp 23–49

Oliveira JT, Santos TC, Martins L, Picciochi R, Marques AP, Castro AG et al (2010) Gellan gum injectable hydrogels for cartilage tissue engineering applications: *in vitro* studies and preliminary *in vivo* evaluation. Tissue Eng Part A 16(1):343–353

Oprea M, Voicu SI (2020) Cellulose composites with graphene for tissue engineering applications. Materials 13(23):5347

Padmanabhan VP, Prakash N, Sankara Narayanan TSN, Kulandaivelu R, Mohammad F, Obulapuram PK et al (2021) Drug delivery and in vitro biological effects of gum ghatti-modified hydroxyapatite nanoporous composites. Mater Chem Phys 263:124385

Pal K, Bharti D, Thakur G, Kim D (2023) Natural gums of plant and microbial origin for tissue engineering applications. In: Advances in biomedical polymers and composites. Elsevier, Amsterdam, pp 497–512

Patel S, Goyal A (2015) Applications of natural polymer gum arabic: a review. Int J Food Prop 18(5):986–998

Patra F, Dey M, Giri TK (2021) Hydrogels based on gum ghatti. In: Plant and algal hydrogels for drug delivery and regenerative medicine. Elsevier, pp 327–356

Pedram Rad Z, Mokhtari J, Abbasi M (2018) Fabrication and characterization of PCL/zein/gum arabic electrospun nanocomposite scaffold for skin tissue engineering. Mater Sci Eng C 93:356–366

Polez RT, Morits M, Jonkergouw C, Phiri J, Valle-Delgado JJ, Linder MB et al (2022) Biological activity of multicomponent bio-hydrogels loaded with tragacanth gum. Int J Biol Macromol 215:691–704

Prajapati VD, Jani GK, Moradiya NG, Randeria NP, Nagar BJ (2013) Locust bean gum: a versatile biopolymer. Carbohydr Polym 94(2):814–821

Prajapati VD, Maheriya PM, Roy SD (2021) Locust bean gum-derived hydrogels. In: Plant and algal hydrogels for drug delivery and regenerative medicine. Elsevier, pp 217–260

Prasad N, Thombare N, Sharma SC, Kumar S (2022) Production, processing, properties and applications of karaya (Sterculia species) gum. Ind Crop Prod 177:114467. ISSN 0926-6690. https://doi.org/10.1016/j.indcrop.2021.114467

Prasadh S, Wong RC-W (2018) Unraveling the mechanical strength of biomaterials used as a bone scaffold in oral and maxillofacial defects. Oral Sci Int 15(2):48–55. ISSN 1348-8643. https://doi.org/10.1016/S1348-8643(18)30005-3

Ragothaman M, Palanisamy T, Kalirajan C (2014) Collagen–poly(dialdehyde) guar gum based porous 3D scaffolds immobilized with growth factor for tissue engineering applications. Carbohydr Polym 114:399–406

Raj V, Lee JH, Shim JJ, Lee J (2021) Recent findings and future directions of grafted gum karaya polysaccharides and their various applications: a review. Carbohydr Polym 258:117687

Rathore HS, Sarubala M, Ramanathan G, Singaravelu S, Raja MD, Gupta S et al (2016) Fabrication of biomimetic porous novel sponge from gum kondagogu for wound dressing. Mater Lett 177:108–111

Rathore HS, Sivagnanam UT, Abraham LS, Prakash D, Panda RC, Senthilvelan T (2022) Green synthesized silver nanoparticles-impregnated novel gum kondagogu–chitosan biosheet for tissue engineering and wound healing applications. Polym Bull 79(9):7215–7227

Rekulapally R, Udayachandrika K, Hamlipur S, Sasidharan Nair A, Pal B, Singh S (2021) Tissue engineering of collagen scaffolds crosslinked with plant based polysaccharides. Prog Biomater 10(1):29–41

Satchanska G, Davidova S, Petrov PD (2024) Natural and synthetic polymers for biomedical and environmental applications. Polymers 16(8). https://doi.org/10.3390/polym16081159

Sahoo DR, Biswal T (2021) Alginate and its application to tissue engineering. SN Appl Sci 3(1):30

Santo VE, Frias AM, Carida M, Cancedda R, Gomes ME, Mano JF et al (2009) Carrageenan-based hydrogels for the controlled delivery of PDGF-BB in bone tissue engineering applications. Biomacromolecules 10(6):1392–1401

Sarika PR, Cinthya K, Jayakrishnan A, Anilkumar PR, James NR (2014) Modified gum arabic cross-linked gelatin scaffold for biomedical applications. Mater Sci Eng C 43:272–279

Sekar MP, Budharaju H, Zennifer A, Sethuraman S, Vermeulen N, Sundaramurthi D et al (2021) Current standards and ethical landscape of engineered tissues—3D bioprinting perspective. J Tissue Eng 12:204173142110276

Selvakumar G, Lonchin S (2020) Fabrication and characterization of collagen-oxidized pullulan scaffold for biomedical applications. Int J Biol Macromol 164:1592–1599

Serio F, da Cruz AF, Chandra A, Nobile C, Rossi GR, D'Amone E et al (2021) Electrospun polyvinyl-alcohol/gum arabic nanofibers: biomimetic platform for in vitro cell growth and cancer nanomedicine delivery. Int J Biol Macromol 188:764–773

Sharma G, Sharma S, Kumar A, Al-Muhtaseb AH, Naushad M, Ghfar AA et al (2018) Guar gum and its composites as potential materials for diverse applications: a review. Carbohydr Polym 199:534–545

Singh RS, Kaur N, Kennedy JF (2015) Pullulan and pullulan derivatives as promising biomolecules for drug and gene targeting. Carbohydr Polym 123:190–207

Singh RS, Kaur N, Rana V, Kennedy JF (2016) Recent insights on applications of pullulan in tissue engineering. Carbohydr Polym 153:455–462

Sinha D, Seth S, Das A, Bhattacharyya UK, Ghosh K, Nandi A et al (2022) Tuned gum ghatti and pectin for green synthesis of novel wound dressing material: engineering aspects and in vivo study. J Drug Deliv Sci Technol 76:103730

Smith AM, Shelton RM, Perrie Y, Harris JJ (2007) An initial evaluation of gellan gum as a material for tissue engineering applications. J Biomater Appl 22(3):241–254

Stevens LR, Gilmore KJ, Wallace GG, in het Panhuis M (2016) Tissue engineering with gellan gum. Biomater Sci 4(9):1276–1290

Su C, Chen Y, Tian S, Lu C, Lv Q (2022) Natural materials for 3D printing and their applications. Gels 8(11):748

Sun J, Tan H (2013) Alginate-based biomaterials for regenerative medicine applications. Materials 6(4):1285–1309

Thombare N, Jha U, Mishra S, Siddiqui MZ (2016) Guar gum as a promising starting material for diverse applications: a review. Int J Biol Macromol 88:361–372

Tiwari A, Grailer JJ, Pilla S, Steeber DA, Gong S (2009) Biodegradable hydrogels based on novel photopolymerizable guar gum–methacrylate macromonomers for in situ fabrication of tissue engineering scaffolds. Acta Biomater 5(9):3441–3452

Tortorella S, Inzalaco G, Dapporto F, Maturi M, Sambri L, Vetri Buratti V et al (2021) Biocompatible pectin-based hybrid hydrogels for tissue engineering applications. New J Chem 45(47):22386–22395

Trinetta V, Cutter CN (2016) Pullulan. In: Antimicrobial food packaging. Elsevier. [cited 2024 Feb 7]. pp 385–397. https://linkinghub.elsevier.com/retrieve/pii/B9780128007235000309

Tsai RY, Kuo TY, Hung SC, Lin CM, Hsien TY, Wang DM et al (2015) Use of gum arabic to improve the fabrication of chitosan–gelatin-based nanofibers for tissue engineering. Carbohydr Polym 115:525–532

Valente JFA, Valente TAM, Alves P, Ferreira P, Silva A, Correia IJ (2012) Alginate based scaffolds for bone tissue engineering. Mater Sci Eng C 32(8):2596–2603

Wu S, Xiao R, Wu Y, Xu L (2024) Advances in tissue engineering of gellan gum-based hydrogels. Carbohydr Polym 324:121484

Xu Z, Li Z, Jiang S, Bratlie KM (2018) Chemically modified gellan gum hydrogels with tunable properties for use as tissue engineering scaffolds. ACS Omega 3(6):6998–7007

Yao L (2015) Injectable shear-thinning xanthan gum hydrogel reinforced by mussel-inspired secondary crosslinking. RSC Adv 5(125). https://doi.org/10.1039/C5RA17246B

Zia I, Jolly R, Mirza S, Umar MS, Owais M, Shakir M (2020) Hydroxyapatite nanoparticles fortified xanthan gum–chitosan based polyelectrolyte complex scaffolds for supporting the osteo-friendly environment. ACS Appl Bio Mater 3(10):7133–7146

5 Natural Product-Loaded Chitosan and Collagen-Based Bio-Scaffold and Their Applications

Khushbu Gumber and Akshita Mahajan

Abstract

Biopolymers are rapidly evolving materials that are being investigated within the scientific community due to their diverse range of applications. Chitosan- and collagen-based materials, in particular, are suitable bioactive materials due to their biocompatibility and biodegradability as well as their remarkable features such as formability into different types of shapes and fabricability which diverse variety of bioactive agents. These molecules are widely used in a variety of biological sectors that include tissue engineering, methods for delivering medications, cosmetology, pharmacological, and the food industry, in their pure or composite form. This book chapter outlines the complete structure, categorization, derivatization, and uses of diverse natural products containing chitosan- and collagen-based bio-scaffolds.

Keywords

Collagen · Chitosan · Natural products · Biomaterials · Food industry · Cosmetology · Drug delivery · Tissue regeneration

5.1 Introduction

The increasing global population along with issues related to society and the environment has made sustainable production a critical issue. Particularly, material valorization may be an effective technique for producing high-value products by minimizing waste (Xiong et al. 2019). These days, polymers and polysaccharides are playing a significant role in the area of sustainable growth. Utilizing the

K. Gumber (✉) · A. Mahajan
Department of Chemistry, Chandigarh University, Gharaun, Mohali, Punjab, India

M. Kumar et al. (eds.), *Natural Product Inspired Scaffolds*,
https://doi.org/10.1007/978-981-97-3111-4_5

numerous advantages of innovative polymeric materials, such as blends and composites, has resulted in some unique developments in technology (Weng et al. 2023). They have numerous applications in the areas of pharmaceuticals, cosmetics, food processing, bone formation, healing of tissues, and drug delivery (Sukpaita et al. 2021). Introducing polysaccharides before anything else, chitosan is a polymer of *N*-acetyl-D-glucosamine and D-glucosamine which is produced by deacetylating chitin. It may be extracted from a variety of organisms, including snails, lobster, mold, and crabs. The majority of the extraction process uses the exoskeleton left overs of snails and lobster, which yield around 2000 tonnes of chitosan a year. It is nontoxic, biocompatible, and biodegradable molecule with exceptional antimicrobial and antioxidant properties. Before anything else, let's discuss about polysaccharides. Deacetylating chitin produces chitosan, which is a copolymer mixture of *N*-acetyl-D-glucosamine along with D-glucosamine. Although it may be obtained from seafood, shrimp, lobster, creatures such as insects, or fungal infection, the major source of chitosan is derived from the exoskeleton residues of these animals, producing around 2000 tonnes of chitosan each year. It is a naturally occurring, biocompatible, and nontoxic chemical with significant antioxidant and antimicrobial properties (Irastorza et al. 2021). On the other hand, collagen is a type of protein that is extracted from the bones of animals and the skin (Senadheera et al. 2023). The main forms of commercially available collagen are pigs and cows, although animal-borne spongiform disease (BSE) has caused a decrease in output (Senadheera et al. 2023) and other prions diseases eruption (Pal and Suresh 2016). For this purpose, utilizing fish by-products as a collagen source is an alternative that works over societal, religious, and health limitations (Coppola et al. 2020). The primary distinction that exists between coastal and collagen from mammals is the amount of the hydroxyproline and proline amino acids which are present (Yousefi et al. 2017). Because of its biological compatibility, low antigenic characteristic, high biodegradability, and strength for cell growth, collagen has found widespread use in the food, medicinal products, healthcare, and biomedical engineering sectors (Subhan et al. 2015).

In today's times, modifying and loading materials based on collagen and chitosan marks a new trend. A number of factors, including the surface modification and preparation methods used, and the efficient properties of collagen and the particles of chitosan may vary. It has a major impact on a variety of applications in widely various educational domains (Yanat and Schroën 2021). Biopolymer(s) and other inorganic material(s) were either combined or treated individually with chitosan and collagen in order to produce scaffolds for applications such as tissue engineering and medical regeneration. The loading of different drugs and natural goods into scaffolds made of collagen and chitosan exhibits the scaffolds' many uses in a wide range of businesses, including the food industry and pharmaceutical sectors. Reviewing recent research on the collagen and chitin biopolymers for a range of applications is the main objective of this article.

5.2 Biological Scaffold Designing

Biological scaffolds are designed on the basis of a variety of factors, depending on what they will be used for and their properties. The six different categories are defined below:

5.2.1 Microstructure

The scaffolding' microarchitecture consists of the permeability, size of pore, and interconnections between the pores. For cell relocation and scaffold connection, the pore size must be satisfactory. Furthermore, it provides the vascularization and penetration, which is the mass movement of waste materials and nutrients in and out of the cell and tissue surfaces. For the best tissue entry and vascularization in vivo, the largest hole size should be within 200–500 μm, whereas a more moderate pore size, which can vary within 75–100 μm in in vitro conditions, is preferable (Janmohammadi et al. 2021).

5.2.2 Degradability in Biological System

When scaffolds are grafted into life forms, they should ultimately break down chemically or enzymatically since they are just a temporary support system for developing cells or tissues. Biodegradability is the tendency of the scaffold materials to break down throughout the biological system (Ahmad Khan et al. 2002). The rate at which new bone is formed or tissue regeneration occurs ought to be roughly proportionate to the scaffold's biodegradation rate. "Creeping substitution" is the term that refers to what happens when freshly engineered tissues are successfully incorporated into the host bone to take the place of biomaterial scaffolds (Zhao et al. 2019). The environmentally friendly scaffolding products will either be reused as metabolites in another metabolic pathway or they will exit the body without interfering with neighboring tissues or other organs (Altun et al. 2019; Wang et al. 2023).

5.2.3 The Ability to Be Biocompatible

In order to keep the cells, the scaffold ought to be very biocompatible. To avoid powerful inflammatory reactions which may lead to the body reject itself, there should not be many chronic immunological responses; recovery should happen in a few of days (Fernández-Pérez et al. 2020; Wang et al. 2023).

5.2.4 The Bioactive Property

The capability of scaffolds to collaborate within the cells that surround the designed tissues is referred to as their bioactivity. Instead of causing scarring, bioactive scaffolds work to promote proper cell mobility or separation, tissue regeneration or neoformation, and interaction within the host, as opposed to standard passive biomaterials that usually exhibit minimal or no contact with the cells that surround them (Janmohammadi et al. 2021). These structures can also be used with physical indicators such as topography to improve cell alignment and morphology or with cell-adhesive ligands to encourage cell adhesion. Moreover, bioactive scaffolds can serve as a transporter or storage for growth-stimulating impulses like GFs to improve tissue regeneration (Ways et al. 2018).

5.2.5 Mechanical Characteristics

The scaffold materials used in the anatomical region of implantation should possess fundamental mechanical characteristics that are very similar with the characteristics of the native bones or organs. In addition to provide structural stability and support, it reduces the risk of stress shielding, osteopenia caused on by implants, and refracture that follows. The scaffold must also be strong enough to sustain surgical manipulation during transplant operations. The modulus of elasticity, tensile tension, fracture resistance, lack of energy, and elongation percentage are examples of mechanical qualities (Janmohammadi et al. 2021; Ways et al. 2018; Kumar et al. 2020, 2021a, b, c; Singh et al. 2022).

5.2.6 Technologies for Manufacturing

In terms of science and technology, it's essential to take availability and cost into account in addition to guaranteeing scaffold efficiency so that, when needed, scaling up manufacturing of the scaffolds is possible. Supplying and packaging the scaffolding to the physicians is another important thing to think about. While off-the-shelf availability is typically preferred by doctors to reduce waiting times before implantations, certain tissue types may not allow for this (Bu et al. 2019; Krishani et al. 2023).

5.3 Structural Properties of Chitosan and Collagen

Structure that consists of chitosan as well as collagen is seen in Fig. 5.1. The main amino groups in chitosan's structure are in charge of giving it characteristics like cohesiveness, transfection, in situ gelation, which precisely controlled drug release. It has been claimed that chitosan-based effective drug-delivering scaffolds that react

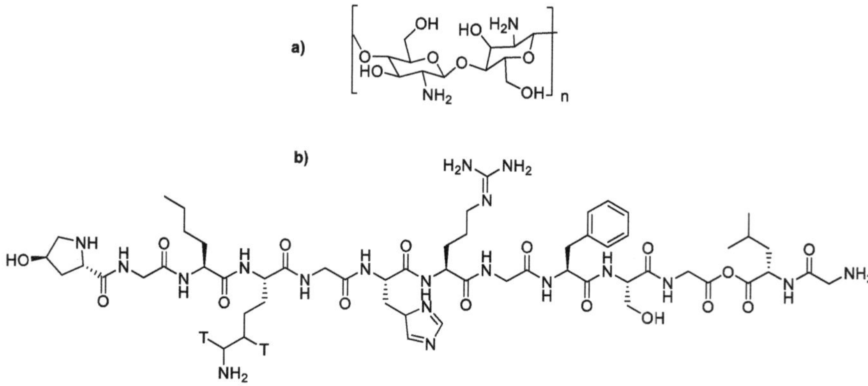

Fig. 5.1 The structure of (**a**) chitosan and (**b**) collagen I α-chain

to environmental cues can deliver medications locally and sustainably in the big bone defect region (Sukpaita et al. 2021).

Because of the formation of covalent cross-links between collagen fibrils, collagen has a stable structure. However, the biodegradability is determined by the type of the protein. The tissues' catabolic mechanisms that involve collagenase enzymolysis break down collagen. Conversely, depending on the need for collagen materials, cross-linking between fibrils might slow down the rate at which collagen degrades (Wang et al. 2023).

5.4 Various Ways to Use Polymeric Materials

5.4.1 Films

The solutions produced by the process of solution casting. The polymer is dissolved in a solvent made from organic materials, and the mixture is subsequently poured inside a mold. Evaporation of the solvent solution occurs by following the casting process (Yeo et al. 2020).

5.4.2 Scaffolds with Pores and Sponges

The most common technique to produce 3D porous structures is lyophilization, which removes the crystals of ice from the frozen solution to create an interconnected porous structure (Irastorza et al. 2021).

5.4.3 Hydrogel-Based

These are networks of polymers that can absorb and hold a lot of water. The hydrophilic segments of polymers in the aqueous phase have an excellent capacity for dissolution; thus, crosslinks are required to keep them from breaking (Hennink and van Nostrum 2012).

5.4.4 Fibers

Their ability to replicate the ultrafine fiber network of the extracellular matrix is enhanced by their linked structures, which facilitate cellular adhesion and proliferation more effectively. However, due to its huge surface area and thin structure, they are not very stable in water. Phase separation and electrospinning are the methods available for creating this kind of material. The latter also offers fibers that are randomly aligned and more closely resemble the extracellular matrix's three-dimensional structure (Ahmed et al. 2018).

5.5 Applications of Chitosan from Food to Biomedicals

Due to chitosan's special qualities—such as biodegradability, non-toxicity, antibacterial activity, and biocompatibility—application in several current sectors has received a lot of attention. Figure 5.2 summarizes the different areas of applications of chitosan.

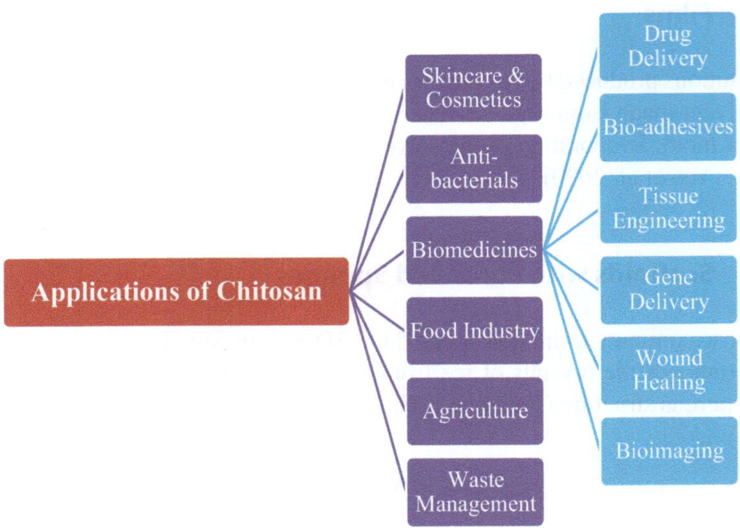

Fig. 5.2 Applications of chitosan in different industries

5.5.1 Biomedical Uses of Chitosan

There are many uses of chitosan in the field of medicines. It is used in the fields of tissue engineering, regeneration of bones, healing wounds, biodegradable bandages, and in many other applications. Because natural products are more biocompatible than synthetic ones, they are better suited for employment in biomedical research. Because lysozyme and other human enzymes can break down chitosan, it is biodegradable. Because chitosan is hydrophilic in nature and has structural characteristics with glycosaminoglycans, it can be used as a scaffold in tissue engineering. The monomeric component of chitosan, *N*-acetylglucosamine, is present in hyaluronic acid, which is an extracellular macromolecule essential for wound healing. It has been reported that chitosan enhances blood clotting and accelerates the tissue regeneration in the process of wound healing. In an earlier study, it was found that the photo-cross-linked electro spun rugs made of polyvinyl pyrrolidine (PVP) and quantized chitosan (QCS) prevented the growth of Gram-positive and Gram-negative bacteria. Results showed that the cross-linked QCS/PVP electrically spun mats could be potential materials for wound dressing. According to Jayakumar et al. (2010) and Sudha et al. (2014), infections caused by bacteria during the infectious stage of burn injuries may retard healing and can lead to dangerous consequences like infection. Minocycline hydrochloride and chitosan have been combined to yield a compound that has antibacterial and wound rehabilitation actions.

5.5.2 The Use of Chitosan in the Food Industry

Natural antioxidant molecules (Lourenço et al. 2019), which contain polyphenols, have gained special attention in the food manufacturing companies (Reddy et al. 2014). Chitosan nanoparticles have gained more interest in recent years due to their potential to improve their functional properties (Shukla et al. 2013). This is due to their increased reactivity and improved relative surface area of the mass due to their broader contact surface, which also increases their antibacterial activity (Badawy et al. 2020).

Chitosan is suitable for the production of antibacterial edible materials for the food sector because of its nontoxic antimicrobial properties. Gram-negative bacteria's endotoxins were less harmful when chitosan reacted (bound or chelated) with them. Chitosan has a significant chelating ability; therefore it might not be necessary to utilize additional chelating agents like EDTA. Chitosan inhibits a wide variety of microorganisms, including yeast, fungi, and viruses, and hence it can be utilized in a variety of antimicrobial materials in laboratory and in living cells settings in various forms (films, solutions, and composite materials). Other chelating chemicals including EDTA aren't essential because of chitosan's strong chelating activity (Qin et al. 2019; Abou Neel et al. 2013). The reactant's chemical as well as nutritional composition, the host, the natural quantity of nutrients, the kind of chitosan, the degree to which of chitosan polymerization, and environmental conditions (such as the substrate's moisture content or water activity, or both) all have an intrinsic effect on these antimicrobial effects. The antibacterial properties of chitosan vary mainly among living host plants; for example, *N*-carboxymethyl chitosan acts more rapidly on fungus and algal cells than on bacteria and has specific fungicidal effects on graminea and vegetable hosts. Furthermore, the growth of *Escherichia coli*, *Fusarium*, *Alternaria*, and *Helminthosporium* is suppressed when chitosan concentrations are above 0.025% (Ahmed and Ikram 2016). Certain microbe development is inhibited by the cationic amino groups found in chitosan, which bind to anionic groups.

Because of its excellent strength and ability to specifically permeabilize to O_2 and CO_2, chitosan is also a good material for food packaging (Liu et al. 2007). These qualities are important for maintaining the quality of food during storage, distribution, and transportation, as such procedures can cause food to spoil chemically and microbiologically (Shin et al. 2019). Active and/or intelligent packaging can be defined as that it is designed to increase the product's shelf life or provide details about its quality. Activated packaging extends the food's shelf life and enhances food quality by preventing food deterioration and infectious microorganisms (Irastorza et al. 2021). Food oxidation is considered as an important factor contributing to food decomposition, which leads to discoloration and sourness that can negatively impact on food quality (Shin et al. 2019). In the food companies, plant-based antioxidant compounds like polyphenols are gaining extra attention (Matos et al. 2017; Schini-Kerth et al. 2011). Chitosan nanoparticles increased attention in recent years due to their capability to improve their functional characteristics (Shukla et al. 2013). This is because their wider contact surface makes them more reactive and improves the mass's relative surface area, which increases antibacterial activity (Eckert et al. 2006). Chitosan's antibacterial properties allows it to protect food against illnesses caused by foodborne bacteria (Shin et al. 2019). This is the most frequent method of action, in which the anionic charges on the outermost layer of the microorganisms interact electrostatically with the a protonated amine of chitosan, resulting in the release of cellular components and finally in cell necrosis (Amato et al. 2018). Therefore, one of the main factors affecting food quality and shelf life is microbial deterioration, which can be prevented by chitosan. Because bacteria have a significant impact on food quality and shelf life, chitosan may be able to stop food from degrading due to bacteria (Gudjónsdóttir et al. 2015). Chitosan can be used as coatings or films in order to accomplish this goal (Lourenço et al. 2019). By immersing what is being eaten in a solution, the coatings—which are tiny layers—are immediately formed on the food's surface (Irastorza et al. 2021). Films are made up of layers which may be used as food pouch or wrapped around food (Islam et al. 2020). In this research, the duration of time that fish sausages covered with chitosan-derived coatings and films were stored was examined. Compared to coated sausages, the sausage wrapping around the film had a stronger texture, a pickled appearance, and a lower pH level and water content; on the other hand, the results indicated that chitosan coatings were invisible and might increase the shelf life of sausages by 15 days. Active and/or intelligent packaging is defined as anything which aims at increasing the product's shelf life or provide information about its quality. Activated packaging increases food shelf life, improves food

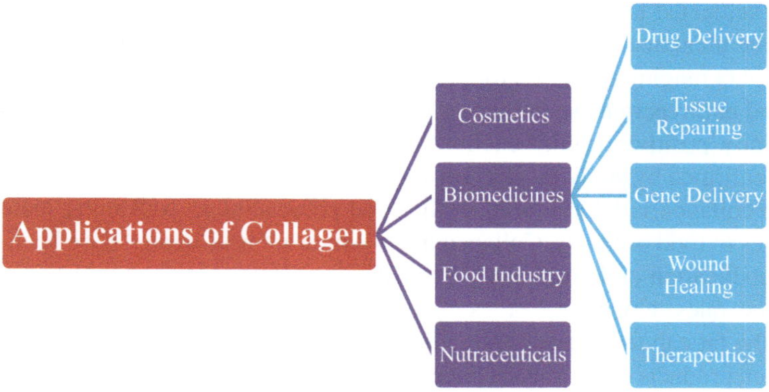

Fig. 5.3 Applications of collagen in various industries

safety, and keeps food from deteriorating or becoming contaminated by foodborne microorganisms (Irastorza et al. 2021).

5.5.3 Utilizing Chitosan in Agriculture

Most of chitosan's antibacterial properties have been shown in living host plants; for example, *N*-carboxymethyl chitosan works much faster on fungus and algae than on bacteria and has distinct fungicidal impacts in plant-based and graminea hosts. Furthermore, at chitosan concentrations above 0.025%, the growth of *Fusarium* species, *Alternaria* species, *Escherichia coli*, bacteria, and *Helminthosporium* is inhibited (Savard et al. 2002). Certain microbes are prevented from growing by the cationic amino groups included in chitosan, which bind to anionic groups.

5.5.4 Collagen in Industry

Collagen also had a wide range of application in below discussed industries (Fig. 5.3).

5.5.5 Collagen in Food and Cosmetic Industry

Collagen is widely utilized in the food and cosmetic sectors due to its many functional qualities and ability to be processed into a broad range of products, such as gel-like substances, porous scaffolding, strands, films, meshes, and micro-/nanoparticles. With molecular weights ranging from 15 to 250 kDa, gelatin is a kind of thermally broken-down collagen that has unique rheological properties. It is widely utilized in the food business as a dietary supplement, microencapsulating agent, and recyclable packaging material (Irastorza et al. 2021). Moreover, gelatin or collagen can be subjected to a variety of chemical hydrolysis processes, such as proteolytic or biochemical treatment, to produce peptides with biological function and molecular weights ranging from 300 to 8000 Da (Vidanarachchi et al. 2014). The antioxidant and antihypertensive properties of collagen-derived peptides, along with other potentially positive health benefits for cosmetic and pharmacological uses, have been confirmed by study results (Irastorza et al. 2021).

The collagen is widely used in the food companies as packaging materials or additives to food products. Collagen fibers which are heat-treated have been utilized as emulsification and particularly in acidic applications, as natural substitutes for synthetic ones. Collagen, for example, improves the rheological properties and keeps the fat cap from developing in oil-in-water emulsions (Irastorza et al. 2021). Collagen products of hydrolysis have therefore been successfully utilized to stop the breaking down of the lipids whose reaction byproducts pose a risk to human health because of their antioxidant properties (Nasri et al. 2013). This antioxidant activity can remove free radicals and is typically linked to the imidazole groups in histidine (Wade and Tucker 1998). In recent years, the growth of environmentally friendly packaging supplies to safeguard, preserve, and extend the shelf life of food has been greatly aided by the use of collagen-based films and coatings. Generally speaking, materials used in food packaging must maintain sensory qualities, be able to function as a barrier against the migration of oxygen and moisture, and prevent the oxidation of fat, color change, and microbial activity (Mihalca et al. 2021; Bhardwaj et al. 2019).

5.5.6 Peptides Used in the Nutraceutical and Cosmetics Industries

Because of its previously mentioned bioactive qualities, collagen has also been utilized more frequently in the production of cosmeceutical goods, such as skin antiaging compositions with providing nourishment, softening, and brilliantly glowing antioxidant, and UV protective capabilities (Kumar et al. 2012). It is stated that collagen that has been hydrolyzed is commonly utilized in cosmetic products due to its greater solubility at neutral pH which provide better results than native collagen (Mihranyan et al. 2012). Hydrolyzed collagen cannot replace the skin's lost collagen, having its small ability to permeate the outermost layer of the skin (Aguirre-Cruz et al. 2020). In this way, oral supplements and injections under the skin are of interest to the cosmetics industry. Because collagen is inexpensive, biodegradable, and easily made, injections of the material have become widely utilized to treat aging symptoms and dermal problems including acne scars (Sadick et al. 2009). Collagen injections into the dermis seem to promote fibroblasts' capacity to synthesize collagen as well as the development of an ideal physiological environment, which in turn boosts cell activity, hydration, and collagen synthesis (Ruszczak 2003). On the other hand, new observations reveal that oral collagen-based treatments improves the physical appearance of the skin. Collagen synthesis and cell density increase are significantly impacted by the quick digestion of liquid or tablet-form activated collagen peptides. Consequently, several studies have demonstrated the advantages of taking every day bioactive gelatin protein supplement in order to improve the skin's hydration, versatility, thickness, and wrinkles (Pu et al. 2023).

Other than cosmetics, according to recent studies, collagen has significant uses in functional foods and medical care applications, such as tablets and beverages (Irastorza et al. 2021). Collagen peptides have been shown to decrease hypertension and hyperlipidemia and also increasing fatty acid metabolism and the breakdown of fat. They can also prevent the liver from producing fatty acids, which reduces the amount of fat that accumulated in the liver (Bai et al. 2017). Additionally, they decreased blood sugar and enhanced insulin sensitivity, which may be related to their antioxidant properties (Aguirre-Cruz et al. 2020). Therefore, in order to prevent or cure diabetes and obesity of any kind, bioactive peptides of collagen are utilized to induce weight loss and improve blood sugar and cholesterol levels (Nath et al. 2017). Previous research indicates that ingesting meals or beverages enhanced with hydrolyzed collagen may be beneficial for treating osteoarthritis, mending wounds, growing new bone, and preserving mineral density (Papadopoulou et al. 2022). Since collagen can improve cardiovascular function, heal broken joints, improve the amount of lean muscle mass, and decrease recovery times, pharmaceutical companies offer nutritional supplements designed to boost nutrition. This industry also has a requirement for collagen (Byrne et al. 2004).

5.5.7 Biomedicine Application of Collagen

There are many biological application of collagen which includes the treatment of heart disease, cancer growth, problems with respiration, healing of wounds, and regeneration of skin. Understanding the significance of collagen using biomanufacturing techniques to generate collagen duplicates (Chan et al. 2020). The use of collagen as a biological ink in three-dimensional (3D) bioprinting techniques used for scaffolding geometry control and cell insertion into biological inks for tissue engineering applications is explained in this article. Collagen may be applied via the extrusion process, the use of inkjet printing, laser-assisted printing, and the stereolithography process printing methods as a bio ink (Blackstone et al. 2021), which explains how to produce collagen with a spatial arrangement for tissue engineering applications using electrospinning technology. Electrospun collagen scaffolds are produced by continuous fiber strands with the aid of an electric field. These fiber strands can be placed in patterns that are spatially structured or randomly oriented. A number of spatial properties, including as fibril sizes, crosslinking intensity levels, and porosity values, can be used to produce electrospun scaffolds that closely resemble the extracellular matrix observed naturally. The electrospun scaffolds can be further cross-linked to boost mechanical strength and resistance to deterioration while maintaining the survival of cells. Another spatial patterning method for creating collagen substrates with microgrooves is soft lithography. In other studies, micropatterning vascular smooth muscle cells reduces cell proliferation and induces smooth muscle differentiation, partly because of nuclear elongation and YAP signals (Nagayama 2021). The combination of these several manufacturing processes allows the manufacture of 3D collagen scaffolds for many biological purposes. In addition to creating collagen scaffolds, several methods have been found to produce hybrid collagen scaffolds that may be used to modify the chemical, mechanical, or cellular behavior of the scaffolds (Walimbe and Panitch 2020), the benefits of incorporating glycosaminoglycans into collagen hydrogels to regulate biological processes such as cell infiltration and blood vessel development. Alginate, heparin treatment, hyaluronic acid, and hyaluronic sulphate are the three major glycosaminoglycans. Patil et al. furthermore examine the current methods for developing hybrid scaffolds that blend collagen with biomaterials derived from both synthetic and natural sources (Patil and Masters 2020). In addition to full-length collagen, collagen-derived polypeptide patterns like DGEA or RGD can be connected to other biological materials. Thanks to developments in collagen synthesis, synthesized collagen and tissue-mimetic peptides have additionally improved. Fertala (2020) explain how regenerated collagen is used for medicine administration, tissue repair, and other applications.

In addition, collagen is utilized to treat a variety of biological issues, including cardiovascular disorders, cancers, healing of wounds, and airway mechanics (Mathew-Steiner et al. 2021), a description of collagen's role in regulating the healing of skin wounds. One of the most significant parts of skin, collagen affects a number of aspects of healing from wounds, such as vascular formation, inflammatory processes, and matrix cells reorganization. Collagen has been studied in clinical studies as a supportive treatment to promote wound healing. Another method for controlling wound healing is to treat with decellularized collagen-rich materials extracted from the epidermis, the umbilical cord, and other source (Yeung and Kelly 2021). Decellularized tissues have been used to study reconstructive surgery, soft tissue repair, burn therapy, and wound care. The authors proceeded over the most recent developments in non-decellularized all-encompassing stem cells, which further aid in tissue regeneration through the activity of the transplanted cells. Another experimental method for healing wounds is the formation of microvascular networks. It has been proven that fibronectin-modified protein hydrogels, which acts as supports for collagen hydrogels, promote further vascularization of co-cultured cells that include cells of smooth muscle and umbilical vein endothelial cells that are produced from human induced pluripotent stem cells. The mechanics of the airways, which comprise the vocal chords, lungs, bronchi vessels, and throat, are greatly affected by collagen. Liu et al. (2021) reviewed the current knowledge on the function of collagen in airway disorders and the corresponding changes in stiffness. The authors discussed about the association between disorders including those with cystic fibrosis, allergies and asthma, and chronic bronchitis and changes in the deposition of collagen, structure, crosslinking, and the mechanical changes that occur from these changes. By modifying airway mechanics, an understanding of collagen remodeling will open up new treatment options for these conditions. Research on the spatial arrangement of collagen in tumors is extremely important from a medical perspective. Ouellette et al. (2021) analyze the area of protein-collagen fiber structuring and how different tumor kinds and illness phases are related to it. The authors talk about the possibility of using collagen fiber patterning as a diagnostic or biomarker for cancer. The lacks in basic understanding are also discussed, including the molecular understanding of how and why collagen patterns spatially inside tumors. Sapudom et al. (2020) study a monocytic cell line's phenotypic modification and infiltration in relation to collagen fibril density in a methodical manner. The authors show that thick collagen matrices induced the synthesis of cytokines associated with inflammation and increased the activity of genes specific to macrophages (Newman et al. 2020). The function of 3D hybrid hydrogels consisting of collagen and elastin-like peptides in promoting adipogenesis in adipo-derived stem cell lines was investigated in the field of adipogenesis. The authors demonstrate how adipogenesis was influenced by the scaffold's mechanical characteristics, degree of cross linkage, and biological composition. According to the authors, fibronectin-functionalized collagen boosted adipogenesis by giving spheroid-like cellular structure to rigid and/or cross-linked hybrid hydrogels. Applications in cardiovascular medicine may involve treating an abdominal aortic aneurysm by stimulating the production of collagen and elastin in the aortic artery wall. Cunnane et al. (2021) discussed the impact of adipose stromal cell-derived extracellular vesicles on smooth muscle cell extracellular matrix remodeling. The extracellular vesicles facilitated smooth muscle migration, proliferating, and elastin as well as collagen deposition, according to the researchers. According to these results, extracellular vesicles may be an effective regulator of smooth muscle cells' and the production of ECM proteins. This might have implications for treating abdominal aortic aneurysms or for the growth of vascular applications for tissue engineering. In conclusion, the human organism has a large amount of collagen, and abnormal collagen processing or deposition is linked to a wide range of diseases. This collection promotes the developments and prospects for collagen to enhance our basic understanding of matrix biology and to use them in the management of several illnesses and wounds. Preclinical studies and even clinical stages of several kinds of illnesses are being intensively explored in the context of collagen. We believe that a better knowledge of how collagen may be targeted for regenerative medicine applications will be made possible in the future by the confluence of technology advancements in personalized medicine and computational biology. These factors lead us to conclude that collagen will continue to be a very active topic of study in the biomedical field for several decades to come.

5.6 Chitosan and Collagen Modified by Natural Products: A New Approach

In recent years, a wide range of physical and chemical techniques have been used to obtain the required characteristics from a broad range of collagen and chitosan materials. Chemical or physical methods can be utilized to modify the mechanical characteristics. These depend on the generation of non-covalent bonds by means of other agents like factors such as temperature or UV radiation. These are cost-effective, simple techniques to avoid extremely harmful chemicals from being introduced. However, the process can be complex sometimes, and it doesn't generate enough cross-linking to meet mechanical specifications (Liu et al. 2019). For that reason, chemical cross-linking is the most often used method of overcoming beyond the drawbacks of biomaterials, including collagen-chitosan scaffolding. In this process, molecules are joined by irreversible covalent bonds formed when a chemical substance is introduced to the material. Both enhanced mechanical qualities and increased stability are given by this method. On the other hand, the scaffold's unaffected cross-linker may enhance the cytotoxicity and result in processing issues due to the reaction itself (Khorshidi et al. 2016). Chemical cross-linking agents vary widely depending on the degree enhancement that is required. Because of its exceptional effectiveness—even at concentrations higher than 10%—it causes cytotoxicity, glutaraldehyde, chemicals used most frequently in the creation of synthetic tissues consisting of collagen and chitosan (Irastorza et al. 2021; Islam et al. 2020; Khan and Alamry 2021). Additionally, tannic acid possesses special antioxidant, antibacterial, anti-inflammatory, and antibacterial qualities (Yeo et al. 2020). *N*-hydroxysuccinimide (EDC-NHS) and *N*-ethyl-*N*′-(3-dimethylamino propyl) carbodiimide combined to maintains the porous structure (Costa et al. 2019). Additional bioconjugation is made possible by citric acid, which makes both biomaterials' binding sites available (Khorshidi et al. 2016). And last, *Gardenia jasminoides* contains the aglycone genipin. This naturally occurring material forms intramolecular and intermolecular unions when combined with primary amines of gelatin and chitosan, resulting in a totally stable structure that maintains its initial permeability, even at low concentrations (Lim et al. 2020). Cell-binding peptides that originate through proteins found in the extracellular matrix can be employed as needed to enable these kinds of interactions between the substance and the cell receptors. Because of its wide distribution throughout the organism, capacity to target multiple cell adhesion receptors, and biological effects on cell anchorage, behavior, and survival, the arginine-glycine-aspartic acid (RGD, which stands) sequence is currently the most useful and widely used peptide sequence in this regard (Jiang et al. 2013). According to reports, the RGD sequence permits the cytoskeleton of cells to be rearranged, which promotes cell adherence and propagation along collagen-chitosan scaffolds. It has been shown in vivo that a low concentration of RGD is required to efficiently improve the cell-material interaction and accelerate the formation of artificial tissues. However, an important challenge that still has to be addressed in scaffold design is the insufficient connection between polymer and cell membranes, which can result in foreign body responses such as irritation and infection. The material of the surface and the cell response designs, which regulate how the cells interact with the substrate, are connected to this problem. Collagen denaturalization can cause conformational modifications and make RGD sequences accessible, even though these sequences are frequently inaccessible in collagen's native shape (Renner et al. 2004). In addition to surface patterns, surface roughness, surface charge, and polarities have all been shown to be important factors in how well cells attached to biomaterials.

Developing a chitosan and collagen product with improved bio-functionality and bulk qualities can be achieved by modifying the material's surface. Coating, oxidation by lower-temperature plasma, a substance, and the inclusion of surfactants are surface modification techniques that are used to blend various derivatives of compounds. By modifying its surface and lyophilizing it, chitosan may also be made into a rigid, porous bio-scaffold. However, its biocompatibility could be affected by combining with other chemicals. Therefore, it is necessary to assess the biocompatibility of different collagen polymers and biomedical-grade chitosan in order to generate a substance that is of excellent quality and biodegradable for a variety of applications. One of its benefits is that they are easily processed into scaffolds, gels, membranes, microscopic particles, small the fibers, nanofibrils, pellets of different sizes, and sponge-like structures. These qualities, along with its biocompatibility, enable it to be applied in several ways, such as tissue engineering, wound healing, and drug administration (Kafi et al. 2019).

One important extra technique to improve chitosan performance is the cross-treatment procedure. The primary variables influencing the characteristics of chitosan are its molecular mass and degrees of degradation (DD), which represent the proportion of deacetylated units (Ahmad Khan et al. 2002). They can also be modified further. For instance, acidic depolymerization can reduce MW and deacetylation can reduce DD. Chitosan is an essential part of hydrogels that are utilized in pharmaceutical and medical applications. In cross-linked hydrogels, a three-dimensional network is formed as a result of cross-linkers connecting polymeric chains. Cross-linkers have a molecular weight (MW) that is much lower compared to the MW of the links that stand between two consecutive crosslinks. The cross-linking density of cross-linked hydrogels is mainly determined by the ratio of cross-linking reagent moles to polymer-repeating unit moles. Furthermore, a suitable number of cross-linking per chain is required to promote network creation. The four structures of chitosan hydrogels which are produced are ionic cross-linked, semi-interpenetrating, hybrid polymer, and chitosan cross-linked with itself (Tsao et al. 2011). Covalent or ionic bonds are the major interactions which form the network, depending on the type of cross-linker.

The incredible properties of collagen and chitosan, especially their versatility, have made them useful for the regeneration of many different types of tissues. Bone regeneration has been one of the most common uses. Collagen-chitosan membranes have been shown to exhibit greater elongation to fracture and a slower rate of degradability than collagen-only scaffolds, both of which are important for bone regeneration (Rahmany and Van Dyke 2013; Luo et al. 2020). In vitro experiments have shown that the collagen-chitosan combination demonstrates osteoid differentiation capability even in the absence of a specific differentiation medium, showing the ability to cause self-differentiation (Kafi et al. 2019). Additionally, the collagen-chitosan complexes have demonstrated optimal properties for in vivo repair of fractured bones. In this study, the models of calvarial bone deficiency, new bone can develop due to electrospun collagen-chitosan membranes. By week 8, large, mature bone tissue had fully repaired the cranial lesion. Similarly, in rabbit cartilage wounds, it has been shown that collagen-chitosan scaffolding cross-linked with carbodiimide promotes cartilage regeneration. As early as 1 month, new cartilage production was seen, and by 6.5 months, the amount and number of cells, together with the compressing and The storage moduli started to take on the appearance of typical cartilage. Recently, hydroxyapatite was included as a bioactive ingredient into a porous collagen-chitosan scaffold (Kafi et al. 2019). Chitosan-collagen scaffolds faithfully replicate the structural organization and functional properties of the extracellular matrix seen in a normal skin. Additionally, the combination has the ability to induce fibroblasts, which might speed up the formation of extracellular matrix components and be used in skin tissue engineering. It also possesses hemostasis and antimicrobial qualities. Due to the fact that cross-linked collagen-chitosan microfiber scaffolds have been demonstrated to be able to regulate water loss through evaporation, the microfiber structures may be more favorable with these qualities. Furthermore, it has been shown that, following 14 days after scaffold application, wounds re-epithelization as a result of directed fibroblast infiltration and collagen remodeling that occurs scaffold disintegration (Irastorza et al. 2021). The thickness of the graft is important in skin tissue engineering since it affects how well the construct is implanted. Research has demonstrated that thinner scaffolds—particularly those that are 0.5 mm thick—promote greater collagen remodeling and organized fibroblast infiltration. In skin defects, these implants completely degraded by week 16, at which point they were replaced by freshly created vasculature, well-organized host tissue, 75.9% of the original skin's tensile strength. Recently, there have been attempts to treat disorders affecting the neural tissue, like spinal cord damage, with a collagen-chitosan composition. In this case, 3D-printed scaffolds may have significant medicinal benefits by gradually re-establishing an environment conducive to axonal regeneration by connecting axons across the fracture and allowing cell migration. Nerve fiber regeneration, neurological function, and locomotor recovery were all attained when the implants were placed in rat models (Kafi et al. 2019).

Chitosan hydrogel sheets are extremely soft and will not irritate the skin when used as dressings for wounds. Furthermore, compared to extracts in their pure form, plant extracts exhibit decreased efficiency and release in hydrogel films and demonstrate incompatibility in hydrogel polymeric formulations as natural antimicrobials. More powerful healers that use a natural approach to treat wounds are plants. *Salix alba*, *Juglans regia*, and gentamycin sulphate-encapsulated hydrogel film demonstrated antibacterial and antifungal properties in both disc diffusion and microbiocidal tests. Given that these characteristics are the most crucial for the best materials for wound healing and give a future location for treating wound care systems (Saghazadeh et al. 2018).

According to a different study, aloe vera's healing properties might be very helpful in the development of active materials for wound dressings. The membranes based on chitosan and aloe vera were developed—the materials for wound dressings. Aloe vera was added to chitosan to increase the antibacterial activity of the resultant membranes (Silva et al. 2013).

Bioglass 45S5 was manufactured together with chitosan (CTS), collagen (Coll), and carbs that were cross-linked using tannic acid (TA). Furthermore, the assessment of the cell viability on the produced scaffolds was conducted both in the presence and absence of glucocorticoid and ascorbic acid. The main goal of this work was to compare how bioglass 45S5 affected both the biological and physicochemical characteristics of scaffolds. The results demonstrated the stability of the biopolymer blends-based scaffolds joined by TA in an aqueous environment. Through the use of a scanning electron microscope, a linked network of pores within a porous scaffold construction could be seen. Bioglass nanoparticles improved the mechanical properties of composite materials and reduced their rate of degradation. Better biological properties were obtained with the incorporation of 20% of tannic acid as opposed to 5%. However, the cells' response to the addition of bioglass 45S5 was not significantly affected (Yeo et al. 2020).

For use as dressings for wounds, scaffolds derived from chitosan (CS) loaded with grape seed extract (GSE) and *Pinus radiata* extract peel (PE) have been produced successfully. The biological and physicochemical properties of GSE and PE were examined in relation to their impact on CS scaffolds. The porous structure of all the scaffolds showed the capacity to absorb over 70 times their own weight when subjected to a solution of phosphate buffer and blood. Over time, the plasma absorption and breakdown rates of the CS scaffolds were enhanced by the addition of GSE and PE. More than 95% of the scaffolds had a clotting capacity, and their surfaces allowed for the attachment of red blood cells. GSE and PE were gradually released from the CS scaffolds. This work suggests a unique synthesis method that, by immobilizing GSE and PE on CS scaffolds, may be used to create innovative material platforms with great potential for wound dressing applications. Studies on the viability of cells in vitro showed that HaCaT cells were not harmed by any of the scaffolds. Crucially, in comparison to the CS scaffold, the combination of GSE along with PE substantially enhanced cell viability. This study proposes a new synthesis approach that might be utilized to build novel material platforms with tremendous promise for wound dressing applications, by immobilizing GSE and PE on CS scaffolds (Borges-Vilches et al. 2023).

The food industry is studying the application of chitosan nanoparticles in food compositions. As an example, when storing beef burgers in the refrigerator, the essential oil of cinnamon has been utilized to stop the burgers from breaking down (Miguel et al. 2017) or those that cover Atlantic horse fish fillets by mixing them with gallic acid (Zarandona et al. 2021). Furthermore, nanoparticles of chitosan can be used to release active ingredients over an extended period of time. In response to consumer concerns over the health and nutritional value of food goods, smart packaging was developed lately to convey data regarding the food being consumed quality of packaged foods. The major portion of smart food packaging techniques has generally been employed to use quality parameters, such as pH-sensitive color-changing films, to determine how fresh fruits, seafood, and meat are (Ahmed et al. 2018). The compounds called anthocyanins, which are extracted from different food wastes, are commonly utilized as pH indicators. The pH-sensitive purple maize anthocyanins have been used in chitosan-silver nanoparticle films. Furthermore, rainbow trout fillets coating using chitosan-*Ferulago angulata* essential oil extended the fish's shelf life by up to 16 days by reducing the rise of thiobarbituric acid reactive chemicals during storage at 4 °C. This was made feasible by the chitosan films' inclusion of anthocyanins made by purple and black vegetable peels (Qin et al. 2019). Apart from improving the antioxidant capacity, bioactive chemicals have the potential to enhance other features of films made of chitosan. The ability of films made from chitosan to break down radicals that are harmful (2,2-diphenyl-1-picrylhydrazil, or DPPH) increased by 43% upon the addition of oak globulus essential oil. Moreover, other naturally occurring antioxidants obtained from fruit waste, such as banana peel extracts, increased the antioxidant capacity of chitosan films while lowering their soluble capacity of the liquid, the amount of moisture, and permeability of water vapor (Islam et al. 2020). Microbiological activity affects food safety in addition to antioxidant ability (Irastorza et al. 2021). Similarly to antioxidants, developed antimicrobial substances like sodium nitrite might have negative impacts on the user's health (Alahakoon et al. 2015). Consequently, the food company seeks for natural alternatives to these substances that don't affect the meal's sensory properties (Irastorza et al. 2021). Essential oils have antibacterial properties along with antioxidant properties while added to chitosan compositions.

Kaczmarek-Szczepańska et al. (2023) published a study on the structure, swelling, and rate of deterioration of collagen-based structures modified by gallic, ferulic, and caffeic acidic substances. It also covered the biological characteristics of the scaffolds, including their antibacterial, hemo- and cytocompatible, and antioxidant activity. When compared to scaffolds made solely of collagen, scaffolds containing phenolic acid had a greater rate of swelling and enzymatic stability. Additionally, the scaffolds' radical scavenging performance ranged from 85% to 91%. Every scaffold was compatible with the surrounding tissues and non-hemolytic. Although a markedly elevated release of LDH was seen in collagen enhanced by ferulic acid, which may have been detrimental to hFOB cells, every material under investigation exhibited antibacterial efficacy against the two kinds of *Staphylococcus aureus* and a variety of *Escherichia coli*. It is likely that phenolic acids, which include substances such gallic, ferulic, and caffeic acid, modify and give collagen-based scaffolds their unique biological characteristics.

5.7 Difficulties and New Opportunities

With the advancement of the application of collagen and chitin scaffolds, novel possibilities for nano-engineering and natural material encapsulation are becoming available in order to enhance the current state of current practice and make it more biocompatible and environmentally friendly. When compared to conventional chitosan coatings, recent research on chitosan nanoparticles in the food sector demonstrate that these particles can enhance practical characteristics because of their greater contact surface. As a result, new approaches to the packaging of food are being tested in an effort to achieve a more well-known surface inclusion due to more noticeable modifications. In the biomedical field, a lot of work is being done to change the properties of chitin and collagen structures according to medicinal requirements. Thus, as enhanced material fabrication techniques enable the rapid and repeatable creation of intricate three-dimensional forms, they have signified a remarkable advancement in the discipline. Furthermore, the development of imaginative and creative 4D printing has allowed for the fusion of conventional and "shrewd" materials. They have the characteristics that allow them to be stimuli-responsive molecules (to react to heat, light, humidity, electromagnetic effects, or pH), changing their characteristics to suit the microenvironment (by producing an electrical flow, changing color or shape, becoming bioactive, or performing predetermined tasks). The ability of substances to have dynamic responses and regulate the spatial and temporal growth of unique molecules is an advantage of 4D printing. Moreover, 4D printing does away with the requirement for additional equipment or post-handling procedures. Consequently, 4D printing may have negative repercussions in the therapeutic field given its incredible promise for secure remote-control therapies like medication administration, biosensors, or regenerative medicine (Irastorza et al. 2021).

References

Abou Neel EA, Bozec L, Knowles JC, Syed O, Mudera V, Day R, Hyun JK (2013) Collagen—emerging collagen based therapies hit the patient. Adv Drug Deliv Rev 65(4):429–456. https://doi.org/10.1016/j.addr.2012.08.010

Aguirre-Cruz G, León-López A, Cruz-Gómez V, Jiménez-Alvarado R, Aguirre-Álvarez G (2020) Collagen hydrolysates for skin protection: oral administration and topical formulation. Antioxidants 9(2):181. https://doi.org/10.3390/antiox9020181

Ahmad Khan T, Khiang Peh K, Seng Ch'ng H (2002) Reporting degree of deacetylation values of chitosan: the influence of analytical methods. J Pharm Pharm Sci 5(3):205–212. www.ualberta.ca/~csps

Ahmed S, Ikram S (2016) Chitosan based scaffolds and their applications in wound healing. Achiev Life Sci 10(1):27–37. https://doi.org/10.1016/j.als.2016.04.001

Ahmed I, Lin H, Zou L, Li Z, Brody AL, Qazi IM, Lv L, Pavase TR, Khan MU, Khan S, Sun L (2018) An overview of smart packaging technologies for monitoring safety and quality of meat and meat products. Packag Technol Sci 31(7):449–471. https://doi.org/10.1002/pts.2380

Alahakoon AU, Jayasena DD, Ramachandra S, Jo C (2015) Alternatives to nitrite in processed meat: up to date. Trends Food Sci Technol 45(1):37–49. https://doi.org/10.1016/j.tifs.2015.05.008

Altun E, Aydogdu MO, Togay SO, Sengil AZ, Ekren N, Haskoylu ME, Oner ET, Altuncu NA, Ozturk G, Crabbe-Mann M, Ahmed J, Gunduz O, Edirisinghe M (2019) Bioinspired scaffold induced regeneration of neural tissue. Eur Polym J 114:98–108. https://doi.org/10.1016/j.eurpolymj.2019.02.008

Amato A, Migneco LM, Martinelli A, Pietrelli L, Piozzi A, Francolini I (2018) Antimicrobial activity of catechol functionalized-chitosan versus Staphylococcus epidermidis. Carbohydr Polym 179:273–281. https://doi.org/10.1016/j.carbpol.2017.09.073

Badawy MEI, Lotfy TMR, Shawir SMS (2020) Facile synthesis and characterizations of antibacterial and antioxidant of chitosan monoterpene nanoparticles and their applications in preserving minced meat. Int J Biol Macromol 156:127–136. https://doi.org/10.1016/j.ijbiomac.2020.04.044

Bai X, Hong W, Cai P, Chen Y, Xu C, Cao D, Yu W, Zhao Z, Huang M, Jin J (2017) Valproate induced hepatic steatosis by enhanced fatty acid uptake and triglyceride synthesis. Toxicol Appl Pharmacol 324:12–25. https://doi.org/10.1016/j.taap.2017.03.022

Bhardwaj A, Alam T, Talwar N (2019) Recent advances in active packaging of agri-food products: a review. J Postharvest Technol 7(1):33–62. http://www.jpht.info

Blackstone B, Gallentine S, Powell H (2021) Collagen-based electrospun materials for tissue engineering: a systematic review. Bioengineering 8:39

Borges-Vilches J, Unalan I, Aguayo CR, Fernández K, Boccaccini AR (2023) Multifunctional chitosan scaffold platforms loaded with natural polyphenolic extracts for wound dressing applications. Biomacromolecules 24(11):5183–5193. https://doi.org/10.1021/acs.biomac.3c00727

Bu Y, Ma J, Bei J, Wang S (2019) Surface modification of aliphatic polyester to enhance biocompatibility. Front Bioeng Biotechnol 7:98. https://doi.org/10.3389/fbioe.2019.00098

Byrne C, Twist C, Eston R (2004) Neuromuscular function after exercise-induced muscle damage theoretical and applied implications. Sports Med 34(1):49–69

Chan WW, Yeo DCL, Tan V, Singh S, Choudhury D, Naing MW (2020) Additive biomanufacturing with collagen inks. Bioengineering 7:66

Coppola D, Oliviero M, Vitale GA, Lauritano C, D'Ambra I, Iannace S, de Pascale D (2020) Marine collagen from alternative and sustainable sources: extraction, processing and applications. Mar Drugs 18(4):214. https://doi.org/10.3390/md18040214

Costa JB, Lima MJ, Sampaio MJ, Neves MC, Faria JL, Morales-Torres S, Tavares APM, Silva CG (2019) Enhanced biocatalytic sustainability of laccase by immobilization on functionalized carbon nanotubes/polysulfone membranes. Chem Eng J 355:974–985. https://doi.org/10.1016/j.cej.2018.08.178

Cunnane E, Ramaswamy A, Lorentz K, Vorp D, Weinbaum J (2021) Extracellular vesicles derived from primary adipose stromal cells induce elastin and collagen deposition by smooth muscle cells within 3D fibrin gel culture. Bioengineering 8:51

Eckert R, Brady KM, Greenberg EP, Qi F, Yarbrough DK, He J, McHardy I, Anderson MH, Shi W (2006) Enhancement of antimicrobial activity against Pseudomonas aeruginosa by coadministration of G10KHc and tobramycin. Antimicrob Agents Chemother 50(11):3833–3838. https://doi.org/10.1128/AAC.00509-06

Fernández-Pérez J, Kador KE, Lynch AP, Ahearne M (2020) Characterization of extracellular matrix modified poly(ε-caprolactone) electrospun scaffolds with differing fiber orientations for corneal stroma regeneration. Mater Sci Eng C 108:110415. https://doi.org/10.1016/j.msec.2019.110415

Fertala A (2020) Three decades of research on recombinant collagens: reinventing the wheel or developing new biomedical products? Bioengineering 7:155

Gudjónsdóttir M, Gacutan MD, Mendes AC, Chronakis IS, Jespersen L, Karlsson AH (2015) Effects of electrospun chitosan wrapping for dry-ageing of beef, as studied by microbiological, physicochemical and low-field nuclear magnetic resonance analysis. Food Chem 184:167–175. https://doi.org/10.1016/j.foodchem.2015.03.088

Hennink WE, van Nostrum CF (2012) Novel crosslinking methods to design hydrogels. Adv Drug Deliv Rev 64(Suppl):223–236. https://doi.org/10.1016/j.addr.2012.09.009

Irastorza A, Zarandona I, Andonegi M, Guerrero P, de la Caba K (2021) The versatility of collagen and chitosan: from food to biomedical applications. Food Hydrocoll 116:106633. https://doi.org/10.1016/j.foodhyd.2021.106633

Islam MM, Shahruzzaman M, Biswas S, Nurus Sakib M, Rashid TU (2020) Chitosan based bioactive materials in tissue engineering applications—a review. Bioact Mater 5(1):164–183. https://doi.org/10.1016/j.bioactmat.2020.01.012

Janmohammadi M, Nourbakhsh MS, Bonakdar S (2021) Electrospun skin tissue engineering scaffolds based on polycaprolactone/hyaluronic acid/L-ascorbic acid. Fibers Polym 22(1):19–29. https://doi.org/10.1007/s12221-021-0036-8

Jayakumar R, Menon D, Manzoor K, Nair SV, Tamura H (2010) Biomedical applications of chitin and chitosan based nanomaterials—a short review. Carbohydr Polym 82(2):227–232. https://doi.org/10.1016/j.carbpol.2010.04.074

Jiang LY, Lv B, Luo Y (2013) The effects of an RGD-PAMAM dendrimer conjugate in 3D spheroid culture on cell proliferation, expression and aggregation. Biomaterials 34(11):2665–2673. https://doi.org/10.1016/j.biomaterials.2013.01.003

Kaczmarek-Szczepańska B, Polkowska I, Małek M, Kluczyński J, Paździor-Czapula K, Wekwejt M, Michno A, Ronowska A, Pałubicka A, Nowicka B, Otrocka-Domagała I (2023) The characterization of collagen-based scaffolds modified with phenolic acids for tissue engineering application. Sci Rep 13(1):9966. https://doi.org/10.1038/s41598-023-37161-6

Kafi MA, Aktar MK, Phanny Y, Todo M (2019) Adhesion, proliferation and differentiation of human mesenchymal stem cell on chitosan/collagen composite scaffold. J Mater Sci Mater Med 30(12):131. https://doi.org/10.1007/s10856-019-6341-8

Khorshidi S, Solouk A, Mirzadeh H, Mazinani S, Lagaron JM, Sharifi S, Ramakrishna S (2016) A review of key challenges of electrospun scaffolds for tissue-engineering applications. J Tissue Eng Regen Med 10(9):715–738. https://doi.org/10.1002/term.1978

Krishani M, Shin WY, Suhaimi H, Sambudi NS (2023) Development of scaffolds from bio-based natural materials for tissue regeneration applications: a review. Gels 9(2):100. https://doi.org/10.3390/gels9020100

Kumar S, Sharma S, Baldi A (2012) Herbal cosmetics: used for skin and hair. 2012;4. https://www.researchgate.net/publication/235944029

Kumar M, Saini JS, Bhunia H (2020) Performance of mechanical joints prepared from carbon-fiber-reinforced polymer nanocomposites under accelerated environmental aging. J Mater Eng Perform 29:7511–7525. https://doi.org/10.1007/s11665-020-05216-8

Kumar M, Saini JS, Bhunia H (2021a) Investigations on MWCNT embedded carbon/epoxy composite joints subjected to hygrothermal aging under bolt preloads. Fibers Polym 22(7):1957–1975. https://doi.org/10.1007/s12221-021-0834-z

Kumar M, Saini JS, Bhunia H, Chowdhury SR (2021b) Behavior of mechanical joints prepared from EB cured CFRP nanocomposites subjected to hygrothermal aging under bolt preloads. Appl Compos Mater 28:271–296. https://doi.org/10.1007/s10443-020-09864-w

Kumar M, Saini JS, Bhunia H, Chowdhury SR (2021c) Aging of bolted joints prepared from electron-beam-cured multiwalled carbon nanotube-based nanocomposites with variable torques. Polym Compos 42(8):4082–4104. https://doi.org/10.1002/pc.26119

Lim KS, Galarraga JH, Cui X, Lindberg GCJ, Burdick JA, Woodfield TBF (2020) Fundamentals and applications of photo-cross-linking in bioprinting. Chem Rev 120(19):10662–10694. https://doi.org/10.1021/acs.chemrev.9b00812

Liu C, Tan Y, Liu C, Chen X, Yu L (2007) Preparations, characterizations and applications of chitosan-based nanoparticles. J Ocean Univ China 6(3):237–243. https://doi.org/10.1007/s11802-007-0237-9

Liu X, Zheng C, Luo X, Wang X, Jiang H (2019) Recent advances of collagen-based biomaterials: multi-hierarchical structure, modification and biomedical applications. Mater Sci Eng C 99:1509–1522. https://doi.org/10.1016/j.msec.2019.02.070

Liu L, Stephens B, Bergman M, May A, Chiang T (2021) Role of collagen in airway mechanics. Bioengineering 8:13

Lourenço SC, Moldão-Martins M, Alves VD (2019) Antioxidants of natural plant origins: from sources to food industry applications. Molecules 24(22):4132. https://doi.org/10.3390/molecules24224132

Luo M, Guo L, Yu M, Wang H (2020) The psychological and mental impact of coronavirus disease 2019 (COVID-19) on medical staff and general public—a systematic review and meta-analysis. Psychiatry Res 291:113190. https://doi.org/10.1016/j.psychres.2020.113190

Mathew-Steiner SS, Roy S, Sen CK (2021) Collagen in wound healing. Bioengineering 8:63

Matos J, Cardoso C, Bandarra NM, Afonso C (2017) Microalgae as healthy ingredients for functional food: a review. Food Funct 8(8):2672–2685. https://doi.org/10.1039/c7fo00409e

Miguel SP, Ribeiro MP, Coutinho P, Correia IJ (2017) Electrospun polycaprolactone/aloe vera_chitosan nanofibrous asymmetric membranes aimed for wound healing applications. Polymers 9(5):183. https://doi.org/10.3390/polym9050183

Mihalca V, Kerezsi AD, Weber A, Gruber-Traub C, Schmucker J, Vodnar DC, Dulf FV, Socaci SA, Fărcaş A, Mureşan CI, Suharoschi R, Pop OL (2021) Protein-based films and coatings for food industry applications. Polymers 13(5):1–24. https://doi.org/10.3390/polym13050769

Mihranyan A, Ferraz N, Strømme M (2012) Current status and future prospects of nanotechnology in cosmetics. Prog Mater Sci 57(5):875–910. https://doi.org/10.1016/j.pmatsci.2011.10.001

Nagayama K (2021) A loss of nuclear—cytoskeletal interactions in vascular smooth muscle cell differentiation induced by a micro-grooved collagen substrate enabling the modeling of an in vivo cell arrangement. Bioengineering 8:124

Nasri R, Younes I, Jridi M, Trigui M, Bougatef A, Nedjar-Arroume N, Dhulster P, Nasri M, Karra-Châabouni M (2013) ACE inhibitory and antioxidative activities of Goby (Zosterissessor ophiocephalus) fish protein hydrolysates: effect on meat lipid oxidation. Food Res Int 54(1):552–561. https://doi.org/10.1016/j.foodres.2013.07.001

Nath S, Ghosh SK, Choudhury Y (2017) A murine model of type 2 diabetes mellitus developed using a combination of high fat diet and multiple low doses of streptozotocin treatment mimics the metabolic characteristics of type 2 diabetes mellitus in humans. J Pharmacol Toxicol Methods 84:20–30. https://doi.org/10.1016/j.vascn.2016.10.007

Newman K, Clark K, Gurumurthy B, Pal P, Janorkar AV (2020) Elastin-collagen based hydrogels as model scaffolds to induce three-dimensional adipocyte culture from adipose derived stem cells. Bioengineering 7:110

Ouellette J, Drifka C, Pointer K, Liu Y, Lieberthal T, Kao W, Kuo J, Loeffler A, Eliceiri K (2021) Navigating the collagen jungle: the biomedical potential of fiber organization in cancer. Bioengineering 8:17

Pal GK, Suresh PV (2016) Sustainable valorisation of seafood by-products: recovery of collagen and development of collagen-based novel functional food ingredients. Innov Food Sci Emerg Technol 37(Part B):201–215. https://doi.org/10.1016/j.ifset.2016.03.015

Papadopoulou SK, Mantzorou M, Kondyli-Sarika F, Alexandropoulou I, Papathanasiou J, Voulgaridou G, Nikolaidis PT (2022) The key role of nutritional elements on sport rehabilitation and the effects of nutrients intake. Sports 10(6):84. https://doi.org/10.3390/sports10060084

Patil VA, Masters KS (2020) Engineered collagen matrices. Bioengineering 7:163

Pu SY, Huang YL, Pu CM, Kang YN, Hoang KD, Chen KH, Chen C (2023) Effects of oral collagen for skin anti-aging: a systematic review and meta-analysis. Nutrients 15(9):2080. https://doi.org/10.3390/nu15092080

Qin Y, Liu Y, Yuan L, Yong H, Liu J (2019) Preparation and characterization of antioxidant, antimicrobial and pH-sensitive films based on chitosan, silver nanoparticles and purple corn extract. Food Hydrocoll 96:102–111. https://doi.org/10.1016/j.foodhyd.2019.05.017

Rahmany MB, Van Dyke M (2013) Biomimetic approaches to modulate cellular adhesion in biomaterials: a review. Acta Biomater 9(3):5431–5437. https://doi.org/10.1016/j.actbio.2012.11.019

Reddy BU, Mullick R, Kumar A, Sudha G, Srinivasan N, Das S (2014) Small molecule inhibitors of HCV replication from pomegranate. Sci Rep 4:5411. https://doi.org/10.1038/srep05411

Renner C, Saccà B, Moroder L (2004) Synthetic heterotrimeric collagen peptides as mimics of cell adhesion sites of the basement membrane. Biopolymers 76(1):34–47

Ruszczak Z (2003) Effect of collagen matrices on dermal wound healing. Adv Drug Deliv Rev 55(12):1595–1611. https://doi.org/10.1016/j.addr.2003.08.003

Sadick NS, Karcher C, Palmisano L (2009) Cosmetic dermatology of the aging face. Clin Dermatol 27(3 Suppl):S3–S12. https://doi.org/10.1016/j.clindermatol.2008.12.003

Saghazadeh S, Rinoldi C, Schot M, Kashaf SS, Sharifi F, Jalilian E, Nuutila K, Giatsidis G, Mostafalu P, Derakhshandeh H, Yue K, Swieszkowski W, Memic A, Tamayol A, Khademhosseini A (2018) Drug delivery systems and materials for wound healing applications. Adv Drug Deliv Rev 127:138–166. https://doi.org/10.1016/j.addr.2018.04.008

Sapudom J, Mohamed WKE, Garcia-Sabaté A, Alatoom A, Karaman S, Mahtani N, Teo JCM (2020) Collagen fibril density modulates macrophage activation and cellular functions during tissue repair. Bioengineering 7:33

Savard T, Beaulieu C, Boucher I, Champagne CP (2002) Antimicrobial action of hydrolyzed chitosan against spoilage yeasts and lactic acid bacteria of fermented vegetables. Journal of food protection, 65(5):828-833.

Schini-Kerth VB, Étienne-Selloum N, Chataigneau T, Auger C (2011) Vascular protection by natural product-derived polyphenols: in vitro and in vivo evidence. Planta Med 77(11):1161–1167. https://doi.org/10.1055/s-0030-1250737

Senadheera TRL, Hossain A, Shahidi F (2023) Marine bioactives and their application in the food industry: a review. Appl Sci 13(21):12088. https://doi.org/10.3390/app132112088

Shin EJ, Joo SR, Kim SH (2019) Cooperation of biopolymer chitosan with hydrogen peroxide for ipso-hydroxylation of arylboronic acids under green conditions. Tetrahedron Lett 60(23):1509–1513. https://doi.org/10.1016/j.tetlet.2019.05.009

Shukla SK, Mishra AK, Arotiba OA, Mamba BB (2013) Chitosan-based nanomaterials: a state-of-the-art review. Int J Biol Macromol 59:46–58. https://doi.org/10.1016/j.ijbiomac.2013.04.043

Silva SS, Popa EG, Gomes ME, Cerqueira M, Marques AP, Caridade SG, Teixeira P, Sousa C, Mano JF, Reis RL (2013) An investigation of the potential application of chitosan/aloe-based membranes for regenerative medicine. Acta Biomater 9(6):6790–6797. https://doi.org/10.1016/j.actbio.2013.02.027

Singh K, Kumar M, Saini JS (2022) Metal inserts for performance improvement in glass fiber-reinforced polymer nanocomposite joints under accelerated aging conditions. Polym Compos 43(3):1448–1461. https://doi.org/10.1002/pc.26465

Subhan F, Ikram M, Shehzad A, Ghafoor A (2015) Marine collagen: an emerging player in biomedical applications. J Food Sci Technol 52(8):4703–4707. https://doi.org/10.1007/s13197-014-1652-8

Sudha PN, Aisverya S, Rose MH, Venkatesan J, Kim S-K (2014) Bionanocomposites of chitosan for multitissue engineering applications. http://www.nibib1.nih.gov/

Sukpaita T, Chirachanchai S, Pimkhaokham A, Ampornaramveth RS (2021) Chitosan-based scaffold for mineralized tissues regeneration. Mar Drugs 19(10):551. https://doi.org/10.3390/md19100551

Tsao CT, Chang CH, Lin YY, Wu MF, Han JL, Hsieh KH (2011) Kinetic study of acid depolymerization of chitosan and effects of low molecular weight chitosan on erythrocyte rouleaux formation. Carbohydr Res 346(1):94–102. https://doi.org/10.1016/j.carres.2010.10.010

Vidanarachchi JK, Ranadheera CS, Wijerathne TD, Udayangani RMC, Himali SMC, Pickova J (2014) Applications of seafood by-products in the food industry and human nutrition. In: Seafood processing by-products: trends and applications, vol 9781461495901. Springer, New York, pp 463–528. https://doi.org/10.1007/978-1-4614-9590-1_23

Wade AM, Tucker HN (1998) Antioxidant characteristics of L-histidine. J Nutr Biochem 9(6):308–315

Walimbe T, Panitch A (2020) Best of both hydrogel worlds: harnessing bioactivity and tunability by incorporating glycosaminoglycans in collagen hydrogels. Bioengineering 7:156

Wang Q, Wang X, Feng Y (2023) Chitosan hydrogel as tissue engineering scaffolds for vascular regeneration applications. Gels 9(5):373. https://doi.org/10.3390/gels9050373

Ways TMM, Lau WM, Khutoryanskiy VV (2018) Chitosan and its derivatives for application in mucoadhesive drug delivery systems. Polymers 10(3):267. https://doi.org/10.3390/polym10030267

Weng J, Durand A, Desobry S (2023) Chitosan-based particulate carriers: structure, production and corresponding controlled release. Pharmaceutics 15(5):1455. https://doi.org/10.3390/pharmaceutics15051455

Xiong X, Yu IKM, Tsang DCW, Bolan NS, Sik Ok Y, Igalavithana AD, Kirkham MB, Kim KH, Vikrant K (2019) Value-added chemicals from food supply chain wastes: state-of-the-art review and future prospects. Chem Eng J 375:121983. https://doi.org/10.1016/j.cej.2019.121983

Yanat M, Schroën K (2021) Preparation methods and applications of chitosan nanoparticles; with an outlook toward reinforcement of biodegradable packaging. React Funct Polym 161:104849. https://doi.org/10.1016/j.reactfunctpolym.2021.104849

Yeo J, Lee J, Yoon S, Kim WJ (2020) Tannic acid-based nanogel as an efficient anti-inflammatory agent. Biomater Sci 8(4):1148–1159. https://doi.org/10.1039/c9bm01384a

Yeung DA, Kelly NH (2021) The role of collagen-based biomaterials in chronic wound healing and sports medicine applications. Bioengineering 8:8

Yousefi M, Ariffin F, Huda N (2017) An alternative source of type I collagen based on by-product with higher thermal stability. Food Hydrocoll 63:372–382. https://doi.org/10.1016/j.foodhyd.2016.09.029

Zarandona I, López-Caballero ME, Montero MP, Guerrero P, de la Caba K, Gómez-Guillén MC (2021) Horse mackerel (Trachurus trachurus) fillets biopreservation by using gallic acid and chitosan coatings. Food Control 120:107511. https://doi.org/10.1016/j.foodcont.2020.107511

Zhao B, Zhao Z, Ma J, Ma X (2019) Modulation of angiogenic potential of tissue-engineered peripheral nerve by covalent incorporation of heparin and loading with vascular endothelial growth factor. Neurosci Lett 705:259–264. https://doi.org/10.1016/j.neulet.2019.01.017

Natural Product-Loaded PCL and PLA-Based Bio-Scaffold and Their Applications

6

Shilpa Chandel, Parijat Pandey, Sapna Saini, Ravinder Sharma, Bharat Parashar, and Kiran Yadav

Abstract

Tissue engineering is a multidisciplinary field that aims to create functional tissues by combining cells, biomaterials, and biochemical factors. In the rapidly advancing field of tissue engineering, bio-scaffolds stand as key protagonists. They act as a structural framework for guiding cells to repair or replace damaged tissues. This dynamic synergy is particularly pronounced with the incorporation of natural products into bio-scaffolds, emerging as a focal point in the domain of tissue engineering and regenerative medicine. Natural products, known for their inherent biocompatibility and therapeutic potential, enrich the bio-scaffold landscape, elevating its capabilities in fostering optimal tissue regeneration. This book chapter not only navigates the synthesis and characterization of bio-scaffolds based on polycaprolactone (PCL) and polylactic acid (PLA) but also extends its exploration to the profound implications of incorporating a variety of natural products. The emphasis on design principles, fabrication techniques, and the controlled release mechanisms elucidates the meticulous engineering behind these bio-scaffolds. Furthermore, the chapter illuminates the vast spectrum of applications, spanning from wound healing and bone regeneration to drug delivery and cancer therapy, underscoring the versatility of natural product-loaded PCL and PLA-based bio-scaffold technology.

S. Chandel · R. Sharma · B. Parashar · K. Yadav (✉)
Faculty of Pharmaceutical Sciences, ICFAI University, Kalujhanda, Baddi, Himachal Pradesh, India

P. Pandey
Department of Pharmaceutical Sciences, Gurugram University, Gurugram, Haryana, India

S. Saini
Janta College of Pharmacy, Butana, Sonepat, Haryana, India

M. Kumar et al. (eds.), *Natural Product Inspired Scaffolds*, https://doi.org/10.1007/978-981-97-3111-4_6

Keywords
Poly(lactic acid) (PLA) · Poly(ε-caprolactone) (PCL) · Bio-scaffold · Natural products · Biocompatibility

6.1 Introduction

Tissue engineering is a multidisciplinary field that aims to regenerate, repair, or replace damaged tissues or organs by integrating principles from biology, materials science, and engineering (Mark and Atala 2014). Central to the success of tissue engineering is the concept of bio-scaffolds—three-dimensional structures that provide a supportive framework for cells to adhere, proliferate, and differentiate, ultimately leading to the formation of functional tissues (Dawson et al. 2011). Bio-scaffolds serve as a temporary framework, mimicking the extracellular matrix (ECM) of the native tissue. The design must facilitate proper cell attachment, migration, and the biochemical signaling necessary for tissue regeneration. Functionality encompasses the ability to support cell activities, such as proliferation and differentiation, leading to the formation of tissue with native-like properties (Salerno et al. 2012; Hollister 2005; Liu and Bhatia 2004).

An optimal scaffold design includes a controlled pore size and interconnectivity. Pore architecture influences nutrient diffusion, waste removal, and cell infiltration (Shi et al. 2002). Achieving a balance between porosity and interconnectivity is crucial to support cell viability and tissue integration (Zhang and Zhang 2001). Bio-scaffolds must exhibit high biocompatibility to avoid adverse immune responses. Compatibility with the host tissue ensures a harmonious integration, preventing inflammation or rejection. Surface modifications may be employed to enhance biocompatibility and promote favorable cell-scaffold interactions (Huebsch and Mooney 2009; Tutar et al. 2019; Van et al. 2011).

The integration of natural products into polymeric bio-scaffolds represents a dynamic frontier in biomedical engineering, continually advancing to address the evolving demands of regenerative medicine and drug delivery. Natural products, sourced from botanical, marine, and microbial origins, boast a rich pharmacological repertoire. Concurrently, poly(ε-caprolactone) (PCL) and poly(lactic acid) (PLA) have risen to prominence as biocompatible and biodegradable polymers, forming the foundation for cutting-edge biomaterials (Zhao et al. 2021).

Additionally, innovative strategies for the sustained release of natural compounds from these scaffolds, including stimuli-responsive delivery systems, have expanded the therapeutic potential. The synergistic interaction between polymeric matrices and natural products has paved the way for personalized medicine, where tailored bio-scaffolds can address specific patient needs.

6.2 Natural Product-Loaded Bio-Scaffolds: An Overview

6.2.1 Natural Products in Biomedicine

Natural products in biomedicine refer to compounds derived from natural sources such as plants, animals, fungi, and microorganisms, which exhibit medicinal properties and potential therapeutic benefits. Throughout history, various cultures have relied on natural remedies for healing purposes. In modern biomedical research, natural products continue to be a valuable source of novel drug candidates, providing a rich reservoir of chemical diversity for drug discovery and development.

These natural products encompass a wide range of chemical classes, including alkaloids, flavonoids, terpenoids, polyphenols, and peptides, among others. Each class offers unique chemical structures and biological activities, making them suitable candidates for targeting various disease pathways and biological processes.

One of the significant advantages of natural products in biomedicine is their ability to interact with specific biological targets with high affinity and selectivity, often resulting in potent pharmacological effects. Artemisinin, for example, has not only been critical in malaria treatment but has also paved the way for the development of semisynthetic derivatives with improved therapeutic efficacy and fewer side effects (Krishani et al. 2023; Soni et al. 2022).

Paclitaxel, derived from the Pacific yew tree, is another compelling example of a natural product with profound biomedical implications. This compound has demonstrated remarkable anticancer activity by disrupting microtubule function, leading to cell cycle arrest and apoptosis. Paclitaxel has become a mainstay in the treatment of various cancers, illustrating how natural products can serve as the foundation for the development of potent anticancer drugs (Slichenmyer and Von Hoff 1991).

The anti-inflammatory and antioxidant properties of curcumin, found in turmeric, have attracted significant attention in the medical community. Studies have explored its potential in managing conditions such as rheumatoid arthritis and inflammatory bowel disease. Curcumin's multifaceted mechanisms of action, including modulation of inflammatory pathways and inhibition of oxidative stress, highlight the versatility of natural products in addressing complex and interconnected disease processes (Yadav and Kumar 2014; Yadav et al. 2023).

Beyond these examples, the exploration of marine organisms has led to the discovery of compounds with antibiotic, antiviral, and anticancer properties. For instance, marine sponges have yielded bioactive compounds like discodermolide, which shows promise in cancer treatment. The biodiversity of our planet offers an extensive reservoir of natural products, each with the potential to unlock novel therapeutic avenues (Calcabrini et al. 2017).

The integration of natural products into modern medicine involves not only the isolation of bioactive compounds but also the development of sustainable harvesting methods, synthetic modifications, and advanced delivery systems. As technology advances, researchers are increasingly able to harness the full potential of natural products, driving innovation and contributing to the development of new drugs and therapies that address the evolving challenges of healthcare. The exploration of the vast and intricate world of natural products remains a dynamic and promising frontier in biomedical research (Li et al. 2015; Liu et al. 2023).

6.2.2 Bio-Scaffolds in Tissue Engineering

Bio-scaffolds form the cornerstone of tissue engineering, providing a scaffold for cells to adhere, proliferate, and differentiate in a manner that closely mimics the native tissue environment. Central to tissue engineering is the concept of scaffolds, which serve as three-dimensional frameworks to support cell growth, tissue formation, and eventually, integration into the host organism. Bio-scaffolds play a crucial role in providing mechanical support, guiding tissue regeneration, and influencing cellular behavior (Chaudhary and Garg 2015; Chen 2011).

6.2.3 Fundamental Principles of Scaffold Design

Biocompatibility Scaffolds must be biocompatible to avoid triggering immune responses. They should not induce toxicity, inflammation, or rejection when interacting with the host tissue (Huebsch and Mooney 2009).

Biodegradability Ideally, scaffolds should degrade at a rate compatible with tissue regeneration. As the tissue grows, the scaffold gradually degrades, allowing the newly formed tissue to take its place (Huebsch and Mooney 2009; Tutar et al. 2019).

Mechanical Properties The mechanical characteristics of the scaffold should mimic those of the target tissue. This is essential for providing the necessary physical support during the regeneration process.

6.2.4 Structural Characteristics of Scaffolds

Porosity An interconnected porous structure is crucial for facilitating the exchange of nutrients, oxygen, and waste products between cells and the surrounding environment. It also aids in cell migration and tissue ingrowth (Shi et al. 2002).

Interconnectivity Proper interconnectivity between pores is essential for cell migration, nutrient diffusion, and vascularization. This ensures that cells throughout the scaffold receive the necessary signals for proliferation and differentiation.

Surface Topography The surface features of scaffolds, such as roughness and patterning, influence cell adhesion, alignment, and differentiation. Mimicking the natural extracellular matrix (ECM) architecture enhances cellular interactions (Shi et al. 2002).

6.2.5 Influence on Cellular Behavior

Cell Adhesion and Proliferation The scaffold's surface properties, including chemistry and topography, impact cell adhesion and proliferation. Integrin-binding sites on the scaffold surface promote cell adhesion, allowing cells to spread and proliferate.

Differentiation Scaffolds can be engineered to release specific biochemical cues, such as growth factors or signaling molecules, to guide cell differentiation along desired pathways. This is crucial for generating functional tissues with specific cell types.

Extracellular Matrix (ECM) Mimicry Scaffolds often aim to replicate the native ECM to provide cues for cell behavior. This involves incorporating ECM components like collagen, fibronectin, and laminin into the scaffold structure (Salerno et al. 2012).

Mechanical Signals The mechanical properties of the scaffold influence cell behavior. Stiffness or elasticity can direct stem cell differentiation, with certain mechanical cues promoting the development of specific cell lineages.

By understanding the fundamental principles of scaffold design, optimizing structural characteristics, and leveraging their influence on cellular behavior, researchers can create scaffolds that effectively support tissue regeneration and contribute to the advancement of regenerative medicine (Chan and Leong 2008).

6.3 Polycaprolactone (PCL) and Polylactic Acid (PLA) as Bio-Scaffold Materials

The selection of appropriate biomaterials for tissue engineering and regenerative medicine is crucial for the success of implantable devices and scaffolds. Polycaprolactone (PCL) and polylactic acid (PLA) have emerged as promising candidates for bio-scaffold materials due to their unique properties, biocompatibility, and tunable degradation characteristics.

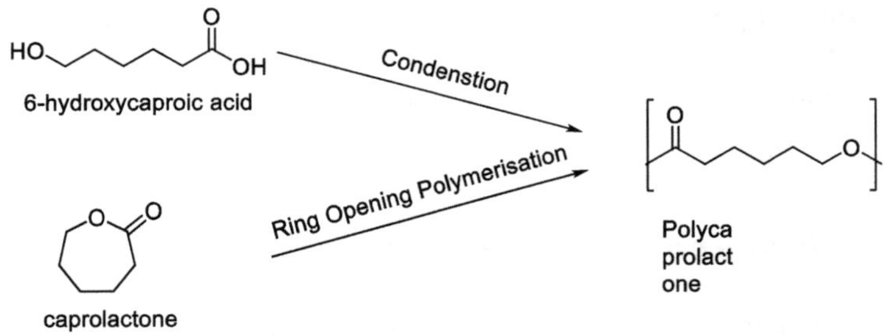

Fig. 6.1 Synthesis of polycaprolactone (PCL)

6.3.1 Polycaprolactone (PCL)

Polycaprolactone (PCL, Fig. 6.1) is biodegradable polyester with exceptional biocompatibility. This property is attributed to the hydrophobic nature of PCL, making it suitable for various biomedical applications without triggering adverse immune responses. The biodegradability of PCL allows for controlled degradation over an extended period, making it advantageous for applications requiring sustained support (Homaeigohar and Boccaccini 2022; Tate et al. 2008; Magnusson et al. 2011).

PCL possesses good mechanical strength, providing structural integrity to bioscaffolds. Its flexibility and ease of processing make it an attractive choice for constructing intricate three-dimensional structures tailored to specific tissue engineering needs. This property allows PCL to adapt to the mechanical demands of various tissues within the human body (Gu et al. 2019; Patil and Masters 2020).

6.3.1.1 Limitations and Methods to Overcome

PCL's relatively slow degradation rate may be considered a limitation for applications requiring faster tissue regeneration. While advantageous in certain scenarios, this property necessitates careful consideration based on the specific requirements of the intended biomedical application (Labet and Thielemans 2009). Below are potential approaches to overcome.

Blend with Faster-Degrading Polymers Combining PCL with polymers that have faster degradation rates can help achieve a more balanced degradation profile, catering to applications requiring quicker tissue regeneration (Ma 2008).

Surface Modification Altering the surface of PCL through techniques like plasma treatment or coating with bioactive agents can influence its degradation rate, allowing for controlled adjustments.

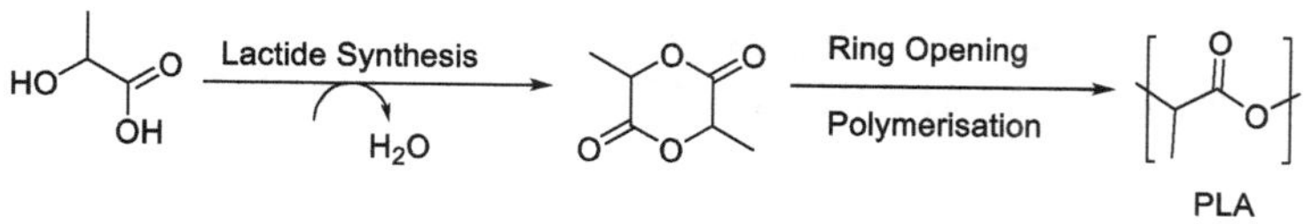

Fig. 6.2 Synthesis of polylactic acid (PLA)

6.3.2 Polylactic Acid (PLA)

Polylactic acid (PLA, Fig. 6.2), derived from renewable resources such as corn starch or sugarcane, exhibits excellent biocompatibility. PLA is known for its sustainable origin, aligning with the growing demand for environmentally friendly biomedical materials. The biocompatibility of PLA makes it suitable for various tissue engineering applications (Assad and Jackson 2019; Bruinink and Luginbuehl 2012).

PLA offers tunable degradation rates, allowing for customization based on the specific needs of the intended application. Its mechanical strength is comparable to certain natural tissues, contributing to its utility in bio-scaffold design. The ability to tailor degradation rates makes PLA versatile for applications requiring controlled tissue regeneration (Chen et al. 2011; Saleem et al. 2024).

6.3.2.1 Limitations and Methods to Overcome

PLA, while versatile, can exhibit brittleness, especially in its early stages of degradation. Additionally, there is a risk of acidification during PLA degradation, which may pose challenges in certain tissue engineering contexts (Woodruff and Hutmacher 2010; Roberto et al. 2016). Below are potential approaches to overcome drawbacks.

1. Polymer blending: Blending PLA with other polymers, such as PCL, can mitigate brittleness, enhancing the overall mechanical properties of the scaffold.
2. Addition of plasticizers: Incorporating plasticizers into PLA formulations can improve flexibility and reduce brittleness.
3. Buffering agents: Addressing the potential acidification issue can be achieved by incorporating buffering agents into the scaffold to maintain a stable pH during degradation.

6.3.3 Surface Modification for Enhanced Biocompatibility

Surface modification strategies are pivotal for improving the biocompatibility of PCL and PLA bio-scaffolds. Various techniques, including plasma treatment, biomolecule coating, and incorporation of bioactive agents, aim to enhance cell adhesion, minimize immune responses, and optimize the interaction between the scaffold and host tissue (Yang et al. 2001; Holzwarth and Ma 2011).

6.4 Fabrication and Characterization of Natural Product-Loaded PCL and PLA Bio-Scaffolds

Incorporating natural products into PCL and PLA bio-scaffolds requires techniques to ensure even distribution within the plastic structure. The method chosen depends on the characteristics of the natural product, the desired bio-scaffold properties, and its intended use.

There are two main methods for incorporating bioactive molecules into the scaffold. The first is the top-down approach, where bone extracellular matrix (ECM) from osteocytes or stem cells are positioned onto the scaffold, forming composite structure that combines the biomaterial scaffold with the ECM. The second is the bottom-down approach, which includes amending the scaffold by incorporating growth factors or cytokines, and attaching peptides through covalent bonding. Additionally, these biologically active compounds are inserted inside the nanoparticles or microspheres which are intended controlled drug release within extracellular matrix (Zhu 2010; Kesireddy and Kasper 2016).

6.4.1 Fabrication Methods

6.4.1.1 Solvent-Based Fabrication and Particle Extraction Method

Solvent-based fabrication is a method that includes dissolving a polymer into a suitable solvent, followed by the addition of an insoluble salt (Fig. 6.3). The solvent then undergoes evaporation, forming a composite of both salt and polymer. The composite is subsequently submerged in water, leading to the gradual dissolution of the salt and resulting in a significantly sponge-like structure (Hutmacher 2000).

6.4.1.2 Lyophilisation Method

This method includes dissolving polymer into the solvent and then pouring it into a mold and finally cooling it below its freezing point to solidify the solvent. The solidified solvent gets evaporated through sublimation process, resulting in the formation of a scaffold having solid structure and numerous interconnected pores. Necessary temperature for solidification step falls within the range of −20 to 80 °C (Roseti et al. 2017). During the cooling process, solutes separate within the ice phase, creating small porous structures. It's possible to adjust the pore sizes by varying and regulating the freezing step (Aranaz et al. 2014).

6.4.1.3 Thermal-Induced Phase Separation (TIPS)

This is the method where the polymer solution undergoes the phase division mechanism, leading to the formation of two phases—one with a low concentration of polymer and another with a high polymer concentration. The phase containing a lower polymer concentration is then eliminated, leaving behind the phase with a higher polymer concentration, which is subsequently solidified. Through this solidification process, a microscale fibroid structure is created, characterized by substantial porosity. This resulting structure serves as a scaffold (Figoli 2014; Li et al.

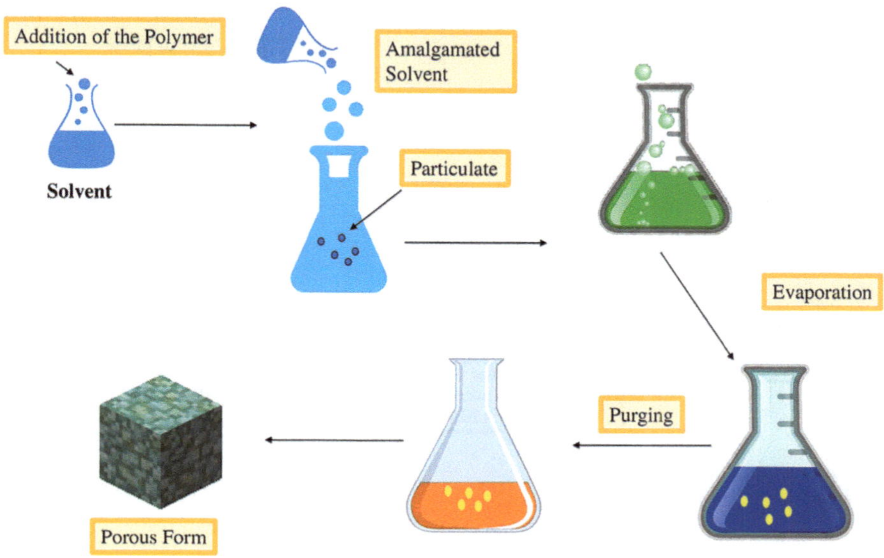

Fig. 6.3 Solvent-based fabrication and particle extraction method

2016). The phase division mechanism plays a crucial role in shaping the unique characteristics of the scaffold for its intended applications in tissue engineering and regenerative medicine.

6.4.1.4 Gas Foaming Method

Gas foaming technique utilizes a nonreactive gas as a foaming agent, applying pressure within pre-molded biodegradable polymers dissolved in solvents like water or fluoroform under elevated temperatures. Gas exerts pressure until it saturates polymer with effervescence, resulting in the formation of a porous structure characterized by 85% of the porosity and pore sizes ranging from 30 to 700 μm (Gorth and Webster 2011; Eltom et al. 2019). A detailed depiction of the method can be found in Fig. 6.4.

6.4.1.5 Powder Foaming Method

The powder foaming method is ideal for producing glass and ceramic scaffolds with notable porosity. This approach involves the dispersion of ceramic particles in a liquid medium, like water or ethanol, to create a slurry. Subsequently, this slurry suspension is employed in crafting green bodies, denoting loosely bonded clay materials presented as compressed powder structures before undergoing the sintering process (Chen et al. 2011). In this ceramic suspension, a surfactant and fillers like gelatine, sucrose, and polymeric microbeads are incorporated. Throughout the sintering process, these fillers are exposed either to combustion or evaporation process inside the slurry, which forms pores as an end product (Thavornyutikarn et al. 2014).

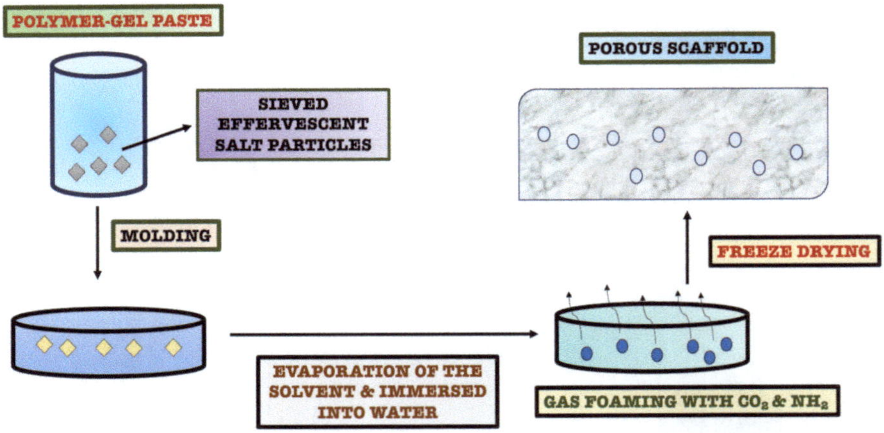

Fig. 6.4 Gas foaming method

6.4.1.6 Electrospinning Method

This method is highly adaptable, allowing the processing of diverse matters to achieve intended porousness and structure in scaffolds, having diameter spanning from microns to nanometers (Gorth and Webster 2011; Ovinuchi 2023). The method involves the use of a high-voltage electric field to produce extremely fine fibers from polymer solutions that are electrically charged. This technique enables the creation of diverse fiber patterns characterized by exceptional porosity. Morphology and diameter of the resulting fibers are determined by crucial factors including viscosity of solution, polymer molecular weight, polymer charge denseness, and strength of electric field. Adjustment of these variables allows for the customization of features in the resulting fibers to meet specific requirements (Pham et al. 2006; Adeli-Sardou et al. 2019). Figure 6.5 provides a comprehensive visual representation of the method.

6.4.1.7 Sol-Gel Method

Sol-gel method is indicative of a chemical technique, specifically wet chemical method, employed for the production of copious nanocomposites, notably metal oxide nanocomposites. This method utilizes molecular precursor, commonly a metal alkoxide, that undergoes dissolution in water or alcohol and is subsequently converted into a gel through the hydrolysis or alcoholises process, facilitated by heating and stirring (Bokov et al. 2021). This method involves the polymerization of metal alkoxides, achieved by forming a sol after the inclusion of surfactant, followed by condensation and gelation processes. Aforementioned methodology is apt for creating glass or ceramic-based scaffolds in diverse configurations, including ceramic fibers, flimsy coatings, nanostructured inorganic membranes, spherical or ultrafine powders, highly porous aerogel materials, and monolithic ceramics and glasses (Raucci et al. 2010).

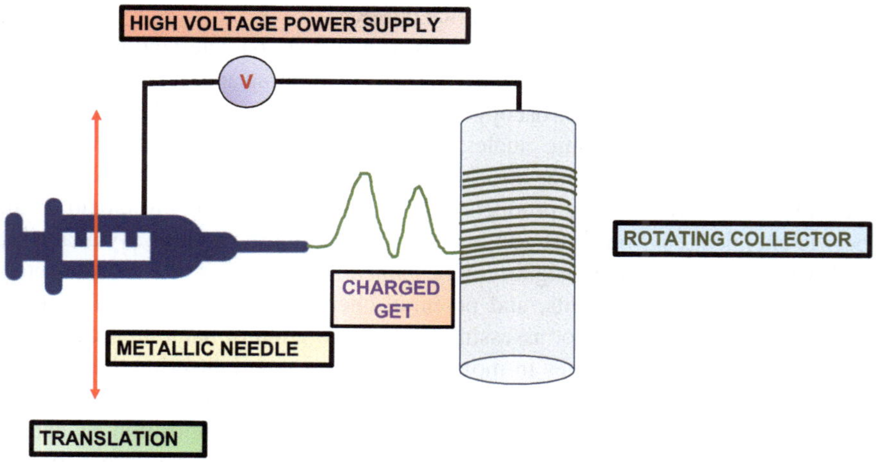

Fig. 6.5 Electrospinning method

6.4.1.8 Computer-Aided Drug Designing Models

In response to the limitations posed by traditional approaches in controlling porosity and pore size within tissue engineering, innovative methods incorporating computer-assisted design and manufacturing have been adopted. This includes the utilization of inkjet printing of polymer powders and fused deposition modelling (FDM) (Kumar et al. 2022). In the case of inkjet printing, the process commences with the digital design of a three-dimensional structure using computer-aided design (CAD) software. Subsequently, successive layers of polymer powder are deposited onto a build platform, and a liquid binder is selectively sprayed using an inkjet print head, solidifying specific areas according to the digital model. This layer-by-layer process continues until the entire scaffold is formed, providing high precision and control over porosity. On the other hand, FDM involves the heating and extrusion of a thermoplastic polymer filament, which is deposited layer by layer to construct the desired object. The molten polymer solidifies as it is extruded, forming a continuous structure. Both techniques offer advanced capabilities in tailoring the architecture of tissue engineering scaffolds to meet specific biological requirements, showcasing the intersection of technology and medical science in the pursuit of optimal regenerative solutions (Rani et al. 2023; Salerno and Netti 2021).

6.4.2 Characterization Techniques

6.4.2.1 Chemical Composition Characterization

The method for recognizing and measuring the chemical composition within a material can be achieved through spectroscopic methods such as FTIR, NMR, and chromatography.

6.4.2.1.1 Nuclear Magnetic Resonance (NMR)

Nuclear magnetic resonance (NMR) serves as a robust characterization technique for natural product-loaded bio-scaffolds in the field of tissue engineering and regenerative medicine. NMR spectroscopy operates on the principle of exploiting the magnetic properties of atomic nuclei, providing detailed information about the molecular structure and composition of a given sample. In the context of bio-scaffolds loaded with natural products, NMR enables researchers to elucidate the specific chemical environments and interactions within the material. By analyzing the NMR spectra, valuable insights can be gained into the types of molecules present, their spatial arrangements, and potential chemical modifications. This technique is particularly useful for assessing the incorporation of natural products into bio-scaffolds, tracking changes in molecular structures, and ensuring the integrity of the loaded therapeutic agents. NMR serves as a nondestructive, noninvasive tool, offering a comprehensive understanding of the chemical composition of natural product-loaded bio-scaffolds, thus contributing to the advancement of precision and efficacy in regenerative medicine applications (Madhav et al. 2019; Pavia et al. 2014; Liu et al. 2016; Brown 2012).

6.4.2.1.2 Fourier's Transform Infrared Spectroscopy (FTIR)

Fourier transform infrared spectroscopy (FTIR) is employed for the chemical composition characterization of bio-scaffolds loaded with natural products. FTIR is a powerful analytical technique that measures the absorption of infrared radiation by a material, providing detailed information about its chemical composition. In the context of bio-scaffolds, FTIR enables the identification of specific functional groups present in both the scaffold material and the loaded natural products. By analyzing the distinct peaks and patterns in the infrared spectrum, researchers can gain insights into the molecular structure, bonding, and interactions within the bio-scaffold. This method proves invaluable in understanding the chemical composition of natural product-loaded bio-scaffolds, facilitating the optimization of these materials for enhanced biocompatibility and therapeutic efficacy in tissue engineering and regenerative medicine applications (Liu et al. 2012).

6.4.2.2 Structural and Morphological Analysis

Identification of the molecular and crystalline structure can be done through the techniques such as X-ray diffraction (XRD) and electron microscopy.

6.4.2.2.1 X-Ray Diffraction (XRD)

X-ray diffraction (XRD) is a powerful technique used for the characterization of materials, including polymer based bio-scaffolds. This characterization technique plays a crucial role in analyzing polymer-based bio-scaffolds, uncovering their crystalline structure, identifying distinct phases, measuring crystallinity, identifying structural alterations, ensuring quality control, and examining interactions among components. These findings are essential for enhancing the design and functionality of bio-scaffolds across diverse applications, such as tissue engineering and drug delivery (Solechan et al. 2022; Sun et al. 2014; Hasan et al. 2018).

6.4.2.2.2 Scanning Electron Microscopy (SEM)

Scanning electron microscopy (SEM) emerges as a pivotal tool in the thorough characterization of polymer-based bio-scaffolds. This technique offers profound insights into critical facets, including surface morphology, pore architecture, particle dispersion, interface features, and overall structural quality. Such meticulous observations serve as instrumental guidance in fine-tuning the design and guaranteeing the efficacy of bio-scaffolds, particularly in the intricate realm of biomedical applications (Hassanajili et al. 2019; Xia et al. 2018).

6.4.2.2.3 Transmission Electron Microscopy (TEM)

Transmission electron microscopy (TEM) emerges as a pivotal characterization technique for natural product-loaded bio-scaffolds in the realm of tissue engineering and regenerative medicine. This sophisticated method offers unparalleled high-resolution imaging capabilities, allowing researchers to delve into the intricate details of the bio-scaffold's nanoscale structure. TEM facilitates a comprehensive understanding of the distribution and morphology of natural product particles within the bio-scaffold. By providing detailed visualizations at the submicron level, TEM enables the examination of the interactions between the loaded natural products and the scaffold material. Moreover, TEM is instrumental in revealing the fine features, porosity, and overall architecture critical for the effectiveness of the bio-scaffold in promoting tissue regeneration. This technique, through its ability to capture nanoscale details, contributes significantly to advancing the design, optimization, and quality assessment of natural product-loaded bio-scaffolds, ultimately enhancing their potential for therapeutic applications in regenerative medicine (Zhai et al. 2015; Wang et al. 2016).

6.4.2.2.4 Porosity and Pore Size Analysis

Mercury intrusion porosimetry: Mercury intrusion porosimetry is typically performed using a specialized instrument called a mercury porosimeter or mercury intrusion porosimeter. These instruments are designed to measure the intrusion of mercury into the pores of a material under controlled pressure conditions. The process involves gradually increasing the pressure in the chamber, causing mercury to intrude into the pores of the material. The volume of mercury intruded at different pressure points is then measured, and this information is used to calculate parameters such as total porosity and pore size distribution. Mercury intrusion porosimetry is a powerful technique that provides detailed information about the porosity and pore size distribution of PLA and PCL bio-scaffolds. This information is essential for optimizing scaffold design to meet the specific requirements of tissue engineering applications (Shum et al. 2005).

6.4.2.3 Stability Studies

6.4.2.3.1 Differential Scanning Calorimetry (DSC)

Differential scanning calorimetry (DSC) is a thermal analysis technique that measures the heat flow associated with thermal transitions in a material. It provides valuable information about the thermal properties and behavior of a material as it is heated or cooled. In the context of scaffold materials, DSC can be employed to assess the thermal stability of the scaffold (Enrione et al. 2013).

DSC operates on the principle that any heat flow into or out of a sample can be detected and measured. As the sample undergoes physical or chemical changes, such as phase transitions or reactions, it either absorbs or releases heat.

The sample is first cooled to a low temperature, then heated at a constant rate, while the heat flow is continuously monitored. During the process, if the sample undergoes any phase transition or thermal event (e.g., melting, crystallization, or chemical reactions), it will either absorb or release heat. The DSC instrument records the temperature at which these events occur and the associated heat flow.

Peaks in the DSC thermogram represent thermal events, and the area under the peaks corresponds to the heat absorbed or released during those events. In the context of bio-scaffolds, DSC can be used to investigate key thermal events such as the following:

Denaturation of proteins: Many bio-scaffolds are composed of proteins. DSC can detect the denaturation temperature, providing insights into the stability of the protein structure within the scaffold.

Phase transitions: Polymers used in bio-scaffold construction may undergo phase transitions, such as melting or crystallization. DSC helps identify these transitions and provides information on the material's thermal stability (Younis et al. 2023).

6.4.2.3.2 Thermogravimetric Analysis (TGA)

Thermogravimetric analysis (TGA) is another thermal analysis technique that complements DSC. TGA measures the weight change of a material as a function of temperature or time. The sample is subjected to a controlled temperature program in an inert or reactive atmosphere. The sample is heated at a constant rate, and as it undergoes thermal decomposition, the weight loss is recorded. The TGA instrument monitors the change in mass as a function of temperature or time (Ebnesajjad 2006).

Peaks or steps in the TGA curve correspond to different stages of decomposition or volatilization of the scaffold material. By analyzing the TGA curve, researchers can determine the temperatures at which significant weight loss occurs and gain insights into the thermal stability and composition of the material. Bio-scaffolds often consist of complex organic and biological components. TGA helps in determining the following:

Thermal stability: TGA can identify the temperature at which significant weight loss occurs, indicating the thermal stability of the bio-scaffold.

Decomposition characteristics: TGA provides information about the stages and temperatures at which components within the bio-scaffold decompose.

6.4.2.4 Degradation Analysis of Bio-Scaffolds

Degradation analysis is a critical aspect of studying bio-scaffolds, particularly in the context of tissue engineering and regenerative medicine. The degradation characteristics of a bio-scaffold determine its ability to provide temporary structural support and guide tissue regeneration. Two important methods for degradation analysis in the context of bio-scaffolds are environmental testing and enzymatic degradation studies.

6.4.2.5 Environmental Testing for Bio-Scaffolds

Environmental testing involves subjecting the bio-scaffold to conditions that mimic physiological environments to assess how well it maintains its structural integrity over time. Bio-scaffolds implanted in the body are exposed to various physiological conditions, including temperature, humidity, and the presence of bodily fluids. Environmental testing simulates these conditions in a controlled laboratory setting.

Bio-scaffold samples are exposed to conditions that replicate the physiological environment. This may involve incubation in a bioreactor or exposure to a specific temperature and humidity regime. Changes in the structural and mechanical properties of the bio-scaffold are monitored over time. Techniques such as microscopy, spectroscopy, and mechanical testing can be used to assess any alterations in the scaffold (Wang et al. 2021). Degradation analysis, through environmental testing studies, is essential for assessing the long-term performance and biocompatibility of bio-scaffolds in biomedical applications. These studies contribute to the development of scaffolds that can effectively guide tissue regeneration while maintaining structural integrity over the desired period (Annor et al. 2012).

6.4.2.6 Biological Assessment of Bio-Scaffolds

The biological assessment of bio-scaffolds is a crucial step in understanding how well these materials interact with biological systems. This assessment involves evaluating various aspects related to cell behavior, cytocompatibility, and in vivo performance. Three important components of biological assessment in the context of bio-scaffolds are cell adhesion and proliferation, cytocompatibility assays, and in vivo studies.

6.4.2.6.1 Cell Adhesion and Proliferation for Bio-Scaffolds

Cell adhesion and proliferation studies assess how well cells adhere to the surface of the bio-scaffold and whether they can proliferate and grow on it. For effective tissue regeneration, it is essential that cells attach to the bio-scaffold and proliferate to form new tissue. This is particularly important for scaffolds used in tissue engineering applications. Cells relevant to the targeted tissue (e.g., osteoblasts for bone tissue) are seeded onto the bio-scaffold. The scaffold is then cultured, and the adhesion and proliferation of cells are monitored over time. Techniques such as microscopy, cell counting, and viability assays are used to assess the extent of cell adhesion and proliferation on the bio-scaffold (Wu et al. 2011; Iijima et al. 2017).

6.4.2.6.2 Cytocompatibility Assays for Bio-Scaffolds

Cytocompatibility assays evaluate the impact of the bio-scaffold on cell viability, function, and overall cellular response. Bio-scaffolds should not induce cytotoxic effects on cells and should support normal cell function. Cytocompatibility assays help ensure that the bio-scaffold is conducive to cell survival and activity. Cells are exposed to extracts or direct contact with the bio-scaffold. Various parameters, including cell viability, proliferation, and functionality, are assessed. Assays such as MTT assays, live/dead staining, and assessments of specific cell functions are conducted to determine the cytocompatibility of the bio-scaffold (Switha et al. 2023; Zhang et al. 2021).

1. In Vivo Studies for Bio-Scaffolds

   In vivo studies involve implanting the bio-scaffold into living organisms to assess its performance within a biological environment. Understanding how the bio-scaffold functions in a living organism is critical for predicting its potential for clinical use. In vivo studies provide insights into the scaffold's integration, degradation, and overall performance. Bio-scaffold implants are placed in animal models (e.g., rodents, rabbits) to observe tissue responses, host interactions, and any potential adverse effects. Histological analysis, imaging, and assessments of tissue regeneration are performed to evaluate the performance of the bio-scaffold within the living organism.

   (a) Tissue integration: The biological assessment helps determine how well the bio-scaffold integrates with host tissues.
   (b) Biocompatibility: Assessing cell behavior and cytocompatibility ensures that the bio-scaffold is biologically compatible and does not elicit adverse reactions.
   (c) Clinical relevance: In vivo studies provide critical data on the potential clinical effectiveness and safety of the bio-scaffold.

Biological assessment of bio-scaffolds through cell adhesion and proliferation studies, cytocompatibility assays, and in vivo studies is essential for ensuring their effectiveness, safety, and suitability for applications in tissue engineering, regenerative medicine, and other biomedical fields (Mohd et al. 2023; Molina et al. 2021).

6.5 Release Mechanisms of Natural Products from Bio-Scaffolds

The methods by which natural products are released from bio-scaffolds may differ and be influenced by the distinctive features of the scaffold and the properties of the incorporated natural product. Choosing the suitable mechanism relies on the specific attributes of the bio-scaffold, the nature of the natural products, and the intended therapeutic or functional objectives.

6.5.1 Diffusion-Controlled Release

This method is a strategy employed for the controlled and gradual release of natural products from bio-scaffolds. In this mechanism, natural products are initially encapsulated within the bio-scaffold during its fabrication. The release process is governed by the principle of diffusion, where a concentration gradient exists between the encapsulated natural products within the scaffold and the external environment. As a result, the natural products diffuse from regions of higher concentration within the scaffold to areas of lower concentration in the surrounding (Aulton 2007; Wen and Park 2010). The rate of diffusion is influenced by factors such as the characteristics of the scaffold material, the size of the encapsulated molecules, and the properties of the surrounding medium. This method often leads to sustained and predictable release kinetics, allowing for a controlled and steady delivery of natural products over an extended period (Rambhia and Ma 2019). The material properties of the bio-scaffold, including porosity and permeability, play a crucial role in shaping the diffusion-controlled release, offering a versatile approach for applications in drug delivery systems and various biomedical fields (Frenning 2011).

6.5.2 Degradation-Controlled Release

The degradation-controlled release method is employed for the precise and controlled liberation of natural products from bio-scaffolds. In this approach, bio-scaffolds are intentionally engineered to be biodegradable, enabling the gradual release of encapsulated natural products as the scaffold undergoes systematic degradation over a specified temporal duration (Lee and Mooney 2001; Rivest et al. 2007; Siepmann and Göpferich 2001). The selection of the bio-scaffold's material is guided by its susceptibility to biological processes or environmental conditions that promote breakdown. As the scaffold degrades, it unveils the encapsulated natural products, facilitating their gradual release into the neighboring environment. The release rate is dependent upon the degradation kinetics of the scaffold material and can be customized by considering factors such as polymer composition, molecular weight, and crosslinking density (Melissinaki et al. 2011; Seyednejad et al. 2012). Degradation-controlled release provides a methodology for sustained and regulated delivery of natural products, particularly advantageous in contexts where a timed and gradual release is favorable, such as in tissue engineering or drug delivery systems. The synchronization of natural product release with the biodegradation of the scaffold amplifies the precision and efficacy of this release mechanism (Simmons et al. 2004; Dado and Levenberg 2009; Siepmann and Siepmann 2008).

6.5.3 Stimuli-Responsive Release

The stimuli-responsive method represents an advanced strategy for the controlled release of natural products from bio-scaffolds (Uto et al. 2019; Marklein and

Burdick 2010; Brown and Anseth 2017). In this approach, bio-scaffolds are tailored to respond to distinct external stimuli, including variations in pH, temperature, or the presence of specific ions (Badeau and DeForest 2019; Ooi et al. 2017). These responsive characteristics initiate the release of encapsulated natural products in a precise and regulated manner (Galdopórpora et al. 2019; Mousavi et al. 2020). The selection or engineering of the bio-scaffold material is conducted with careful consideration to induce alterations in its structural or chemical properties upon exposure to the designated stimuli. As the external conditions align with the predetermined triggers, the bio-scaffold undergoes modifications that facilitate the controlled release of natural products (Mura et al. 2013). This method enables on-demand release, serving as a dynamic and adaptable platform for applications where precise timing and responsiveness are critical, such as in drug delivery systems necessitating targeted and triggered release in response to specific physiological conditions. The stimuli-responsive method introduces a level of sophistication to the controlled release of natural products, offering flexibility and accuracy in therapeutic and biomedical contexts (Gonçalves and Paiva 2017; Pattni and Torchilin 2015).

6.6 Applications of Natural Product-Loaded PCL and PLA Bio-Scaffolds

6.6.1 Enhanced Healing Properties

Injuries can result from mechanical, chemical, microbiological, or thermal sources, leading to wounds. The healing process involves a coordinated interplay of cells, matrix components, and various biological factors aimed at restoring tissue integrity. The phases of wound healing, including homeostasis, inflammation, proliferation/granulation, and remodeling/maturation, form a sequential and overlapping series of events (Negut et al. 2020).

The utilization of scaffold structures as innovative carriers for cell and drug delivery has emerged as a promising avenue to enhance wound healing. These scaffolds contribute to the differentiation of endothelial and epithelial cells, along with the production of angiogenic growth factors, particularly in cutaneous wounds. Currently, scaffolds find applications across diverse areas of tissue engineering, addressing issues ranging from nasal and auricular malformations to bone formation, cartilage development, periodontal regeneration, artificial corneas, heart valves, ligament replacement, tendon repair, and tumor treatment.

Scaffolds based on PCL and PLA polymers have been extensively studied for their efficacy in promoting wound healing. To enhance the wound-healing effects, electrospun (ES) nanofiber membranes incorporating various bioactive or antibacterial agents have been developed, as illustrated in Fig. 6.6. Drawing on ancient practices, certain plant extracts, rich in biologically active substances such as flavonoids, phenols, and polysaccharides, with antimicrobial, anti-inflammatory, and antioxidant properties have been employed to expedite the wound healing process.

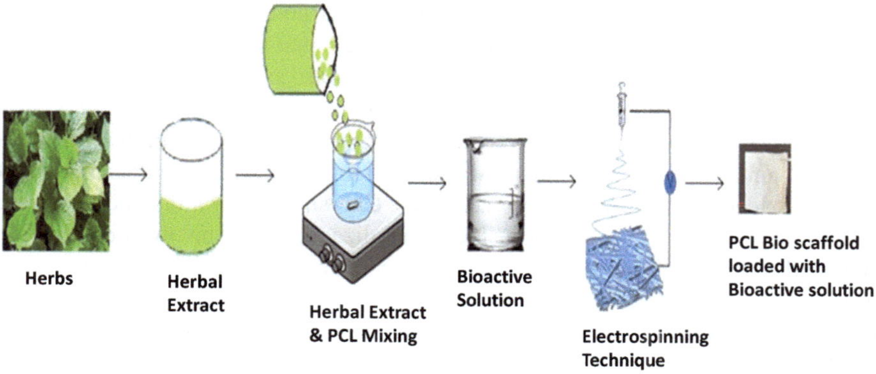

Fig. 6.6 Fabrication of electrospun nanofibrous PCL scaffold loaded with herbal bioactive

Wound healing dressing composed of PCL nanofibers loaded with pomegranate peel extract (PPE) using the electrospinning (ES) method was developed. The resulting nanofibrous membrane underwent analysis through scanning electron microscopy (SEM), confirming its uniform formulation and the successful incorporation of PPE, as corroborated by FTIR. The prepared membrane demonstrated sustained release kinetics for up to 40 h. Notably, this membrane exhibited significant antioxidant, antimicrobial, and proliferative properties, and particularly benefiting L929 cells. Consequently, the membrane holds promising potential for successful utilization as a wound healing material (Wang et al. 2023).

Similarly, El Fawal et al. (2021) employed the electrospinning (ES) technique to create a PCL/gelatin scaffold loaded with oregano oil for wound healing applications. Characterization through X-ray diffraction (XRD), FTIR, and SEM confirmed the scaffold's structure. The resulting scaffold demonstrated robust antimicrobial, cytotoxic, and antioxidant activities.

In another study, researchers utilized the electrospinning (ES) method to prepare a composite scaffold of PCL and PLA loaded with black seed (*Nigella sativa* (NS)) for wound healing. Optimization of manufacturing parameters, including feed rate, voltage, and PCL and PLA concentrations, was achieved through response surface methodology. The researchers reported that the fabricated nanofibrous scaffold displayed excellent antimicrobial potential, cell viability, and cell proliferation without inducing toxicity (Sharifi et al. 2020).

6.6.2 Antibacterial Effects

In 2019, Gamez et al. successfully engineered antimicrobial PCL mats (scaffolds) via the electrospinning (ES) method, incorporating thymol and carvacrol natural polyphenolic compounds. These mats demonstrated a sustained release of bioactive compounds and exhibited robust antimicrobial potential against both gram-positive (*Staphylococcus aureus*) and gram-negative (*Escherichia coli* S17) bacteria.

Additionally, the researchers substantiated that the prolonged release was attributed to the presence of hydrogen bonds between the PCL mat fibers and the bioactives.

Polyphenolic compounds and flavonoids, recognized as valuable secondary plant metabolites, are known for their potent anti-inflammatory and antimicrobial properties. In 2021, Alipour et al. undertook the development and exploration of the antioxidant, antimicrobial, anti-inflammatory, and mineralization characteristics of chrysin-loaded PCL and gelatin bio-scaffolds. Chrysin, a natural flavonoid sourced from honey, was used in the study. The mineralization study involved dental pulp stem cells (DPSCs). The scaffold exhibited significant antimicrobial activity against *Acinetobacter baumannii*, *Pseudomonas aeruginosa*, *Staphylococcus aureus*, and *Enterococcus faecalis*. Proliferation of dental stem cells was assessed through the MTT assay. Notably, the presence of chrysin-loaded scaffolds led to a substantial decrease in TNF and DCF levels in inflamed DPSCs. The findings strongly support the potential use of chrysin-loaded PCL scaffolds in various dental applications where anti-inflammatory and antimicrobial actions are required.

6.6.3 Osteoinductive Properties

Osteoinduction involves the recruitment of immature cells and the stimulation of these cells to develop into preosteoblasts. In situations requiring bone healing, such as fractures, osteoinduction plays a pivotal role, serving as a prerequisite biological mechanism for new bone formation after injury. In the context of repairing skeletal defects, particularly traumatic bone cracks, advanced tissue engineering techniques have become indispensable. This innovative approach involves the application of biodegradable scaffolds to defective sites, incorporating seed cells, growth factors, and scaffold materials. As the scaffold material breaks down, the released factors initiate cell proliferation and differentiation, facilitating the repair of defects. Scaffolds fabricated from biodegradable materials such as PLC, PLA, and gelatin not only provide essential support for new tissue growth but also exhibit significant osteoinductivity (Agarwal et al. 2021; Albrektsson and Johansson 2001).

Utilizing natural compounds with angiogenic activities holds promise for successful tissue regeneration and osteoinductive effects. Ma et al. formulated a PCL-collagen scaffold with inherent osteoinductive properties. They created a PCL/COL (collagen) scaffold loaded with the berberine bioactive compound through an electrospinning process. The physicochemical properties of the electrospun scaffolds were thoroughly characterized. Additionally, the impact of BBR/PCL/COL scaffolds on the osteogenic differentiation of DPSCs in vitro and their ability to repair bone defects in vivo were assessed. In the in vivo study, rats were selected, and bone defects were created on two sides of the sagittal suture using a trephine. Subsequently, scaffolds (PCL/COL, BBR/PCL/COL) with a diameter of 6 mm were implanted to cover the bone defects, while the control group was left empty as a negative control. Results from the in vivo study demonstrated that the BBR/PCL/COL electrospun scaffold accelerated the bone defect repair process (Ma et al. 2021).

6.6.4 Scaffold for Growth Factors

The significance of growth factors (GFs) in cell adhesion, proliferation, differentiation, and gene expression underscores their importance in tissue regeneration. Incorporating GFs into scaffolds enhances the success rate of tissue regeneration. When growth factors are integrated or immobilized on scaffolds, they shield the factors from direct contact with water, restrict their diffusion, and consequently prolong their activity in vivo. Ideal scaffolds should possess the capability to secrete and slowly release growth factors, mirroring the natural tissue's requirements during clinical applications. This functionality aims to replicate the biological model of natural tissue, achieving the simulation of scaffold functions (De Witte et al. 2018; Shen and Hu 2021).

The term "growth factor" encompasses several families of intracellular signaling proteins, each with defined roles such as cell recruitment, promotion of migration, and differentiation. Specific growth factors play vital roles in the formation, maturation, and maintenance of tissues throughout the body. In the context of bone generation and repair, certain growth factors stimulate pathways specific to bone formation, fostering osteoinduction. Others, such as vascular endothelial growth factor (VEGF) and platelet-derived growth factor (PDGF), promote angiogenesis—the sprouting of new blood vessels from existing ones (Blackwood et al. 2012).

In 2022, Qin et al. produced electrospun PCL nanofibers loaded with growth factors (Fig. 6.7). The methodology involved the PEGylation of nanofibers using azide-alkyne groups that bonded with nanocapsules containing growth factors. The resulting scaffold exhibited satisfactory cell adhesion and growth properties. Additionally, through the specific immobilization of pH-responsive nanocapsules containing bone morphogenetic protein 2 (BMP-2), controlled release of active BMP-2 from the PCL nanofibers was achieved over a span of 21 days. When bone mesenchymal stem cells were cultured on this nanofiber scaffold, enhanced ossification was observed, correlating with the time-dependent release of BMP-2. This

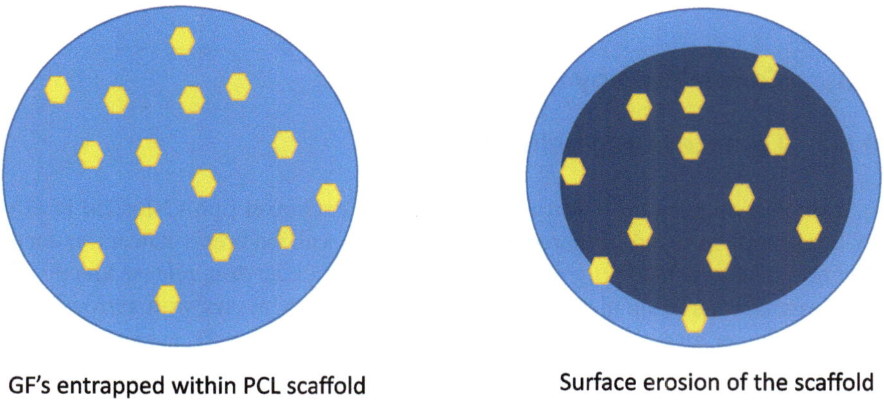

Fig. 6.7 Schematic representation of release of growth factors (GFs) from PCL scaffolds

established surface modification has the potential to be applied as a universal approach to hydrophobic nanomaterials, enabling prolonged and sustainable release of multiple signaling proteins for tissue engineering.

6.7 Controlled Drug Release

PCL, an FDA-approved biodegradable and bioresorbable polymer, serves as a suitable material for various medical applications. It is utilized for both long-term drug delivery, as seen in birth control implants, and short-term release of substances such as antibiotics, anti-inflammatory agents, bone morphogenetic proteins, immune modulators, and naturally occurring biologically active compounds (Volokhova et al. 2021). The porous and interconnected scaffolds, made from materials like PCL and PGA, enhance cell penetration and proliferation on their surfaces. These scaffolds, employed in tissue engineering, enable controlled release through either the diffusion of drugs by polymer degradation or the diffusion of impregnated drugs into scaffold walls (Rambhia and Ma 2015).

In a study by Adeli-Sardou et al. (2019), PCL/gelatin scaffolds loaded with lawsone were prepared using the electrospinning method. The resulting scaffolds demonstrated prolonged drug release of lawsone over a 20-day period. In vitro bioactivity evaluation on human gingiva fibroblast cells (HGF) using the MTT assay revealed that the scaffold loaded with 1% lawsone significantly increased cell attachment and proliferation. Additionally, RT-qPCR analysis indicated a noteworthy increase in transforming growth factor β (TGF-B1), collagen (COL1), and epidermal growth factor (EGF). Specifically, there was a significant increase in TGF-B1 and COL1 gene expression in PCL/Gel/Law 0.5% and 1% mats. In an in vivo wound healing activity assessment using a rat wound model, the lawsone-loaded scaffolds, particularly PCL/Gel/Law 1%, exhibited the most substantial impact on healing by enhancing reepithelialization after 14 days. This suggests that the incorporation of 1% lawsone in the core of PCL/Gel electrospun nanofibers possesses excellent characteristics and can be utilized as a wound dressing patch in medicine.

6.8 Cancer Therapy

6.8.1 Localized Drug Delivery

In a study conducted by Nwazojie et al. in 2023, paclitaxel EphA2-loaded PLGA-PCL microspheres were prepared for targeted and localized breast cancer treatment. The observed sustained drug release followed non-Fickian drug release kinetics. In vitro assessments using alamar blue and flow cytometry assays with various drug-loaded polymer formulations revealed cell death and cytotoxicity, evidenced by cell inhibition and late apoptosis in triple-negative breast cancer (TNBC) cells (MDA-MB 231). In vivo studies on groups of mice induced with subcutaneous TNBC demonstrated the effectiveness of the localized release of EphA2-conjugated drugs, leading to the complete elimination of residual tumors after local surgical resection.

6.8.2 Immunomodulatory Effects

The substantial surface area of the biodegradable scaffold promotes superior cell attachment and proliferation. Researchers have highlighted the immunomodulatory effects of PCL scaffolds loaded with natural bioactives. In a study conducted by Abdulrahman et al. in 2022, a PCL scaffold loaded with curcumin and vitamin D was fabricated and evaluated for treating bone defects, demonstrating anti-inflammatory and immunomodulatory effects. Vitamin D, known as an immunomodulator, has been reported to enhance various immune cell functions, including macrophage activity, cytokine production, and phagocytosis. Their findings indicated that the PCL/Vit D-Cur scaffold not only enhanced osteoblast cell proliferation and osteogenic gene expression but also effectively contributed to the upregulation and regulation of M2-related genes, such as IL-10 and IL-1ra. This process played an anti-inflammatory role by producing anti-inflammatory cytokines.

6.9 Challenges

The effective delivery of drugs using natural product-loaded PCL and PLA-based bio-scaffolds is contingent upon addressing the challenges inherent in this approach. Tackling these challenges is imperative to ensure the successful and efficient delivery of drugs through these biomaterials.

Biocompatibility Optimization Achieving optimal biocompatibility remains a challenge, especially when integrating natural products into PCL- and PLA-based bio-scaffolds. Variations in individual responses and potential immunogenic reactions must be carefully addressed to ensure the safety and effectiveness of these biomaterials in diverse patient populations.

Controlled Release Precision The precise control of release mechanisms, particularly in drug delivery applications, poses a challenge. Striking a balance between sustained release and avoiding burst release requires further refinement to meet the specific therapeutic needs of different applications.

Long-Term Stability and Durability Ensuring the long-term stability and durability of natural product-loaded bio-scaffolds is critical for their success in clinical applications. Addressing issues such as material degradation over time and maintaining structural integrity under physiological conditions is essential for sustained therapeutic effects.

Standardization of Fabrication Modification techniques to further enhance biocompatibility.

Methods Standardizing fabrication methods is a challenge due to the variety of techniques employed in creating these bio-scaffolds. Consistency in production is crucial for reproducibility and regulatory approval, necessitating the establishment of standardized protocols for fabrication.

In Vivo Performance Validation The translation of promising in vitro results to in vivo success which poses a significant challenge. Comprehensive studies that evaluate the performance of natural product-loaded PCL and PLA bio-scaffolds in relevant animal models are necessary to bridge the gap between laboratory findings and clinical applications.

6.10 Future Directions

Future directions should explore advanced surface functionality and the integration of smart and responsive elements for precise control over drug release. A personalized medicine approach, incorporating patient-specific factors, could optimize treatment outcomes, while the incorporation of advanced imaging techniques can provide real-time insights into bio-scaffold performance. Additionally, the development of multifunctional bio-scaffolds capable of addressing multiple therapeutic aspects simultaneously holds promise for broadening their applicability across various medical fields. Overall, these challenges and future directions underscore the dynamic nature of research in this field, shaping the trajectory of natural product-loaded bio-scaffold applications in biomedicine.

6.11 Conclusion

This chapter has provided a comprehensive exploration of the intersection between natural product-loaded bio-scaffolds and their applications in various biomedical fields. The initial sections introduced the key components of the study, emphasizing the significance of both natural products and bio-scaffolds in the realm of biomedicine. The subsequent focus on polycaprolactone (PCL) and polylactic acid (PLA) as bio-scaffold materials delved into their individual properties and the methods employed for enhancing their biocompatibility through surface modifications.

The fabrication and characterization processes were discussed in detail, covering various aspects such as structural and morphological analysis, stability, degradation, and biological assessments. The elucidation of release mechanisms, including diffusion-controlled, degradation-controlled, and stimuli-responsive release, added depth to the understanding of how natural products interact within these bio-scaffolds.

Furthermore, the exploration of applications demonstrated the versatility of natural product-loaded PCL and PLA bio-scaffolds. From wound healing with enhanced properties and antibacterial effects to bone regeneration with osteoinductive properties and drug delivery facilitating controlled release and targeted therapies, these bio-scaffolds exhibit immense potential in diverse biomedical scenarios, including cancer therapy with localized drug delivery and immunomodulatory effects.

References

Adeli-Sardou M, Yaghoobi MM, Torkzadeh-Mahani M, Dodel M (2019) Controlled release of lawsone from polycaprolactone/gelatin electrospun nano fibers for skin tissue regeneration. Int J Biol Macromol 124:478–491. https://doi.org/10.1016/j.ijbiomac.2018.11.237

Agarwal T, Tan SA, Onesto V, Law JX, Agrawal G, Pal S, Lim WL, Sharifi E, Moghaddam FD, Maiti TK (2021) Engineered herbal scaffolds for tissue repair and regeneration: recent trends and technologies. Biomed Eng Adv 2:100015

Al-Bishari AM, Al-Shaaobi BA, Al-Bishari AA, Al-Baadani MA, Yu L, Shen J, Cai L, Shen Y, Deng Z, Gao P (2022) Vitamin D and curcumin-loaded PCL nanofibrous for engineering osteogenesis and immunomodulatory scaffold. Front Bioeng Biotechnol 10:975431

Albrektsson T, Johansson C (2001) Osteoinduction, osteoconduction and osseointegration. Eur Spine J 10 Suppl 2(Suppl 2):S96–S101. https://doi.org/10.1007/s005860100282. PMID: 11716023; PMCID: PMC3611551

Alipour M, Pouya B, Aghazadeh Z, SamadiKafil H, Ghorbani M, Alizadeh S, Aghazadeh M, Dalir Abdolahinia E (2021) The antimicrobial, antioxidative, and anti-inflammatory effects of polycaprolactone/gelatin scaffolds containing chrysin for regenerative endodontic purposes. Stem Cells Int 2021:3828777

Annor AH, Tang ME, Pui CL, Ebersole GC, Frisella MM, Matthews BD, Deeken CR (2012) Effect of enzymatic degradation on the mechanical properties of biological scaffold materials. Surg Endosc 26(10):2767–2778

Aranaz I, Gutierrez MC, Ferrer ML, Monte FD (2014) Preparation of chitosan nanocomposites with a macroporous structure by unidirectional freezing and subsequent freeze-drying. Mar Drugs 12(11):5619–5642

Assad M, Jackson N (2019) Biocompatibility evaluation of orthopedic biomaterials and medical devices: a review of safety and efficacy models. Encyclopaedia Biomed Eng 2(1):281–309

Aulton ME (2007) Aulton's pharmaceutics: the design and manufacture of medicines. Churchill Livingstone, London

Badeau BA, DeForest CA (2019) Programming stimuli-responsive behavior into biomaterials. Annu Rev Biomed Eng 21:241–265. https://doi.org/10.1146/annurev-bioeng-060418-052324

Bhadran A, Shah T, Babanyinah GK, Polara H, Taslimy S, Biewer MC, Stefan MC (2023) Recent advances in polycaprolactones for anticancer drug delivery. Pharmaceutics 15(7):1977

Blackwood KA, Bock N, Dargaville TR, Ann Woodruff M (2012) Scaffolds for growth factor delivery as applied to bone tissue engineering. Int J Polym Science 2012:174942

Bokov D, Turki A, Chupradit S, Suksatan W, Ansari MJ, Shewael IH, Valiev GH, Kianfar A (2021) Nanomaterial by sol-gel method: synthesis and application. Adv Mater Sci Eng 5102014:1–20

Brown SP (2012) Applications of high-resolution 1H solid-state NMR. Solid State Nucl Magn Reson 41:1–27

Brown TE, Anseth KS (2017) Spatiotemporal hydrogel biomaterials for regenerative medicine. Chem Soc 46:6532

Bruinink A, Luginbuehl R (2012) Evaluation of biocompatibility using in vitro methods: interpretation and limitations. Adv Biochem Eng Biotechnol 126:117–152. https://doi.org/10.1007/10_2011_111
Calcabrini C, Catanzaro E, Bishayee A, Turrini E, Fimognari C (2017) Marine sponge natural products with anticancer potential: an updated review. Mar Drugs 15(10):310
Chan B, Leong KW (2008) Scaffolding in tissue engineering: general approaches and tissue specific considerations. Eur Spine J 17(4):467–479
Chaudhary C, Garg T (2015) Scaffolds: a novel carrier and potential wound healer. Crit Rev Ther Drug Carrier Syst 32(4):277–321. https://doi.org/10.1615/critrevtherdrugcarriersyst.2015011246. PMID: 26080925
Chen QZ (2011) Foaming technology of tissue engineering scaffolds—a review. Bubble Sci Eng Technol 3(2):34–47
Chen J, Zhang YG, Li S, Jia J, Fang H, Zhang Z (2011) Effects of in situ and physical mixing on mechanical and bioactive behaviors of nano hydroxyapatite–chitosan scaffolds. J Biomater Sci Polym 22:2097–2106
Dado D, Levenberg S (2009) Cell–scaffold mechanical interplay within engineered tissue. Semin Cell Dev Biol 20:656
Dawson JI, Kanczler J, Aarvold A, Smith J, Oreffo ROC (2011) Tissue engineering: bone. In: Comprehensive biotechnology, vol 2
De Witte T-M, Fratila-Apachitei LE, Zadpoor AA, Peppas NA (2018) Bone tissue engineering via growth factor delivery: from scaffolds to complex matrices. Regen Biomater 5(4):197–211. https://doi.org/10.1093/rb/rby013
Ebnesajjad S (2006) 4—Surface and material characterization techniques. In: Surface treatment of materials for adhesion bonding. William Andrew Publishing, pp 43–75
El Fawal G, Hong H, Mo X, Wang H (2021) Fabrication of scaffold based on gelatin and polycaprolactone (PCL) for wound dressing application. J Drug Deliv Sci Technol 63:102501
Eltom A, Zhong G, Muhammad A (2019) Scaffold techniques and designs in tissue engineering functions and purposes: a review. Adv Mater Sci Eng 2019(4):1–13
Enrione J, Díaz-Calderón P, Weinstein-Oppenheimer CR, Sánchez E, Fuentes MA, Brown DI, Herrera H, Acevedo CA (2013) Designing a gelatin/chitosan/hyaluronic acid biopolymer using a thermophysical approach for use in tissue engineering. Bioprocess Biosyst Eng 36(12):1947–1956
Figoli A (2014) Thermally induced phase separation (TIPS) for membrane preparation. In: Encyclopedia of membranes, pp 1–2
Frenning G (2011) Modelling drug release from inert matrix systems: from moving-boundary to continuous-field descriptions. Int J Pharm 418(1):88–99
Galdopórpora JM, Morcillo MF, Ibar A, Perez CJ, Tuttolomondo MV, Desimone MF (2019) Development of silver nanoparticles/gelatin thermoresponsive nanocomposites: characterization and antimicrobial activity. Curr Pharm Des 25:4121–4129
Gamez E, Mendoza G, Salido S, Arruebo M, Irusta S (2019) Antimicrobial electrospun polycaprolactone-based wound dressings: an in vitro study about the importance of the direct contact to elicit bactericidal activity. Adv Wound Care (New Rochelle) 8(9):438–451. https://doi.org/10.1089/wound.2018.0893. Epub 2019 Aug 9. PMID: 31737424; PMCID: PMC6855275
Gonçalves GAR, Paiva RDMA (2017) Gene therapy: advances, challenges and perspectives. Einstein (São Paulo) 15:369–375. https://doi.org/10.1590/s1679-45082017rb4024
Gorth DJ, Webster T (2011) Matrices for tissue engineering and regenerative medicine. In: Biomaterials for artificial organs, pp 270–286
Gu L, Shan T, Ma YX, Tay FR, Niu L (2019) Novel biomedical applications of crosslinked collagen. Trends Biotechnol 37(5):464–491
Hasan A, Soliman S, El H, Tseng F, Yalcin YT, Marei HC (2018) Fabrication and in vitro characterization of a tissue engineered PCL-PLLA heart valve. Sci Rep 8:8187

Hassanajili S, Pour A, Oryan A, Khozani TT (2019) Preparation and characterization of PLA/PCL/HA composite scaffolds using indirect 3D printing for bone tissue engineering. Mater Sci Eng C 104:109960

Hollister S (2005) Porous scaffold design for tissue engineering. Nat Mater 4:518

Holzwarth JM, Ma PX (2011) 3D nanofibrous scaffolds for tissue engineering. J Mater Chem 21:10243–10251

Homaeigohar S, Boccaccini AR (2022) Nature-derived and synthetic additives to poly(ε-caprolactone) nanofibrous systems for biomedicine; an updated overview. Front Chem 19:9

Huebsch N, Mooney DJ (2009) Inspiration and application in the evolution of biomaterials. Nature 462:426–432

Hutmacher DW (2000) Scaffolds in tissue engineering bone and cartilage. Biomaterials 21(24):2529–2543

Iijima K, Tsuji Y, Kuriki I, Kakimoto A, Nikaido Y, Ninomiya R, Iyoda T, Fukai F, Hashizume M (2017) Control of cell adhesion and proliferation utilizing polysaccharide composite film scaffolds. Colloids Surf B: Biointerfaces 160:228–237

Kesireddy V, Kasper FK (2016) Approaches for building bioactive elements into synthetic scaffolds for bone tissue engineering. J Mater Chem B 4(42):6773–6786

Krishani M, Shin WY, Suhaimi H, Sambudi NS (2023) Development of scaffolds from bio-based natural materials for tissue regeneration applications: a review. Gels 9(2):100

Kumar R, Kumar M, Tyagi R, Singh R (2022) Electrospinning based nanofibers for 3D printing applications. In: Encyclopedia of materials: plastics and polymers, pp 253–263. https://doi.org/10.1016/B978-0-12-820352-1.00247-9

Labet M, Thielemans W (2009) Synthesis of polycaprolactone: a review. Chem Soc Rev 38:3484–3504

Lee KY, Mooney DJ (2001) Hydrogels for tissue engineering. Surgery 101:1869

Li C, Li Q, Mei Q, Lu T (2015) Pharmacological effects and pharmacokinetic properties of icariin, the major bioactive component in Herba Epimedii. Life Sci 126:57–68

Li Z, Xie MB, Li Y, Ma Y, Li Y, Dai FY (2016) Recent progress in tissue engineering and regenerative medicine. J Biomater Tissue Eng 6(10):755–766

Liu T, Bhatia V (2004) Three-dimensional tissue fabrication. Adv Drug Deliv Rev 56:1635

Liu Y, Thibodeaux D, Gamble G (2012) Characterization of attenuated total reflection infrared spectral intensity variations of immature and mature cotton fibers by two dimensional correlation analysis. Appl Spectrosc 66(2):198–207

Liu S, Zhu Y, Meng W, He Z, Feng W, Zhang C, Giesy JP (2016) Characteristics and degradation of carbon and phosphorus from aquatic macrophytes in lakes: insights from solid-state 13C NMR and solution 31P NMR spectroscopy. Sci Total Environ 543:746–756

Liu Y, Yang H, Xiong J, Zhao J, Guo M, Chen J, Zhao X, Chen C, Zhou Y (2023) Icariin as an emerging candidate drug for anticancer treatment: current status and perspective. Biomed Pharmacother 157:113991

Ma PX (2008) Biomimetic materials for tissue engineering. Adv Drug Deliv Rev 60:184–198

Ma L, Yu Y, Liu H, Sun W, Lin Z, Liu C, Miao L (2021) Berberine-releasing electrospun scaffold induces osteogenic differentiation of DPSCs and accelerates bone repair. Sci Rep 11(1):1027

Madhav H, Singh N, Jaiswar G (2019) Thermoset, bioactive, metal–polymer composites for medical applications. In: Materials for biomedical engineering. Elsevier, pp 105–143

Magnusson JP, Saeed AO, Fernandez-Trillo F, Alexander C (2011) Synthetic polymers for biopharmaceutical delivery. Polym Chem 2:48–59

Mark EF, Atala A (2014) Principles of tissue engineering. Principles of Tissue Engineering (Fourth Edition) 2014, 83–123

Marklein RA, Burdick JA (2010) Controlling stem cell fate with material design. Adv Mater 22:175

Melissinaki V, Gill AA, Ortega I (2011) Direct laser writing of 3D scaffolds for neural tissue engineering applications. Biofabrication 3:045005

Mohd N, Razali M, Fauzi MB, Abu Kasim NH (2023) In vitro and in vivo biological assessments of 3D-bioprinted scaffolds for dental applications. Int J Mol Sci 24(16):12881

Molina MIE, Malollari KG, Komvopoulos K (2021) Design challenges in polymeric scaffolds for tissue engineering. Front Bioeng Biotechnol Sec Biomater 9:617141
Mousavi ST, Harper GR, Municoy S, Ashton MD, Townsend D, Alsharif GHK, Oikonomou VK, Firlak M, Au-Yong S, Murdock BE (2020) Electroactive silk fibroin films for electrochemically enhanced delivery of drugs. Macromol Mater Eng 305:2000130. https://doi.org/10.1002/mame.202000130
Mura S, Nicolas J, Couvreur P (2013) Stimuli-responsive nanocarriers for drug delivery. Nat Mater 12:991–1003. https://doi.org/10.1038/nmat3776
Negut I, Dorcioman G, Grumezescu V (2020) Scaffolds for wound healing applications. Polymers (Basel) 12(9):2010. https://doi.org/10.3390/polym12092010. PMID: 32899245; PMCID: PMC7563417
Nwazojie CC, Obayemi JD, Salifu AA, Borbor-Sawyer SM, Uzonwanne VO, Onyekanne CE, Akpan UM, Onwudiwe KC, Oparah JC, Odusanya OS, Soboyejo WO (2023) Targeted drug-loaded PLGA-PCL microspheres for specific and localized treatment of triple negative breast cancer. J Mater Sci Mater Med 34(8):41
Ooi HW, Hafeez S, Blitterswijk CA, Moroni L, Baker MB (2017) Hydrogels that listen to cells: a review of cell-responsive strategies in biomaterial design for tissue regeneration. Mater Horizons 4:1020–1040. https://doi.org/10.1039/C7MH00373K
Ovinuchi E (2023) A perspective on the synergistic use of 3D printing and electrospinning to improve nanomaterials for biomedical applications. Nano Trends 4:100025. https://doi.org/10.1016/j.nwnano.2023.100025
Patil VA, Masters KS (2020) Engineered collagen matrices. Bioengineering (Basel) 7(4):163
Pattni BS, Torchilin VP (2015) Targeted drug delivery systems: strategies and challenges. In: Devarajan PV, Jain S (eds) Targeted drug delivery: concepts and design. Springer, Cham, pp 3–38
Pavia DL, Lampman GM, Kriz GS, Vyvyan JR (2014) Introduction to spectroscopy. Cengage Learning, Boston
Pham QP, Sharma U, Mikos AG (2006) Electrospinning of polymeric nanofibers for tissue engineering applications: a review. Tissue Eng 12(5):1197–1211
Qin X, Wu Y, Liu S, Yang L, Yuan H, Cai S, Flesch J, Li Z, Tang Y, Li X, Zhuang Y, You C, Liu C, Yu C (2022) Surface modification of polycaprolactone scaffold with improved biocompatibility and controlled growth factor release for enhanced stem cell differentiation. Front Bioeng Biotechnol 9:802311. https://doi.org/10.3389/fbioe.2021.802311. PMID: 35071210; PMCID: PMC8782149
Rambhia KJ, Ma PX (2015) Controlled drug release for tissue engineering. J Control Release 219:119–128
Rambhia KJ, Ma PX (2019) Controlled drug release for tissue engineering. J Control Release 219:119–128. https://doi.org/10.1016/j.jconrel.2015.08.049
Rani P, Yadav V, Pandey P, Yadav K (2023) Recent patent-based review on the role of three-dimensional printing technology in pharmaceutical and biomedical applications. Pharm Pat Anal 12(4):159–175
Raucci MG, Guarino V, Ambrosio L (2010) Hybrid composite scaffolds prepared by sol–gel method for bone regeneration. Compos Sci Technol 70(13):1861–1868
Rivest C, Morrison DWG, Ni B, Rubin J, Yadav V, Mahdavi A (2007) Microscale hydrogels: fabrication, applications and mechanical properties. Mech Mater Struct 2:756
Roberto S, Francesco L, Luigi B, Salvatrice R, Giulio G (2016) Integration of PCL and PLA in a monolithic porous scaffold for interface tissue engineering. J Mech Behav Biomed Mater 63:303–313
Roseti L, Parisi V, Petretta M, Cavallo C, Desando G, Bartolotti I, Grigolo B (2017) Scaffolds for bone tissue engineering: state of the art and new perspectives. Mater Sci Eng C 78:1246–1262
Saleem MM, Kumar M, Malhi GS (2024) Predictive analysis of fused deposition modeling of PLA material through machine learning. AIP Conf Proceed 2986(1)
Salerno A, Netti PA (2021) Review on computer-aided design and manufacturing of drug delivery scaffolds for cell guidance and tissue regeneration. Front Bioeng Biotechnol 9:682133

Salerno A, Maio ED, Iannace S, Netti P (2012) Tailoring the pore structure of PCL scaffolds for tissue engineering prepared via gas foaming of multi-phase blends. J Porous Mater 19:181
Seyednejad H, Gawlitta D, Bruin A, Nostrum CF, Vermonden T, Dhert WJA, Hennik WE (2012) In vivo biocompatibility and biodegradation of 3D-printed porous scaffolds based on a hydroxyl-functionalized poly(e-caprolactone). Biomaterials 33:4309
Sharifi M, Bahrami SH, Nejad NH, Milan PB (2020) Electrospun PCL and PLA hybrid nanofibrous scaffolds containing Nigella sativa herbal extract for effective wound healing. J Appl Polym Sci 137(46):49528
Shen H, Hu X (2021) Growth factor loading on aliphatic polyester scaffolds. RSC Adv 11(12):6735–6747. https://doi.org/10.1039/d0ra10232f. PMID: 35423177; PMCID: PMC8694921
Shi G, Cai Q, Wang C, Wang LNS, Bei J (2002) Fabrication and biocompatibility of cell scaffolds of poly(L-lactic acid) and poly(L-lactic-co-glycolic acid). Polym Adv Technol 13:227
Shum AWT, Li J, Mak AFT (2005) Fabrication and structural characterization of porous biodegradable poly(dl-lactic-co-glycolic acid) scaffolds with controlled range of pore sizes. Polym Degrad Stab 87(3):487–493
Siepmann J, Göpferich A (2001) Mathematical modeling of bioerodible, polymeric drug delivery systems. Adv Drug Deliv Rev 48:229
Siepmann J, Siepmann F (2008) Mathematical modeling of drug delivery. Int J Pharm 364:328–343
Simmons A, Hyvarinena J, Odell RA, Martinc DJ, Gunatillake PA, Noble KR, Poole-Warrena LA (2004) Long-term in vivo biostability of poly(dimethylsiloxane)/poly(hexamethylene oxide) mixed macrodiol-based polyurethane elastomers. Biomaterials 25:4887
Slichenmyer WJ, Von Hoff DD (1991) Taxol: a new and effective anti-cancer drug. Anti-Cancer Drugs 2(6):519–530. PMID: 1687206
Solechan S, Suprihanto A, Widyanto SA, Triyono J, Fitriyana DF, Siregar JP, Cionita T (2022) Investigating the effect of PCL concentrations on the characterization of PLA polymeric blends for biomaterial applications. Materials 15:7396
Soni R, Shankar G, Mukhopadhyay P, Gupta V (2022) A concise review on Artemisia annua L.: a major source of diverse medicinal compounds. Ind Crop Prod 184:115072
Sun H, Han B, Kong J, Meng J, Zhu LF (2014) Microstructure, thermal properties and rheological behavior of PLA/PCL blends for melt-blown nonwovens. Polymer 38:477–483
Switha D, Basha SK, Kumari VS (2023) In vitro cytocompatibility evaluation of nanostarch reinforced polyaniline-polyvinyl alcohol conductive bionanocomposites for skin tissue engineering application. J Umm Al-Qura Univ Appl Sci 9:252–259
Tate MLK, Falls TD, McBride SH, Atit R, Knothe UR (2008) Mechanical modulation of osteochondroprogenitor cell fate. Int J Biochem Cell Biol 40:2720–2738
Thavornyutikarn B, Chantarapanich N, Sitthiseripratip K, Thouas GA, Chen Q (2014) Bone tissue engineering scaffolding: computer-aided scaffolding techniques. Prog Biomater 3(2):61–102
Tutar R, Motealleh A, Khademhosseini A, Kehr NS (2019) Functional nanomaterials on 2D surfaces and in 3D nanocomposite hydrogels for biomedical applications. Adv Funct Mater 29:1904344. https://doi.org/10.1002/adfm.201904344
Uto K, Arakawa CK, DeForest CA (2019) Next-generation biomaterials for culture and manipulation of stem cells. Cold Spring Harbor Perspect Biol 12:035691
Van VS, Dubruel P, Schacht E (2011) Biopolymer-based hydrogels as scaffolds for tissue engineering applications: a review. Biomacromolecules 12:1387–1408
Volokhova AA, Kudryavtseva VL, Spiridonova TI, Kolesnik I, Goreninskii SI, Sazonov RV, Remnev GE, Tverdokhlebov SI (2021) Controlled drug release from electrospun PCL nonwoven scaffolds via multi-layering and e-beam treatment. Mater Today Commun 26:102134
Wang W, Caetano GF, Chiang WH, Braz AL, Blake JJ, Paulo MACF, Bártolo J (2016) Morphological, mechanical and biological assessment of PCL/pristine graphene scaffolds for bone regeneration. Int J Bioprint 2:95–105
Wang P, Sun Y, Shi X, Shen H, Ning H, Liu H (2021) Bioscaffolds embedded with regulatory modules for cell growth and tissue formation: a review. Bioact Mater 6(5):1283–1307

Wang Y, Wang X, Zhou D, Xia X, Zhou H, Wang Y et al (2023) Preparation and characterization of polycaprolactone (PCL) antimicrobial wound dressing loaded with pomegranate peel extract. ACS Omega 8(23):20323–22033. https://doi.org/10.1021/acsomega.2c08180
Wen H, Park K (2010) Introduction and overview of oral controlled release formulation design. Wiley, New York
Woodruff MA, Hutmacher DW (2010) The return of a forgotten polymer-polycaprolactone in the 21st century. Prog Polym Sci 351:1217–1256
Wu SC, Chang WH, Dong GC, Chen KY, Chen YS, Yao CH (2011) Cell adhesion and proliferation enhancement by gelatin nanofiber scaffolds. J Bioact Compat Polym 26(6):565–577
Xia Y, Sun J, Zhao L, Zhang F, Liang XJ, Guo Y, Weir MD, Reynolds MA, Gu N, Xu HHK (2018) Magnetic field and nano-scaffolds with stem cells to enhance bone regeneration. Biomaterials 183:151–170
Yadav D, Kumar N (2014) Nanonization of curcumin by antisolvent precipitation: process development, characterization, freeze drying and stability performance. Int J Pharm 477(1–2):564–577
Yadav K, Jaishwal S, Yadav D, Kumar S, Sharma C, Saroha K, Nanda S (2023) Development and in-vitro evaluation of curcumin proniosomes for enhanced stability and antimicrobial activity. Eur Chem Bull 12(13):1942–1958
Yang SF, Leong KF, Du ZH, Chua CK (2001) The design of scaffolds for use in tissue engineering. Part 1. Traditional factors. Tissue Eng 7:679–689
Younis H, Khan HU, Maheen S, Saadullah M, Shah S, Ahmad N, Alshehri S, Majrashi MAA, Alsalhi A, Siddique R et al (2023) Fabrication, characterization and biomedical evaluation of a statistically optimized gelatin scaffold enriched with co-drugs loaded into controlled-release silica nanoparticles. Molecules 28:5233
Zhai P, Chen XB, Schreyer DJ (2015) PLGA/alginate composite microspheres for hydrophilic protein delivery. Mater Sci Eng C Mater 56:251–259
Zhang Y, Zhang M (2001) Synthesis and characterization of macroporous chitosan/calcium phosphate composite scaffolds for tissue engineering. J Biomed Mater Res 55:304
Zhang Y, Tan W, Wu M et al (2021) Characterization and cytocompatibility of 3D porous biomimetic scaffold derived from rabbit nucleus pulposus tissue in vitro. J Mater Sci Mater Med 32:8
Zhao D, Zhu T, Cui JLL, Zhang Z, Zhuang X, Ding J (2021) Poly(lactic-co-glycolic acid)-based composite bone-substitute materials. Bioact Mater 6(2):346–360. https://doi.org/10.1016/j.bioactmat.2020.08.016
Zhu JM (2010) Bioactive modification of poly(ethylene glycol) hydrogels for tissue engineering. Biomaterials 31(17):4639–4656

Applications of Bioscaffolds in Plastic Surgery

7

Ram Kumar, Ashish Baldi, Navneesh, Pawan Kumar, Babita Dhiman, Sanjeev Kumar, and Sushant

Abstract

Plastic surgery encompasses a wide array of specialties, including liposuction, facial cosmetic, breast surgery, maxillofacial surgery, body contouring, and skin grafting procedures. Given the critical need for biological safety in plastic surgery, biomaterials are extensively utilized due to their exceptional biocompatibility and biodegradability. A diverse range of biomaterials is currently employed in plastic surgery, offering versatile applications. The recent surge in three-dimensional (3D) printing technology has enabled the production of highly precise and personalized bioscaffolds with microporous structures, promising significant advancements in biomaterials. This chapter delves into the principles and processes of 3D bioprinting technology, elucidating the merits and drawbacks of various bioprinting techniques. Additionally, it provides an overview of

R. Kumar · A. Baldi
University College of Pharmacy, Guru Kashi University, Talwandi Sabo, Bathinda, Punjab, India

Navneesh
Department of Pharmaceutical Chemistry, ISF College of Pharmacy, Moga, Punjab, India

P. Kumar
Department of Pharmaceutical Science and Technology, Maharaja Ranjit Singh Punjab Technical University, Bathinda, Punjab, India

B. Dhiman
Electronics and Communication Engineering, UIC, Chandigarh University, Mohali, Punjab, India

S. Kumar
University Institute of Pharma Sciences, Chandigarh University, Gharuan, Punjab, India

Sushant (✉)
Faculty of Pharmaceutical Sciences, The ICFAI University, Baddi, Himachal Pradesh, India

M. Kumar et al. (eds.), *Natural Product Inspired Scaffolds*,
https://doi.org/10.1007/978-981-97-3111-4_7

the bioprinting materials currently available and examines the rationale behind dynamic 3D bioprinting (4D bioprinting), achieved by modulating the combination strategy of bioprinting materials. The field of microsurgery has experienced remarkable growth over the past decades, propelled by advancements in surgical instruments, magnification technology, perforator dissection techniques, and vascular imaging. Consequently, this chapter discusses plastic surgery biomaterials and the latest advancements in their 3D manufacturing technology, emphasizing the integration of simulation models in laboratory settings for early skills development and assessment in microsurgery within residency programs.

Keywords
Scaffolds · Tissue · Bioprinting · Biomaterials · Plastic surgery

7.1 Introduction

The term "plastic surgery" was first introduced by German surgeon Karl Ferdinand von Greffy (1787–1840) in 1818. As the nineteenth century came to a close, the emergence of feminism in the wake of the first industrial revolution played a pivotal role in advancing cosmetic surgery. The early twentieth century witnessed significant developments in plastic surgery, particularly due to the aftermath of the two world wars, which resulted in a surge of tissue defects and deformities. This prompted oral and maxillofacial surgeons as well as otolaryngologists to embark on extensive research, catalyzing unprecedented progress in the field of plastic surgery (Peng et al. 2020).

Plastic surgery constitutes a specialized field utilizing surgical methods and tissue transplantation to mend, rebuild, and improve human tissues and organs. Presently, the field is marked by continuous advancements in theories and methodologies within plastic and aesthetic medicine, accompanied by the swift evolution of biomaterials and cutting-edge technologies. With the improvement of living standards and an escalating emphasis on beauty, there is an expanding concern for the safety of cosmetic materials used in plastic surgery. This heightened awareness underscores the intersection of innovation, safety, and the growing societal interest in aesthetic enhancements (Yang et al. 2023; Al-Himdani et al. 2017).

The clinical landscape of plastic surgery is witnessing a notable expansion in the application of local minimally invasive techniques and non-surgical methods. Recovery periods for individuals undergoing plastic surgery typically span between 1 and 3 months, contingent on the nature of the specific procedure. As of 2023, a discernible trend toward a more natural aesthetic is evident, with a preference for subtle enhancements such as understated breast implants and Brazilian Butt Lifts (BBLs) gaining popularity. Demand remains high for procedures like liposuction and non-surgical facial rejuvenation, reflecting a collective pursuit of aesthetic refinement with a focus on achieving a harmonious and natural appearance (Egro and Coleman 2020; Devgan et al. 2019).

Plastic surgeons employ a diverse array of both surgical and non-surgical techniques to address a spectrum of functional and aesthetic concerns, with the overarching goal of enhancing the appearance and functionality of various body parts. These professionals specialize in tailoring interventions to meet individual needs, whether the focus is on reconstructive procedures to restore form and function or cosmetic enhancements aimed at refining and rejuvenating physical features. This comprehensive approach underscores the versatility of plastic surgery in addressing a wide range of concerns, emphasizing both the artistic and functional dimensions of the discipline (Giele and Cassell 2016; Carro and Tokgöz 2023).

In 2023, a report revealed a substantial spike in cosmetic surgeries almost 30% higher in 2021 compared to 2019 and nearly 50% more than in 2020. This surge reflects a growing societal inclination toward cosmetic enhancements and evolving attitudes toward beauty standards. Plastic surgery encompasses two main branches: reconstructive surgery (RS) and cosmetic surgery (CS). RS aims to restore both form and function in body parts affected by conditions like congenital deformities, trauma, or disease, encompassing procedures such as breast reconstruction and cleft lip repair. Conversely, cosmetic surgery is dedicated to enhancing specific features for aesthetic reasons, addressing concerns related to symmetry, proportion, and overall attractiveness. Common cosmetic procedures include rhinoplasty, facelift, breast augmentation, liposuction, and tummy tuck (Fig. 7.1) (Abbas and Dawood 2022; Hamilton III et al. 2004).

In the realm of construction, scaffolds play a pivotal role as fundamental frameworks for supporting building materials. Over the last 50 years, a significant progress has been achieved in the field of regenerative medicine through the innovation of biodegradable scaffolds. These scaffolds serve as crucial frameworks, guiding the migration of host cells and facilitating the gradual process of revascularization. Governed by cellular enzymes, the scaffold undergoes a meticulous breakdown, synchronizing with the formation of new tissues. This orchestrated dance between degradation and synthesis ensures an optimal balance, ultimately resulting in the replacement of the scaffold by thriving the living tissue. This dynamic process guarantees sustained structural support over time, marking a noteworthy advancement in regenerative medicine (Panayi and Orgill 2019).

The intricate preparation of biomaterial scaffolds involves a nuanced classification into natural and synthetic materials. The trajectory of advancements in both basic research and clinical applications is intricately linked to the origin of these biomaterials. Scaffolds, crucial in fostering cell growth and organization, serve as integral frameworks in the healing process. Common examples of implantable biomaterial scaffolds encompass hyaluronic acid (HA), collagen, and polymethyl methacrylate (PMMA). While biomaterials offer benefits such as accessibility, non-invasive donor procedures, and customization, they also present challenges like foreign body reactions, inflammation, deformation, and the need for repeated injections, distinguishing them from autologous fat transplantation. This duality underscores the complexity and considerations in the selection and application of biomaterial scaffolds (Jiang et al. 2020; Van Nieuwenhove et al. 2017; Spector and Lim 2016). An ideal biomaterial should possess the following features such as being

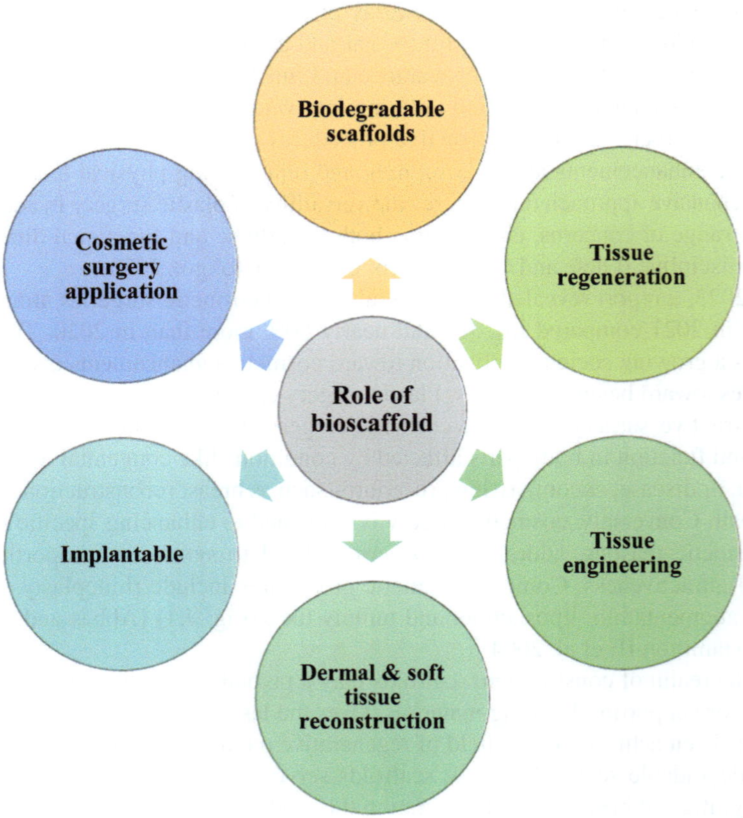

Fig. 7.1 The role of bioscaffold

compatible, non-allergic, and non-carcinogenic, not inducing any foreign body reaction, possessing the ability to attach with the host, easy to store, etc. (Peng et al. 2020).

7.2 Evaluation of Plastic Surgery Techniques

The history of plastic surgery is rich and spans thousands of years, with evidence of early procedures found in various cultures. Here's a brief overview of the historical context and evolution of plastic surgery:

7.2.1 Ancient Civilizations

In ancient civilizations, particularly during the Vedic periods in India, foundational surgical knowledge was established by revered scholars and practitioners. The

Rigveda, the earliest account of ancient Indian civilization, references the Ashwini Kumaras as chief surgeons, laying the groundwork for legendary surgical operations. Charaka Samhita, compiled by Charaka, serves as a comprehensive compendium of medical science, illuminating the principles of Ayurveda and the contributions of early medical practitioners like Acharya Atreya and Acharya Agnivesh. Sushruta, often hailed as the father of surgery, significantly advanced the field by providing detailed descriptions of surgical techniques under eight heads, including excision, scarification, and puncturing. His remarkable contributions encompass reconstructive procedures for various defects, such as rhinoplasty, otoplasty, and lobuloplasty. Sushruta's writings reflect essential surgical principles like precision, hemostasis, and perfection, showcasing a level of understanding that has transcended the ages (Selin 2013; Kaur and Saraf 2011; Singh and Sharma 2008).

7.2.2 Middle Ages

The enduring legacy of the Islamic Golden Age, spanning over a millennium, remains profoundly relevant in contemporary medical and surgical practices. The foundational principles established during this era continue to play a crucial role in the ongoing development and refinement of medical skills. A prime illustration is the enduring commitment to advancing neurosurgical expertise through hands-on approaches such as laboratory work, anatomical studies, and cadaver dissection. Among the influential figures of this golden age, the Persian physician Al-Zahrawi, often hailed as the "father of surgery," significantly contributed to the field by meticulously documenting surgical procedures in his comprehensive medical encyclopedia. His enduring legacy underscores the timeless impact of this historical period on the evolution of medical knowledge and practices (Annajjar and Ebrahim 2021).

7.2.3 Renaissance and the Sixteenth Century

Among the earliest medical procedures are accounts of ancient skin translocation methods, with autologous skin translocation, or skin grafting, standing as a historical cornerstone in addressing burn wounds. The journey of skin grafting, marked by cultural, religious, and anecdotal fluctuations, gained a widespread acceptance in the 1800s as a safe and effective treatment for wound management, particularly for burns. The nineteenth and twentieth centuries witnessed remarkable strides in skin grafting, propelled by medical advancements and the impact of global conflicts. This era saw the development of specialized surgical instruments tailored for skin grafting, affording surgeons greater precision in managing tissue depth and consistency, thereby significantly enhancing overall outcomes in the practice of skin grafting (Ozhathil et al. 2021).

7.2.4 The Nineteenth Century

The nineteenth century witnessed the formalization of plastic surgery as a distinct medical specialty, a transformation underscored by the exigencies of the First World War. The unprecedented impact of modern weaponry led to a surge in soldier casualties with severe injuries, necessitating innovative approaches to reconstruction. ENT surgeon Harold D. Gillies, serving in the Royal Army Medical Corps, emerged as a pioneering figure. Applying his expertise, Gillies revolutionized techniques for facial reconstruction, particularly in nose reconstruction and skin grafts. By the war's end, he had treated around 11,000 casualties, laying the groundwork for the specialization of plastic and reconstructive surgery. Gillies' groundbreaking contributions during this critical period continue to resonate in contemporary surgical practices, showcasing the enduring impact of his innovative techniques (Dumont 2015).

7.2.5 Modern-Day Surgery

In the latter half of the twentieth century and continuing into the present day, plastic surgery has undergone remarkable advancements, achieving a widespread public acceptance as a means to attain the desired aesthetic outcomes for both men and women. The accessibility of plastic surgery has increased, transforming the appearance and lives of numerous individuals. Plastic surgeons have pioneered innovative techniques, offering solutions to mitigate the effects of aging on the face and body, as well as reshaping features like noses and breasts. The evolving landscape of plastic surgery reflects its transformative impact on personal aesthetics and the broader societal embrace of cosmetic enhancements (Hausamen 2001).

7.2.6 Plastic and Cosmetic Surgery Principles in the Twentieth Century

The mid-twentieth century witnessed a notable surge in the interest and popularity of cosmetic surgery, driven by significant advancements in anesthesia and surgical techniques. Concurrently, societal changes played a pivotal role in the increasing acceptance of elective procedures for aesthetic purposes. This convergence of medical progress and evolving cultural attitudes created a conducive environment for individuals to seek and embrace cosmetic surgery as a means to enhance their aesthetic appearance. The era marked a transformative shift in the perception of elective procedures, reflecting a growing recognition of the intersection between medical innovation and the pursuit of aesthetic ideals (Molloy and Ardill 1980).

7.2.7 Technological Advancements: Late Twentieth Century to Present

In the latter part of the twentieth century, there was a swift embrace of technological progress, highlighted by the advent of lasers, endoscopic surgery, and minimally invasive techniques. These groundbreaking innovations ushered in a revolution in both reconstructive and cosmetic plastic surgeries. The precision and less invasive nature of these technologies not only transformed surgical procedures but also significantly contributed to enhanced outcomes and patient experiences. This era marked a pivotal juncture in the field, where advanced technologies became integral tools, reshaping the landscape of plastic surgery and expanding the possibilities for both reconstructive and cosmetic interventions (Treacy 2022).

7.3 Role of Bioscaffold

Three-dimensional (3D) scaffolds frequently find application in tissue engineering (TE) endeavors, fostering an environment supportive of cell integration and growth factor utilization aimed at the regeneration or substitution of impaired tissues or organs (Khoshnazar et al. 2023). In plastic surgery, bioscaffolds are used for various purposes which are discussed in Fig. 7.1.

7.3.1 Tissue Regeneration

A scaffold needs to exhibit a mechanical strength that mirrors the characteristics absent in the tissue it intends to bolster temporarily. Moreover, it must have ample strength and mechanical stability to endure implantation and withstand the usual stresses experienced in living tissues. The mechanical attributes of a material are closely linked to how it reacts to loading and deformation. This relationship is rooted in two fundamental concepts: firstly, loading causes deformation in a material, and secondly, the interaction between loading and deformation dictates the material's stiffness. This understanding forms the foundation for designing scaffolds that effectively mimic and support natural tissue mechanics (Guimarães et al. 2020).

7.3.2 Tissue Engineering

For effective wound healing and skin regeneration, tissue-engineered scaffolds must possess the essential ability to absorb a substantial amount of fluid while maintaining a moist environment in the wound bed. Commonly used cotton pads typically have a limited fluid-holding capacity of approximately six to seven times their dry weight. An exceptional absorbent dressing or scaffold is imperative for surgical wounds to minimize the need for frequent pad replacements during the intervention and for chronic wounds that generate significant exudates (Shyna et al. 2020). Scaffold porosity plays a critical role in facilitating oxygen and nutrient transport to the wound site. The optimal porosity for wound dressing is highlighted within the range of 60–90% (Wu et al. 2012). Polymers are excellent candidates in tissue engineering (TE) applications due to their versatile and flexible nature, offering a broad spectrum of mechanical, chemical, and physical properties. These materials exhibit strong biocompatibility, are lightweight, and show resilience against biochemical degradation (Krishani et al. 2023). Also fiber-reinforced plastic (FRP) composite joints may be advantageous in making bioscaffold (Singh et al. 2019, 2022; Kumar et al. 2021a, b).

7.3.3 Dermal and Soft Tissue Reconstruction

Bioscaffolds play a crucial role in reconstructive surgery, addressing skin and soft tissue defects. They prove especially beneficial when additional structural support is required or when replacing a sizable area of tissue (Hassan et al. 2023).

7.3.4 Implantable Devices

Medical implants are devices or tissues positioned inside or on the body. Bioscaffolds serve as components in implantable devices, such as breast or cosmetic implants, enhancing tissue integration and minimizing complications (Pial et al. 2022).

7.3.5 Cosmetic Surgery Applications

In cosmetic surgery, bioscaffolds find application in procedures like breast augmentation, facial rejuvenation, and body contouring. They contribute to achieving natural and enduring results by fostering tissue integration (Jarvis et al. 2020; Zielińska and Nowak 2017).

7.3.6 Biodegradable Scaffolds

The scaffold in tissue engineering should match the natural regeneration rate of damaged tissues and be biodegradable. Its body-neutral nature eliminates the requirement for additional surgeries to remove the scaffold once its purpose is fulfilled (Zielińska and Nowak 2017).

7.4 Advantages of Bioscaffold in Plastic Surgery

1. *Tissue Regeneration*: Certain biomaterials are crafted to support tissue regeneration by serving as a scaffold for cell growth and differentiation (Fig. 7.2) (Chan and Leong 2008).
2. *Biocompatibility*: Ideal biomaterials for plastic surgery must be biocompatible, ensuring they do not induce immune responses or adverse reactions when introduced into the body (Huzum et al. 2021).
3. *Customization*: Bioscaffold technology offers potential customization, enabling surgeons to tailor implants or scaffolds to match the patient's specific anatomy and needs. Conventional implants struggle to be customized according to the recipient's skeletal anatomy and defect characteristics (Meng et al. 2021).
4. *Reduced Scarring*: Advanced materials play a role in enhancing wound healing and minimizing scarring, thereby improving the aesthetic outcomes of plastic surgery procedures (Xiao et al. 2023).
5. *Stability and Longevity*: If bioscaffold ensures stability and durability, it has the potential to yield longer-lasting results in plastic surgery procedures (Peng et al. 2020).

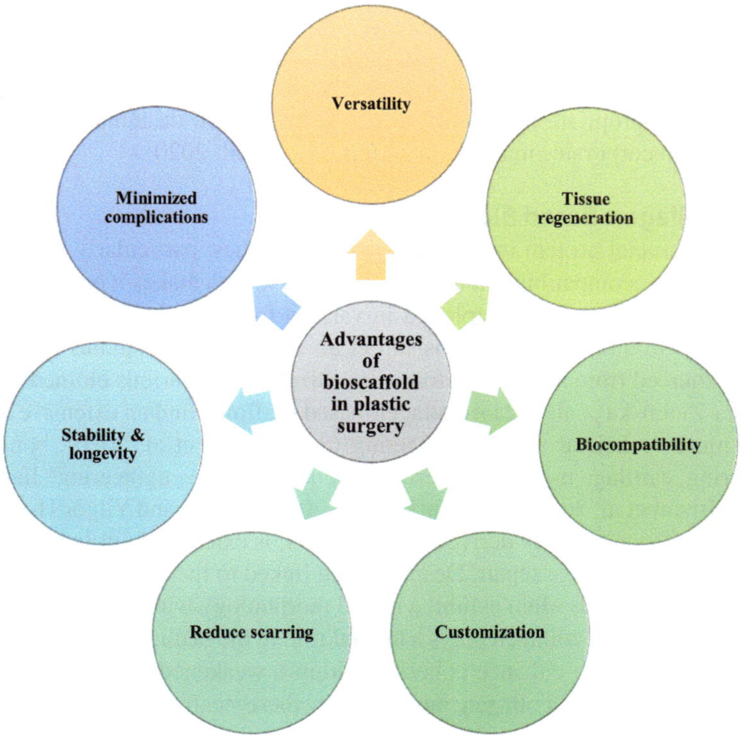

Fig. 7.2 Advantages of bioscaffold in plastic surgery

6. *Minimized Complications*: Cutting-edge materials can contribute to lowering the risk of complications, such as infection or rejection, resulting in smoother and safer surgical outcomes (Mohammadi et al. 2023).
7. *Versatility*: With its diverse design and properties, bioscaffold holds the potential for versatile applications across various types of plastic surgery procedures (Feng et al. 2023).

7.5 Biomaterial and Fabrication Techniques in Plastic Surgery

The choice of scaffold material should be tailored to the patient's condition. Plastic materials used in cosmetic procedures can be categorized into natural biomaterials and synthetic materials based on their origin. Synthetic materials encompass both organic polymers and inorganic non-metallic materials.

7.5.1 Natural Biomaterial

Bio-derived natural materials exhibit biocompatibility, safety, and lack of toxic byproducts upon biodegradation. Consequently, it is strongly advisable to utilize these materials in scaffold fabrication (see Table 7.1). Thus far, no single material has demonstrated the complete array of scaffold requirements. Thus, the amalgamation of two or more materials is advocated to achieve the desired properties. Selecting an appropriate fabrication method is crucial in designing a dependable scaffold that incorporates multiple materials (Peng et al. 2020).

7.5.1.1 Collagen-Based Bioscaffold

Collagen, a crucial protein in animal connective tissues, particularly in mammals, exhibits high biocompatibility and biodegradability. This makes it a versatile application in the medical field, employed in various forms, including scaffolds, drug carriers, and wound dressings (Krishani et al. 2023). Recent studies indicate that collagen sourced from marine life forms is utilized across various biomedical fields (Grabska-Zielińska et al. 2021). Collagen-based scaffolds find an extensive application in musculoskeletal tissue engineering (Szychlinska et al. 2020), bone tissue engineering, cartilage tissue engineering, and neural tissue engineering (Jiang et al. 2020; Krishani et al. 2023; Szychlinska et al. 2020; Bayrak and Yilgor Huri 2018).

Type I collagen, notably derived from animals, is extensively utilized in clinical applications for cartilage repair. Despite being linked to the fibrocartilaginous phenotype, these scaffolds often exhibit a mixed morphology with both type I and type II collagens. Type III collagen plays a crucial role in the healing response, acting as a covalent modifier to enhance cohesion within a weakened type II collagen network. Cross-linking of collagen molecules in these scaffolds strengthens fibrils, stabilizes structure, prolongs resorption time, and improves biocompatibility. Commonly used type I/type III collagen membranes for articular cartilage repair

Table 7.1 Applications of natural biological materials

Natural biological materials	About	Form	Applications	Advantages	Limitations
Collagen	Animal connective tissues such as bovine chondrocytes	Injection	Dermal filler for wrinkles, scars, crow feet, etc.	Biocompatible, biodegradable, immunologically low, good mechanical properties, retain moisture	Possibility of rejection reaction
Hyaluronic acid		Injection	Dermal filler for wrinkles, nose, eyebrows, acne, etc.	Rejection rate minimal	Swelling (mild)
Fibrin glue/bio-protein glue	Animal blood	Injection	Burn wound reconstruction, facelift surgeries, skin graft	Biocompatible	Narrow range of applications
Decellularized tissue	Human tissue	Biological patch	Skin transplantation	Biocompatible	Expensive
Silicone	Organic polymer	Prosthesis, breast implants, rhinoplasty	Nasal deformities, maxillofacial abnormalities, wrinkles, facial lines	Biocompatible, resistance to heat and cold, good physical as well as mechanical strength	Hydrophobicity, rupturing of silicone implants
Hydroxyapatite	Calcium-phosphate mineral	Prosthesis	Treating bone defects	Biocompatible, biodegradable	Brittle, less toughness

feature a bilayer structure, with the synovia-facing layer being compact, smooth, and low-porosity.

While the pure collagen scaffold boasts excellent biological properties, its mechanical characteristics and structural stability are lacking. Intermolecular cross-linking of collagen, achievable through physical treatment or chemical agents, proves instrumental in modifying the scaffold's properties. Commonly employed physical treatments include ultraviolet (UV) irradiation, gamma radiation, and dehydrothermal treatment (DHT). These methods enhance the mechanical properties of the collagen scaffold, simultaneously diminishing its solubility and absorption without introducing toxicity (Wang et al. 2015; Takitoh et al. 2015; Maslennikova et al. 2015).

Collagen is an ideal biomaterial for soft tissue augmentation as it is having good cell adhesion property, is immunologically compatible, has moisture retention, and is resistant toward proteolysis (Özgentas et al. 1994). Preclinical and clinical reports suggest the use of type I and tye III collagen (injectable) as dermal fillers. The clinical trials on 123 patients with skin deficiencies showed satisfactory results with very few side effects. Thus, type I and III collagen can be addressed as safe biomaterial for soft tissue grafting procedures. Various collagen-based products are available in the market for the correction of facial wrinkles, acne scars, and other small areas of skin defects. Collagen matrix can also be used in periodontal plastic surgeries (Xu et al. 2019; Atieh et al. 2016).

7.5.1.2 Hyaluronic Acid-Based Bioscaffold

Hyaluronic acid, alternatively referred to as hyaluronan (HA), was first discovered by Meyer and Palmer in 1934, when they isolated it from bovine vitreous humor. This glycosaminoglycan, found in the extracellular tissue, is obtained from bacteria and cockscomb. With its versatile properties, hyaluronic acid has become integral to biomaterials science, finding utility in various fields including tissue culture scaffolds and cosmetic products. Its unique physical and biochemical attributes, whether in solution or hydrogel forms, render it highly desirable for technologies aimed at bodily repair. Tissue engineering strategies typically involve three main approaches, such as conductive, inductive, and cellular, with hyaluronic acid being utilized as a scaffold material alongside synthetic options like polyethylene glycol (Chircov et al. 2018). In recent years, there has been a growing interest in developing and utilizing hyaluronic acid (HA)-based scaffolds for tissue engineering purposes. These scaffolds are designed to regulate cell behavior and create optimal tissue engineering products. HA's histocompatibility, capacity for chemical modification, and biodegradability make it an ideal material for scaffolds in tissue engineering applications. Moreover, the hydrogel properties of HA mimic the water content found in the human tissue, which facilitates the exchange of oxygen, nutrients, and metabolic waste (Li et al. 2016; Hunt and Grover 2010). Hyaluronic acid (HA) is highly valued in cosmetic dermatology and skincare due to its remarkable properties, including consistency, biocompatibility, hydrophilicity, limited immunogenicity, and distinctive viscoelasticity. Its potential as a biomaterial for tissue engineering is promising. However, unaltered HA is swiftly absorbed by the body, making it unsuitable for tissue engineering purposes. To address this issue, chemical modification of HA is pursued through two primary methods: esterification and cross-linking. Esterification involves binding HA with specific hydrophobic groups, reducing its polyanion characteristics. Under certain conditions, HA's carboxyl group can be esterified to produce HYAFF, an esterified derivative of HA (Laurent 1998; Aigner et al. 1998; Zhu et al. 2017).

7.5.1.3 Bio-Protein Glue

It is also known as fibrin glue/sealant. The bio-protein glue is mainly made up of fibrinogen, thrombin, and $CaCl_2$ with other components having a similar behavior of the last stage of coagulation. It provides excellent adhesion in very less time. Various studies showed the safety of bio-protein glue except for few allergic responses. It can be used in facelift surgeries, skin graft, etc. (Marchac and Sándor 1994; Flemming 1992). Skin grafting on 16 patients was conducted by Chittoria and co-workers using a fibrin glue suggesting good adhesion in less surgical time. It also revealed the safety and efficacy of fibrin glue for skin grafting (Reddy et al. 2017). A study for facelift surgery was performed using the aerosolized fibrin glue in 48 patients which showed the promising results in face lifting (Fezza et al. 2002). Despite its various uses and safety, there is lack of clinical trials on the efficacy of fibrin glue in plastic surgery (Mooney et al. 2009).

7.5.1.4 Decellularized Tissue

Decellularized extracellular matrix is a new class of biomaterial obtained by decellularization process (chemically/physically/biologically) in order to remove antigens which cause rejection in tissue transplantation (Cebotari et al. 2010; Cheng et al. 2019). It consists of glycosaminoglycan, growth factors, and surface ligands. It is mainly used in transplantation of the skin. Most importantly, it does not cause any rejection reaction. Its use has been studied in the skin, nose, ears, face, and limb remodeling (Shang et al. 2023).

7.5.1.5 Silicone

Silicone is an organic polymer and is used medically owing to its various properties such as non-toxicity, resistance toward heat and cold, resistance toward aging, and good physical as well as mechanical properties. Since the 1950s, silicone implants have been utilized in cosmetic surgeries (Ohlwiler 1967). It is useful in correcting nasal deformities (Deva et al. 1998; Gubisch and Kotzur 1998; Kim et al. 2011), maxillofacial abnormalities (Harris 1964; Ahmed et al. 2010), wrinkeles, facial lines (Pollack 1990), etc. Due to its soft texture and biocompatibility, it is best suited for prostheses for facial deformities. Its implants are also used for breast implantation; however, few studies report the chances of local inflammation. The silicone implants suffer from limitations due to its poor antibacterial potential, rupturing of silicone implant, and hydrophobic nature (Hillard et al. 2017; Gorczyca et al. 2007).

7.5.1.6 Hydroxyapatite

The chemical formula of hydroxyapatite is $Ca_{10}(PO_4)_6(OH)_2$ which is a natural calcium-phosphate mineral that occurs in rocks and sea corals (Minelli et al. 2022). As hydroxyapatite is an important component of the bone, it is used in treating bone defects, eye brow projection (Minelli et al. 2022), etc. Its porous granules are significantly useful in improving facial augmentation (Mendelson et al. 2010; Huggins and Mendelson 2017). Hydroxyapatite cement has also been used for craniofacial contour defect correction (Bravo et al. 2024). Owing to its collagen-stimulant property, it has also been used as a biostimulant (Bravo et al. 2024).

7.5.1.7 Chitosan

Chitosan is a biopolymer made up of *N*-acetyl-D-glucosamine and D-glucosamine. It is having unique properties such as the ability to form film, antimicrobial, antioxidant, and good adsorption capacity in addition to properties of biopolymer like biocompatibility, biodegradability, and non-toxicity. Chitosan is a good candidate for scaffold matrix and exhibits various applications in tissue engineering including regeneration, cartilage, and dentistry (Kulka and Sionkowska 1817; Madni et al. 2021).

7.5.2 Synthetic Biomaterial

Synthetic scaffolds, including commonly used materials like polylactic acid, polyglycolic acid, poly-lactic-co-glycolide, polyethylene glycol, poly-4-hydroxybutyrate, polybutylene terephthalate, polycaprolactone, and polydioxanone, are manufactured through a well-controlled process. These yield large quantities of minimally variable, cost-effective products with a prolonged shelf life. Initially in the form of viscous liquids, these scaffolds are either directly injected into the defect or molded to assume the required shape (Panayi and Orgill 2019).

7.5.2.1 Polylactic Acid

Poly-L-lactic acid (PLLA, Table 7.2) stands out as a synthetic, biocompatible, and biodegradable polymer, safely degrading along the metabolic pathway of lactic acid. With a track record of over 30 years in clinical applications, including dissolvable sutures, intrabone implants, and soft tissue implants, PLLA has proven its safety. The physiochemical properties of fillers, including collagen stimulators, are continually evolving and refining, impacting their clinical performance. Chemical properties like pH, charge, water affinity, and physical attributes such as size, shape, texture, and surface area, both in their intact and degraded forms, contribute to the biomaterial's overall performance (Valantin et al. 2003; Lorenc and Lee 2016). PLLA serves as a potent tool for clinicians in addressing long-lasting correction of facial volume loss. Over time, as experience has grown and technical issues have been refined, PLLA has proven to be a safe and effective product, yielding predictable and reproducible results. Similar to other filler agents, a lesser amount of PLLA is needed in younger or fuller-faced patients to achieve desirable outcomes

Table 7.2 Applications of synthetic materials

Synthetic biomaterials	Properties and applications
Polymethyl methacrylate (PMMA)	Synthetic, nondegradable, biocompatible polymer; used as an intradermal filler
Poly-L-lactic acid (PLLA)	A biodegradable thermoplastic polymer finds applications in orthopedics, neurosurgery, and craniofacial surgery, particularly for absorbable plates and screws. Its primary use in soft tissue reconstruction involves expanding and filling the deep dermis or subcutaneous layer, as well as addressing facial fat atrophy
Calcium hydroxyapatite (CH)	Inherent to human bone and tooth enamel, it consists of calcium and phosphate ions, demonstrating robust biocompatibility

(Fitzgerald et al. 2018). In a clinical study by Stein et al., PLLA was injected into the upper arms of 21 patients. Post-injection, significant type III and type I collagens were observed, along with upregulated gene expression for their transcripts. After 9 months, microscopic examination revealed no trace of PLLA, indicating complete degradation (Stein et al. 2015).

PLLA injections have been proved to be safe and efficacious in restoring volume loss in patients undergone facelift with fat grafting (O'Daniel and Kachare 2022).

7.5.2.2 Polycaprolactone: Polycaprolactone (PCL)

It is a biocompatible, biodegradable, and bioresorbable polymer, belonging to the poly-α-hydroxy acid group, akin to polylactic and polyglycolic acids. First synthesized in the early 1930s through the ring-opening polymerization of the cyclic monomer Ɛ-caprolactone, PCL exhibits favourable properties for various applications in the biomedical field. PCL consists of a repeated single-unit sequence of Ɛ-caprolactone $(C_6H_{10}O_2)n$. Distinguished by superior viscoelastic properties compared to other biodegradable polymers, PCL is easily manufacturable and manipulable, enabling the creation of diverse structures such as microspheres, fibers, micelles, films, nanofibers, and foams. Its versatility extends to blending with other polymers to generate copolymers with varied physicochemical properties and biodegradability (Woodruff and Hutmacher 2010; Labet and Thielemans 2009; Benjamin et al. 2019; Sinha et al. 2004; Azimi et al. 2014; Abedalwafa et al. 2013; Mohamed and Yusoh 2016). In tissue engineering, 3D printing has emerged as a rapidly advancing field, and PCL plays a crucial role due to its favorable physicochemical properties, including a low melting point, moldability, and longevity. The 3D printing process involves successive layer deposition using a computerized approach, enabling the creation of structures for repairing and replacing tissues or organs. This innovative therapeutic approach holds a great promise and has evolved into a tangible reality (Medaglini et al. 2015; Yeong and Chua 2014; Christen and Vercesi 2020). Polycaprolactone stimulates collagen, thus acting as dermal filler (Christen and Vercesi 2020) which is formulated in the form of a gel. The reports suggest that the polycaprolactone gel is safe for treating nasolabial folds (Christen and Vercesi 2020).

7.5.2.3 3D Bioprinting

A notable advancement in tissue engineering is the emergence of 3D bioprinting, a concept originally proposed by Mironov et al. as "organ printing." It entails computer-aided, jet-based 3D tissue engineering for living human organs, following the typical 3D printing process chain. This approach utilizes microtissues, often in the form of spheroids, offering advantages such as automation and high cell density compared to traditional scaffold-based methods. The closely placed spheroids undergo self-assembly and fusion, forming the basis for 3D printability in organ printing. While organ printing currently finds application in in vitro drug testing, it holds an immense potential to address organization and complexity issues in engineered tissues and pave the way for an industry-scale robotic tissue fabrication line. Mironov suggests that the first bioprinted organ transplant in the world is likely to be the thyroid gland of a mouse, expected in 2015 (Yeong and Chua 2014; Mironov et al. 2003, 2008, 2009, 2011; An et al. 2013, 2015).

7.5.3 Key Steps for Patient-Specific Bioscaffold

Here is an outline of the key steps (Fig. 7.3) and considerations involved in patient-specific bioscaffold design:

7.5.3.1 Patient Imaging

Medical imaging plays a vital role in visually depicting the diverse tissues and organs within the human body, enabling the observation of both normal and abnormal anatomy and physiology. A range of techniques including X-ray, computed tomography (CT), positron emission tomography (PET), magnetic resonance imaging (MRI), single-photon emission computed tomography (SPECT), digital mammography, and diagnostic sonography, medical professionals are able to perform this essential task. These advanced imaging methods play a pivotal role in diagnosing a wide range of medical conditions, including myocardial diseases, different tissue cancers, neurological disorders, congenital heart disease, abdominal illnesses, complex bone fractures, and various serious medical conditions (Hussain et al. 2022).

7.5.3.2 Data Processing and Modeling

The conversion of imaging data into a 3D digital model is achieved through specialized software. These software tools are categorized into control tools, general computer-aided design (CAD) tools, tools for converting medical data to CAD formats, and a subset of highly specialized research-project tools. The analysis focuses on different geometry representations, evaluating their advantages and disadvantages in the context of heterogeneous volume modeling and bioprinting. The key criterion for assessment is the software's suitability for applications like

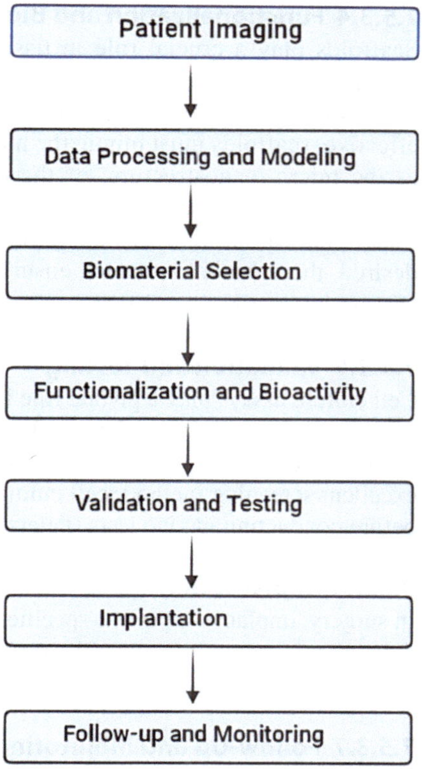

Fig. 7.3 Key steps involved in patient-specific bioscaffold design

heterogeneous volume modeling and bioprinting, emphasizing their commonality in technological requirements (Pakhomova et al. 2020).

7.5.3.3 Biomaterial Selection

Biomaterials are crucial components across multiple healthcare domains, spearheading progress in tissue engineering, medical implants, drug delivery systems, and immunotherapies. Their significance lies in their ability to provide biocompatible structures, like scaffolds and devices, to support various biological functions. Additionally, they can encapsulate and shield biological substances such as cells, chemicals, and proteins, ensuring their efficacy and safety. Moreover, biomaterials enable convenient alterations to their chemical and physicochemical attributes, enhancing their adaptability for diverse applications. This broad category encompasses an extensive range of substances, from naturally occurring biological macromolecules to entirely synthetic coatings. The selection of a suitable biomaterial hinges on factors such as the target tissue and its specific properties, including biocompatibility, mechanical strength, and degradation rate (Mariani et al. 2019).

7.5.3.4 Functionalization and Bioactivity

Scaffolds play a crucial role in tissue engineering, requiring modifications to incorporate essential biological cues influencing cellular functions like adhesion, migration, survival, proliferation, differentiation, and biomineralization. To be effective, scaffolds must mimic the mechanical and biological properties, as well as the micro-/nanostructure, of the native bone and tissue. Additionally, they should be sterilizable to prevent infections at the implantation site. Functionalization should be tailored based on the patient's specific needs and the desired therapeutic outcome, ensuring personalized and effective utilization (Kesireddy and Kasper 2016).

7.5.3.5 Validation and Testing

Cell culture assays play a pivotal role in assessing the biocompatibility of materials or extracts by utilizing isolated cells in vitro. These techniques are valuable for evaluating the potential toxicity or irritancy of materials and chemicals. Offering an excellent screening method, cell culture assays enable the assessment of materials before conducting in vivo tests (Peters et al. 2004).

7.5.3.6 Implantation

In surgery, implant the patient-specific bioscaffold in the targeted area, and monitor the patient's response. Adjust the treatment plan as needed based on the observed outcomes (Herath et al. 2023).

7.5.3.7 Follow-Up and Monitoring

Establish a postoperative monitoring plan to evaluate the effectiveness of the bioscaffold and the overall tissue regeneration process. Continuously assess the patient's progress and feedback, making necessary adjustments to the treatment plan as needed (Colazo et al. 2019).

7.6 Application of Plastic Surgery in Bioscaffolds

Plastic surgery encompasses a spectrum of surgical techniques, including tissue and organ transplantation, aimed at rectifying defects and deformities in human tissues and organs. Additionally, it is utilized to refine the natural human form for both aesthetic enhancement and functional restoration. Plastic surgeons tackle a wide array of tissue reconstruction challenges, grappling with issues such as infection, pain, and donor area irregularity linked with grafting procedures. An emergence of 3D bioprinting has revolutionized plastic surgery by enabling the creation of diverse efficient cells, extracellular matrix (ECM), growth factors, and supportive biodegradable polymer materials. The technology facilitates the fabrication of personalized implants tailored to individual patient needs, with the option for in situ printing directly onto the affected areas to expedite wound healing. These advancements have significantly mitigated surgical complexities, minimized postoperative

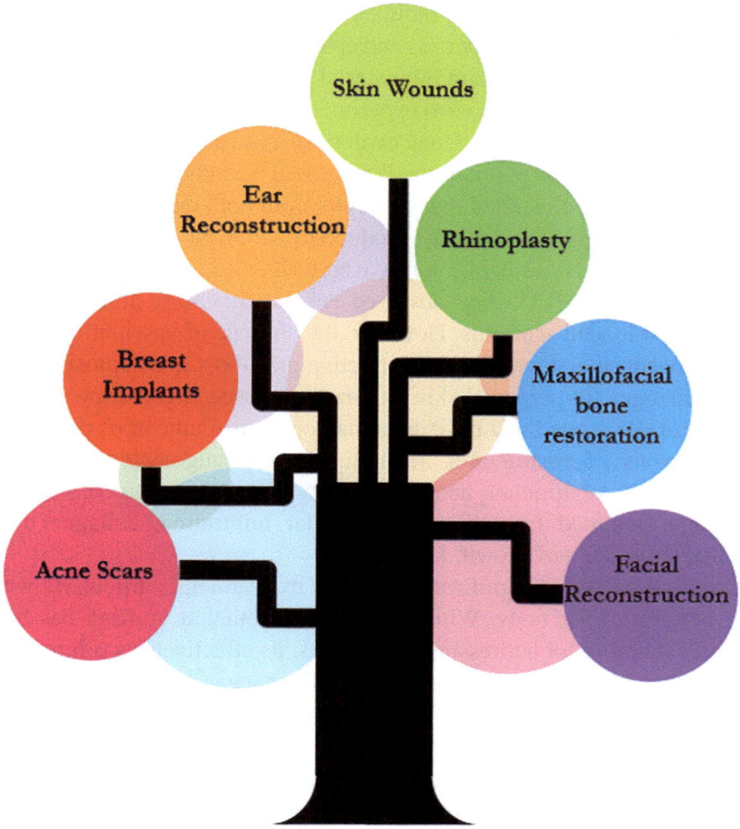

Fig. 7.4 Applications of plastic surgery in bioscaffolds

complications, and ushered in an era of accurate medicine (Giele and Cassell 2016; Carro and Tokgöz 2023). In this context, this part explores the modern research advancements in 3D bioprinting technology across key domains of plastic surgery, including rhinoplasty, ear reconstruction, skin regeneration, breast augmentation, and maxillofacial bone repairs shown in Fig. 7.4.

7.6.1 Skin Wounds

Patients suffering from diabetic ulcers, tumors, severe burns, or other causes often experience significant skin tissue damage, hindering natural skin regeneration. The advent of 3D bioprinting for the skin tissue deals novel avenues for plastic physicians to address skin syndromes and facilitate skin regeneration. GelMA, renowned for its outstanding biocompatibility and light-curing properties, stands out as a widely utilized material in skin tissue engineering (Shi et al. 2018).

Y Shi et al. introduced a novel bioink comprised of GelMA/COL/tyrosinase to facilitate formation of both the dermis and epidermis (Shi et al. 2018). Additionally, CHO has emerged as a promising biomaterial in skin repair. Intini's group investigated CHO biocompatibility in vitro experiments, highlighting its potential for skin regeneration (Intini et al. 2018). Dong et al. developed a 3D bioactive sponge (SIS-PEG) by recross-linking porcine small intestinal submucosa (SIS) (Dong et al. 2019). Their research illustrated that skin cells from both the epidermis and dermis, infused with SIS-PEG, successfully produced revitalized skin with newly grown hair in just 21 days. By utilizing non-woven CHO/GEL and polylactide/COL gel scaffold to infill the MFU, they observed enhanced healing ability compared to single-seed cell-loaded scaffolds. However, the absence of vascular building, inadequate angiogenesis induction, and ineffective graft-host anastomosis remain significant challenges for perpetual skin alternates in tissue engineering. Li's research (Li et al. 2021) illustrated that utilizing a macroporous filamentous protein scaffold with nanofibrous microstructures at multiple scales augmented the capacity of transplanted cells to stimulate neovascularization and dermal reconstitution. This enhancement stemmed from improved cellular infiltration, collagen deposition (COL), and expression of growth factors.

Severe burns present a significant challenge in treatment, particularly when they cover over 60% of the body. While autologous epithelial grafting has become a viable clinical option for addressing these burns, its effectiveness is limited when it comes to repairing deep dermal structures. Roshangar et al. conducted a study where they quarantined adipose-derived stem cells (ADSCs) and introduced them in a 3D gel scaffolds using 3D bioprinting technology. The scaffold's properties and cell viability were evaluated, noting accelerated wound healing in rat models treated with these scaffolds, particularly those incorporating ADSC (Roshangar et al. 2021). Moving forward, the combination of 3D bioprinting and stem cells holds a promise for addressing severe skin injuries. Additionally, research in regenerative medicine extends to restoring skin-related appendages like hair follicles, which is of significant interest due to widespread concerns about hair loss. Despite these advancements, the current research on hair follicle regeneration primarily remains at the cellular level in vitro, lacking robust animal studies to demonstrate its clinical feasibility. Bridging this gap through animal experiments could pave the way for translating laboratory findings into clinical applications. Hence, further exploration and attention from scholars are warranted in this area of research.

7.6.2 Ear Reconstruction

The ear holds a significant importance in social, cultural, and efficient contexts, and individuals with congenital or acquired ear deformities often experience considerable psychological distress. Traditional ear reconstruction techniques have relied on the use of finely crafted rib cartilage, requiring surgical expertise and artistic precision to sculpt an aesthetically pleasing auricular framework. However, the advent of 3D bioprinting technology offers a promising alternative, easing surgical complexities and enabling customized solutions for patients. By utilizing DICOM CT images to replicate the ear's external anatomy and adapting it to individual needs, this technology allows for the creation of tailored ear cartilage products using suitable biomaterial inks or bioinks. A notable milestone was achieved in 2018 by Zhou et al., who pioneered the clinical application of a tissue-engineered ear composed of polyglycolic acid/polylactic acid and chondrocytes, marking a significant advancement in the field (Zhou et al. 2018).

However, the technique's immaturity was evident through postoperative deformation and inflammation. In 2022, Jia et al. introduced a novel approach within the same team to tackle inflammatory responses and operational deformations that can arise in rebuilt structure. They utilized bioactive bioinks derived from auricular chondrocytes and a biomimetic microporous methacrylate-modified decellularized cartilage matrix, incorporating GelMA, polyethylene oxide, and PCL. Through integrated multi-nozzle bioprinting technology, they crafted biologic auricular buildings with accurate shapes and superior mechanical properties (Jia et al. 2022). Meanwhile, Brennan team pursued an alternative 3D ear cartilage printing method. They created numerous ear scaffold prototypes using laser sintering of PCL and then implanted each one under the skin of rats without a thymus. Over the following weeks, they carefully observed for any signs of ulcers, infection, or distortion of the scaffold at the surgical site. At the eighth week, the stents were distant and subjected to analysis via micro-computed tomography and histological tint.

7.6.3 Rhinoplasty

Plastic surgeons face the challenge of reconstructing the nasal cartilage due to various factors such as tumors, injuries, and patient aesthetic preferences. Since he nasal cartilage has limited self-repair capabilities, current rhinoplasty procedures often rely on autologous gristle or artificial implants for both therapeutic and cosmetic resolves. Autologous gristle, in particular, is favored for its minimal risk of immune rejection. Rhinoplasty, akin to ear reconstruction, demands exceptional manual dexterity and artistic finesse from surgeons, making it intricate plastic surgery techniques (Lan et al. 2021). In a study by Lan et al., the impact of culture time on the development of extracellular matrix (ECM) and mechanical aspects of 3D-bioprinted constructions of collagen (COL) hydrogels filled with human nasal chondrocytes was investigated. The findings suggest that 3D bioprinting offers a viable solution for nasal cartilage reconstruction in rhinoplasty, potentially easing surgical complexity. Moreover, researchers such as De Greve et al., Suszynski et al., and Choi et al. have utilized 3D modeling and printing techniques to pretend the post-surgical nasal look and cartilage structure (Choi et al. 2017; Suszynski et al. 2018; De Greve et al. 2022). This innovative approach enables surgeons to visualize and plan surgeries more effectively, contributing to improved outcomes in rhinoplasty procedures.

They have successfully created a bioactive 3D tissue construct using various proportions of dPNCG, chondrocytes, and ADSCs. The findings validate dPNCG as an effective scaffold, fostering an optimal environment for chondrocytes and enabling the printing of functional nasal cartilage structures. However, the realm of cartilage tissue engineering includes the challenge of covalently binding growth factors to printable bioinks that persist. Hauptstein et al. addressed this challenge by developing a two-stage cross-linked hyaluronic acid-based bioink skillful of covalently binding growth factor β_1 (Hauptstein et al. 2022). Leveraging the advantages of this bioink in facilitating the growth and development of printed biological tissues, it holds a promise for future advancements in tissue engineering research.

7.6.4 Breast Implants

As the evolving aesthetic preferences of individuals drive change, there's a rising demand for breast augmentation procedures. Primarily, two methods are utilized: breast prosthesis imbedding and autologous fat imbedding. Autologous fat implantation involves transferring the fat tissue to the breast, but it often leads to complications like tissue fibrosis, cavity formation, necrosis, and fat resorption due to early inflammatory factors and ischemic environments of transplanted fat tissue. On the other hand, breast prosthesis implantation offers a safer and more convenient augmentation option, but issues such as late-stage infections, stiffness, and implant wrapping development can hinder meeting patients' aesthetic expectations (Coombs et al. 2019). Efforts by Dilyana Todorova's and Shan Mou's teams suggest utilizing extracellular vehicles (EVs) of ADSCs to enhance adipose transplantation. However, this technique faces challenges including the limited time for adipocytes to endure hypoxia, insufficient angiogenesis, and restricted nutrient penetration into tissue fluid (Todorova et al. 2017). Saljo et al. delved into assessing enduring viability of 3D-bioprinted adipose tissue derived from lipoaspirate (LAT), analyzing its cellular composition and proteomic profile in vivo over the long term. Their experiments showcased LAT's favorable proteomic profile and the sustained survival of its cellular components, including 3D-bioprinted ADSCs, endothelial progenitor cells, and blood vessels, over an extended period. This outcome underscores the viability of employing 3D bioprinting in adipose tissue engineering (Säljö et al. 2022). Nevertheless, a challenge persists in achieving high survival rates of the adipose tissue within 3D bioscaffolds. Zhou et al. developed and fabricated four variants of polyurethane-based breast scaffold. Each scaffold's fundamental unit cell mirrors the lattice structure, maintaining consistent porosity while exhibiting distinct mechanical properties. Tissue engineering chambers (TECs) present a promising method for producing a substantial volume of adipose tissue. However, their practical clinical claim is impeded by the need for reoperation.

7.6.5 Maxillofacial Bone Restoration

Tissue loss in the intricate maxillofacial region, which encompasses maxillofacial bone, skeletal muscles, gums, and periodontal ligaments, frequently occurs as a result of conditions such as tumors, trauma, or pathology. Restoration typically involves xenograft surgery, allograft and autologous, often requiring joint cure by plastic surgery, maxillofacial and oral. Maxillofacial bone restoration depends on the defect's size and condition. Small defects can be repaired using the periosteum method, leveraging its regenerative and osteogenic potential (Debelmas et al. 2018; Nakahara et al. 2016). However, large defects often necessitate a blend of seed cells, typically a biological scaffold and bone marrow-derived ADSCs. Some studies augment biological scaffolds with bioactive components to enhance stem cell proliferation and differentiation (Mukherjee et al. 2020). In a departure from traditional bone tissue engineering, Shie et al. utilized a novel approach. They created a 3D porous bioceramic scaffold through a combination of printing techniques. Initially, ceramic scaffolds were fabricated using extrusion-based bioprinting, followed by direct printing of stem cells onto the scaffold surface by means of a piezoelectric nozzle. Their experiments validated cell adhesion and hydrophilicity properties of polydopamine-calcium/silicate/polycaprolactone, providing promising insights into advanced bone tissue engineering. Piezoelectric nozzle printing technology offers precise cell printing capabilities, holding a promise for repairing intricate bone tissue in the maxillofacial region (Shie et al. 2019). Conversely, the inflexibility of stiff, solid 3D scaffolds presents challenges in adapting to the intricate structure of maxillofacial bone tissue. Thus, there's a shift toward scaffolds capable of injecting semisolid or gel-like constituents openly into defects, offering a more practical and promising approach. Sadrabadi et al. developed an Alg-based osteoconductive hydrogel biomaterial that showcased its biodegradability, biocompatibility, and ability to promote maxillofacial bone repair in mouse models (Hasani-Sadrabadi et al. 2020). Furthermore, the effectiveness of collagen (COL) gels in promoting soft tissue healing around bone defects has been substantiated. Salamanca et al.'s research on collagenized porcine grafts showcased notable results, indicating enhanced bone regeneration and mitigated bone loss (Salamanca et al. 2020).

References

Abbas AA, Dawood KM (2022) Benzofuran as a promising scaffold for the synthesis of novel antimicrobial agents. Expert Opin Drug Discov 17(12):1357–1376

Abedalwafa M, Wang F, Wang L, Li C (2013) Biodegradable poly-epsilon-caprolactone (PCL) for tissue engineering applications: a review. Rev Adv Mater Sci 34(2):123–140

Ahmed B, Butt AM, Hussain M, Amin M, Yazdanie N (2010) Rehabilitation of nose using silicone based maxillofacial prosthesis. J Coll Physicians Surg Pak 20(1):65–67

Aigner J, Tegeler J, Hutzler P, Campoccia D, Pavesio A, Hammer C et al (1998) Cartilage tissue engineering with novel nonwoven structured biomaterial based on hyaluronic acid benzyl ester. J Biomed Mater Res 42(2):172–181

Al-Himdani S, Jessop ZM, Al-Sabah A, Combellack E, Ibrahim A, Doak SH et al (2017) Tissue-engineered solutions in plastic and reconstructive surgery: principles and practice. Front Surg 4:4

An J, Chua CK, Yu T, Li H, Tan LP (2013) Advanced nanobiomaterial strategies for the development of organized tissue engineering constructs. Nanomedicine 8(4):591–602

An J, Teoh JEM, Suntornnond R, Chua CK (2015) Design and 3D printing of scaffolds and tissues. Engineering 1(2):261–268
Annajjar MJ, Ebrahim R (2021) The history of neurosurgery in the Islamic era in the Middle Ages. J Br Islam Med Assoc 9(3)
Atieh MA, Alsabeeha N, Tawse-Smith A, Payne AGT (2016) Xenogeneic collagen matrix for periodontal plastic surgery procedures: a systematic review and meta-analysis. J Periodontal Res 51(4):438–452
Azimi B, Nourpanah P, Rabiee M, Arbab S (2014) Poly (y (ε-Caprolacto-Caprolactone)) fiber: an overview. J Eng Fibers Fabr:9
Benjamin MA, Sinnott C, Bawa S, Kaufman DI, Guarino K, Addona T (2019) Re-excision rate after partial mastectomy in oncoplastic breast-conserving surgery: a single-institutional experience and review of the literature. Ann Plast Surg 82(4S):S170–S172
Bayrak E, Yilgor Huri P (2018) Engineering musculoskeletal tissue interfaces. Front Mater 5:24
Bravo B, Carvalho R, Bravo L, Penedo L, Elias M (2024) Blending hyaluronic acid and calcium hydroxylapatite for injectable facial dermal fillers: a clinical and ultrasonography assessment. Cosmetics 11(2):61
Carro MA, Tokgöz E (2023) Nonsurgical facial aesthetic procedures. In: Cosmetic and reconstructive facial plastic surgery: a review of medical and biomedical engineering and science concepts. Springer Nature Switzerland, Cham, pp 1–59
Cebotari S, Tudorache I, Jaekel T, Hilfiker A, Dorfman S, Ternes W et al (2010) Detergent decellularization of heart valves for tissue engineering: toxicological effects of residual detergents on human endothelial cells. Artif Organs 34(3):206–210
Chan BP, Leong KW (2008) Scaffolding in tissue engineering: general approaches and tissue-specific considerations. Eur Spine J 17:467–479
Cheng J, Wang C, Gu Y (2019) Combination of freeze-thaw with detergents: a promising approach to the decellularization of porcine carotid arteries. Biomed Mater Eng 30(2):191–205
Chircov C, Grumezescu AM, Bejenaru LE (2018) Hyaluronic acid-based scaffolds for tissue engineering. Romanian J Morphol Embryol 59(1):71–76
Choi YD, Kim Y, Park E (2017) Patient-specific augmentation rhinoplasty using a three-dimensional simulation program and three-dimensional printing. Aesthet Surg J 37(9):988–998
Christen MO, Vercesi F (2020) Polycaprolactone: how a well-known and futuristic polymer has become an innovative collagen-stimulator in esthetics. Clin Cosmet Investig Dermatol 13:31–48
Colazo JM, Evans BC, Farinas AF, Al-Kassis S, Duvall CL, Thayer WP (2019) Applied bioengineering in tissue reconstruction, replacement, and regeneration. Tissue Eng B Rev 25(4):259–290
Coombs DM, Grover R, Prassinos A, Gurunluoglu R (2019) Breast augmentation surgery: clinical considerations. Cleve Clin J Med 86(2):111–122
De Greve G, Malka R, Barnett E, Robotti E, Haug M, Hamilton G et al (2022) Three-dimensional technology in rhinoplasty. Facial Plast Surg 38(5):483–487
Debelmas A, Picard A, Kadlub N, Boisson J (2018) Contribution of the periosteum to mandibular distraction. PLoS One 13(6):e0199116
Deva AK, Merten S, Chang L (1998) Silicone in nasal augmentation rhinoplasty: a decade of clinical experience. Plast Reconstr Surg 102(4):1230–1237
Devgan L, Singh P, Durairaj K (2019) Minimally invasive facial cosmetic procedures. Otolaryngol Clin N Am 52(3):443–459
Dong Y, Liu Z, Qi F, Jin L, Zhang L, Zhu N (2019) Polyethylene-glycol-ornamented small intestinal submucosa biosponge for skin tissue engineering. ACS Biomater Sci Eng 5(5):2457–2465
Dumont BL (2015) Interlocus gene conversion explains at least 2.7% of single nucleotide variants in human segmental duplications. BMC Genomics 16:1–11
Egro FM, Coleman SR (2020) Facial fat grafting: the past, present, and future. Clin Plast Surg 47(1):1–6
Feng Y, Shi Y, Tian Y, Yang Y, Wang J, Guo H et al (2023) The collagen-based scaffolds for bone regeneration: a journey through electrospun composites integrated with organic and inorganic additives. Processes 11(7):2105

Fezza JP, Cartwright M, Mack W, Flaharty P (2002) The use of aerosolized fibrin glue in face-lift surgery. Plast Reconstr Surg 110(2):658–664
Fitzgerald R, Bass LM, Goldberg DJ, Graivier MH, Lorenc ZP (2018) Physiochemical characteristics of poly-L-lactic acid (PLLA). Aesthet Surg J 38(suppl_1):S13–S17
Flemming I (1992) Fibrin glue in face lifts. Facial Plast Surg 8(1):79–88
Giele H, Cassell O (2016) Plastic and reconstructive surgery. Oxford University Press, New York
Gorczyca DP, Gorczyca SM, Gorczyca KL (2007) The diagnosis of silicone breast implant rupture. Plast Reconstr Surg 120(7):49S–61S
Grabska-Zielińska S, Sionkowska A, Carvalho Â, Monteiro FJ (2021) Biomaterials with potential use in bone tissue regeneration—collagen/chitosan/silk fibroin scaffolds cross-linked by EDC/NHS. Materials 14(5):1105
Gubisch W, Kotzur A (1998) Our experience with silicone in rhinomentoplasty. Aesth Plast Surg 22:237–244
Guimarães CF, Gasperini L, Marques AP, Reis RL (2020) The stiffness of living tissues and its implications for tissue engineering. Nat Rev Mater 5(5):351–370
Hamilton GS III, Carrithers JS, Karnell LH (2004) Public perception of the terms "cosmetic," "plastic," and "reconstructive" surgery. Arch Facial Plast Surg
Harris HI (1964) The use of silicones in maxillofacial abnormalities. Plast Reconstr Surg 34(4):419–425
Hasani-Sadrabadi MM, Sarrion P, Pouraghaei S, Chau Y, Ansari S, Li S et al (2020) An engineered cell-laden adhesive hydrogel promotes craniofacial bone tissue regeneration in rats. Sci Transl Med 12(534):eaay6853
Hassan N, Krieg T, Zinser M, Schröder K, Kröger N (2023) An overview of scaffolds and biomaterials for skin expansion and soft tissue regeneration: insights on zinc and magnesium as new potential key elements. Polymers 15(19):3854
Hauptstein J, Forster L, Nadernezhad A, Groll J, Teßmar J, Blunk T (2022) Tethered TGF-β1 in a hyaluronic acid-based bioink for bioprinting cartilaginous tissues. Int J Mol Sci 23(2):924
Hausamen JE (2001) The scientific development of maxillofacial surgery in the 20th century and an outlook into the future. J Craniofac Surg 29(1):2–21
Herath B, Laubach M, Suresh S, Schmutz B, Paige Little J, Yarlagadda PK et al (2023) The development of a modular design workflow for 3D printable bioresorbable patient-specific bone scaffolds to facilitate clinical translation. Virtual Phys Prototyp 18(1):e2246434
Hillard C, Fowler JD, Barta R, Cunningham B (2017) Silicone breast implant rupture: a review. Gland Surg 6:163–168
Huggins RJ, Mendelson BC (2017) Biologic behavior of hydroxyapatite used in facial augmentation. Aesth Plast Surg 41:179–184
Hunt NC, Grover LM (2010) Cell encapsulation using biopolymer gels for regenerative medicine. Biotechnol Lett 32:733–742
Hussain S, Mubeen I, Ullah N, Shah SSUD, Khan BA, Zahoor M et al (2022) Modern diagnostic imaging technique applications and risk factors in the medical field: a review. Biomed Res Int 2022(1):5164970
Huzum B, Puha B, Necoara RM, Gheorghevici S, Puha G, Filip A et al (2021) Biocompatibility assessment of biomaterials used in orthopedic devices: an overview. Exp Ther Med 22(5):1–9
Intini C, Elviri L, Cabral J, Mros S, Bergonzi C, Bianchera A et al (2018) 3D-printed chitosan-based scaffolds: an in vitro study of human skin cell growth and an in-vivo wound healing evaluation in experimental diabetes in rats. Carbohydr Polym 199:593–602
Jarvis T, Thornburg D, Rebecca AM, Teven CM (2020) Artificial intelligence in plastic surgery: current applications, future directions, and ethical implications. Plast Reconstr Surg–Glob Open 8(10):e 3200
Jia L, Hua Y, Zeng J, Liu W, Wang D, Zhou G et al (2022) Bioprinting and regeneration of auricular cartilage using a bioactive bioink based on microporous photocrosslinkable acellular cartilage matrix. Bioact Mater 16:66–81

Jiang JP, Liu XY, Zhao F, Zhu X, Li XY, Niu XG et al (2020) Three-dimensional bioprinting collagen/silk fibroin scaffold combined with neural stem cells promotes nerve regeneration after spinal cord injury. Neural Regen Res 15(5):959–968
Kaur CD, Saraf S (2011) Topical vesicular formulations of Curcuma longa extract on recuperating the ultraviolet radiation–damaged skin. J Cosmet Dermatol 10(4):260–265
Kesireddy V, Kasper FK (2016) Approaches for building bioactive elements into synthetic scaffolds for bone tissue engineering. J Mater Chem B 4(42):6773–6786
Khoshnazar SM, Asadi A, Roshancheshm S, Karimian A, Abdolmaleki A (2023) Application of 3D scaffolds in tissue engineering. Cell Tissue Biol 17(5):454–464
Kim YJ, Go MJ, Hu C, Hong CB, Kim YK, Lee JY et al (2011) Large-scale genome-wide association studies in East Asians identify new genetic loci influencing metabolic traits. Nat Genet 43(10):990–995
Krishani M, Shin WY, Suhaimi H, Sambudi NS (2023) Development of scaffolds from bio-based natural materials for tissue regeneration applications: a review. Gels 9(2):100
Kulka K, Sionkowska A (2023) Chitosan based materials in cosmetic applications: a review. Molecules 28(4):1817
Kumar M, Saini JS, Bhunia H (2021a) Investigations on MWCNT embedded carbon/epoxy composite joints subjected to hygrothermal aging under bolt preloads. Fibers Polym 22(7):1957–1975
Kumar M, Saini JS, Bhunia H, Chowdhury SR (2021b) Behavior of mechanical joints prepared from EB cured CFRP nanocomposites subjected to hygrothermal aging under bolt preloads. Appl Compos Mater 28:271–296
Labet M, Thielemans W (2009) Synthesis of polycaprolactone: a review. Chem Soc Rev 38(12):3484–3504
Lan X, Liang Y, Erkut EJ, Kunze M, Mulet-Sierra A, Gong T et al (2021) Bioprinting of human nasoseptal chondrocytes-laden collagen hydrogel for cartilage tissue engineering. FASEB J 35(3):e21191
Laurent TC (1998) Hyaluronan as a clinical marker of pathological processes. Wenner-Gren Symposium
Li L, He ZY, Wei XW, Wei YQ (2016) Recent advances of biomaterials in biotherapy. Regener Biomater 3(2):99–105
Li X, You R, Zhang Q, Yan S, Luo Z, Qu J, Li M (2021) Engineering vascularized dermal grafts by integrating a biomimetic scaffold and Wharton's jelly MSC-derived endothelial cells. J Mater Chem B 9(32):6466–6479
Lorenc ZP, Lee JC (2016) Composite volumization of the aging face: supra-periosteal space as the foundation for optimal facial rejuvenation. J Drugs Dermatol: JDD 15(9):1136–1141
Madni A, Kousar R, Naeem N, Wahid F (2021) Recent advancements in applications of chitosan-based biomaterials for skin tissue engineering. J Bioresources Bioproducts 6(1):11–25
Marchac D, Sandor GK (1994) Face lifts and sprayed fibrin glue: an outcome analysis of 200 patients. Br J Plast Surg 47(5):306–309
Mariani E, Lisignoli G, Borzì RM, Pulsatelli L (2019) Biomaterials: foreign bodies or tuners for the immune response? Int J Mol Sci 20(3):636
Maslennikova A, Kochueva M, Ignatieva N, Vitkin A, Zakharkina O, Kamensky V et al (2015) Effects of gamma irradiation on collagen damage and remodeling. Int J Radiat Biol 91(3):240–247
Medaglini D, Harandi AM, Ottenhoff TH, Siegrist CA, Consortium VE, Agnandji ST et al (2015) Ebola vaccine R&D: filling the knowledge gaps. Sci Transl Med 7(317):317ps24-317ps24
Mendelson BC, Jacobson SR, Lavoipierre AM, Huggins RJ (2010) The fate of porous hydroxyapatite granules used in facial skeletal augmentation. Aesth Plast Surg 34:455–461
Meng Y, Wang Z, Tao Y, Abdukerim A, Bo Z, Chen W et al (2021) Dark matter search results from the PandaX-4T commissioning run. Phys Rev Lett 127(26):261802
Minelli L, Richa J, Mendelson BC (2022) Aesthetic enhancement of the brow using hydroxyapatite. Aesth Plast Surg 46(3):1201–1210
Mironov V, Boland T, Trusk T, Forgacs G, Markwald RR (2003) Organ printing: computer-aided jet-based 3D tissue engineering. Trends Biotechnol 21(4):157–161

Mironov V, Kasyanov V, Drake C, Markwald RR (2008) Organ printing: promises and challenges. Regen Med 3(1):93–103
Mironov V, Visconti RP, Kasyanov V, Forgacs G, Drake CJ, Markwald RR (2009) Organ printing: tissue spheroids as building blocks. Biomaterials 30(12):2164–2174
Mironov V, Kasyanov V, Markwald RR (2011) Organ printing: from bioprinter to organ biofabrication line. Curr Opin Biotechnol 22(5):667–673
Mohamed RM, Yusoh K (2016) A review on the recent research of polycaprolactone (PCL). Adv Mater Res 1134:249–255
Mohammadi AT, Sanjarian S, Tehrany PM, Khorram R, Vafadar R, Mohseni H et al (2023) Cutting-edge advances in surgery. Nobel Sciences
Molloy AM, Ardill J, Tomkin GH (1980) The effect of metformin treatment on gastric acid secretion and gastrointestinal hormone levels in normal subjects. Diabetologia 19:93–96
Mooney E, Loh C, Pu LL, ASPS/PSEF Technology Assessment Committee (2009) The use of fibrin glue in plastic surgery. Plast Reconstr Surg 124(3):989–992
Mukherjee S, Agarwal M, Bakshi A, Sawant S, Thomas L, Fujii N et al (2020) Chemokine SDF1 mediated bone regeneration using biodegradable Poly (D, L-lactide-co-glycolide) 3D scaffolds and bone marrow-derived mesenchymal stem cells: implication for the development of an "Off-the-Shelf" pharmacologically active construct. Biomacromolecules 21(12):4888–4903
Nakahara K, Haga-Tsujimura M, Saulacic N (2016) Periosteum-induced bone formation by distraction osteogenesis: histologic and microcomputed tomography analysis. Int J Oral Maxillofac Implants 31(4)
O'Daniel TG, Kachare MD (2022) The utilization of poly-L-lactic acid as a safe and reliable method for volume maintenance after facelift surgery with fat grafting. Aesthetic Surg J Open Forum 4:ojac014. US: Oxford University Press
Ohlwiler DA (1967) The use of silicone implants in plastic surgery. Nurs Clin N Am 2(3):495–501
Özgentas HE, Pindur A, Spira M, Liu B, Shenaq S (1994) A comparison of soft-tissue substitutes. Ann Plast Surg 33(2):171–177
Ozhathil DK, Tay MW, Wolf SE, Branski LK (2021) A narrative review of the history of skin grafting in burn care. Medicina 57(4):380
Pakhomova C, Popov D, Maltsev E, Akhatov I, Pasko A (2020) Software for bioprinting. Int J Bioprinting 6(3)
Panayi AC, Orgill DP (2019) Current use of biological scaffolds in plastic surgery. Plast Reconstr Surg 143(1):209–220
Peng W, Peng Z, Tang P, Sun H, Lei H, Li Z et al (2020) Review of plastic surgery biomaterials and current progress in their 3D manufacturing technology. Materials 13(18):4108
Peters K, Unger RE, Kirkpatrick CJ, Gatti AM, Monari E (2004) Effects of nano-scaled particles on endothelial cell function in vitro: studies on viability, proliferation and inflammation. J Mater Sci Mater Med 15:321–325
Pial MMH, Tomitaka A, Pala N, Roy U (2022) Implantable devices for the treatment of breast cancer. J Nanotheranostics 3(1):19–38
Pollack SV (1990) Acne Keloidalis. J Dermatol Surg Oncol 16(6)
Reddy KS, Chittoria RK, Babu P, Marimuthu SK, Kumar SH, Subbarayan EK et al (2017) Effectiveness of fibrin glue in adherence of skin graft. J Cutan Aesthet Surg 10(2):72–75
Roshangar L, Rad JS, Kheirjou R, Khosroshahi AF (2021) Using 3D-bioprinting scaffold loaded with adipose-derived stem cells to burns wound healing. J Tissue Eng Regen Med 15(6):546–555
Salamanca E, Hsu CC, Yao WL, Choy CS, Pan YH, Teng NC, Chang WJ (2020) Porcine collagen–bone composite induced osteoblast differentiation and bone regeneration in vitro and in vivo. Polymers 12(1):93
Säljö K, Apelgren P, Stridh Orrhult L, Li S, Amoroso M, Gatenholm P, Kölby L (2022) Long-term in vivo survival of 3D-bioprinted human lipoaspirate-derived adipose tissue: proteomic signature and cellular content. Adipocyte 11(1):34–46
Selin H (ed) (2013) Encyclopaedia of the history of science, technology, and medicine in non-western cultures. Springer Science & Business Media

Shang Y, Wang G, Zhen Y, Liu N, Nie F, Zhao Z et al (2023) Application of decellularization-recellularization technique in plastic and reconstructive surgery. Chin Med J 136(17):2017–2027
Shi Y, Xing TL, Zhang HB, Yin RX, Yang SM, Wei J, Zhang WJ (2018) Tyrosinase-doped bioink for 3D bioprinting of living skin constructs. Biomed Mater 13(3):035008
Chiu YC, Shie MY, Lin YH, Lee AKX, Chen YW (2019) Effect of strontium substitution on the physicochemical properties and bone regeneration potential of 3D printed calcium silicate scaffolds. Int J Mol Sci 20(11):2729
Shyna S, Nair PD, Thomas LV (2020) A nonadherent chitosan-polyvinyl alcohol absorbent wound dressing prepared via controlled freeze-dry technology. Int J Biol Macromol 150:129–140
Singh B, Sharma N (2008) Mechanistic implications of plastic degradation. Polym Degrad Stab 93(3):561–584
Singh J, Kumar V, Kim KH, Rawat M (2019) Biogenic synthesis of copper oxide nanoparticles using plant extract and its prodigious potential for photocatalytic degradation of dyes. Environ Res 177:108569
Singh K, Kumar M, Saini JS (2022) Metal inserts for performance improvement in glass fiber-reinforced polymer nanocomposite joints under accelerated aging conditions. Polym Compos 43(3):1448–1461
Sinha VR, Bansal K, Kaushik R, Kumria R, Trehan A (2004) Poly-ϵ-caprolactone microspheres and nanospheres: an overview. Int J Pharm 278(1):1–23
Spector M, Lim TC (2016) Injectable biomaterials: a perspective on the next wave of injectable therapeutics. Biomed Mater 11(1):014110
Stein P, Vitavska O, Kind P, Hoppe W, Wieczorek H, Schürer NY (2015) The biological basis for poly-L-lactic acid-induced augmentation. J Dermatol Sci 78(1):26–33
Suszynski TM, Serra JM, Weissler JM, Amirlak B (2018) Three-dimensional printing in rhinoplasty. Plast Reconstr Surg 141(6):1383–1385
Szychlinska MA, Calabrese G, Ravalli S, Dolcimascolo A, Castrogiovanni P, Fabbi C et al (2020) Evaluation of a cell-free collagen type I-based scaffold for articular cartilage regeneration in an orthotopic rat model. Materials 13(10):2369
Takitoh T, Bessho M, Hirose M, Ohgushi H, Mori H, Hara M (2015) Gamma-cross-linked nonfibrillar collagen gel as a scaffold for osteogenic differentiation of mesenchymal stem cells. J Biosci Bioeng 119(2):217–225
Todorova D, Simoncini S, Lacroix R, Sabatier F, Dignat-George F (2017) Extracellular vesicles in angiogenesis. Circ Res 120(10):1658–1673
Treacy P (2022) The evolution of aesthetic medicine: the evolution of a new field of medicine by a pioneer voted the top aesthetic doctor in the world. Austin Macauley Publishers
Valantin MA, Aubron-Olivier C, Ghosn J, Laglenne E, Pauchard M, Schoen H et al (2003) Polylactic acid implants (New-Fill)® to correct facial lipoatrophy in HIV-infected patients: results of the open-label study VEGA. AIDS 17(17):2471–2477
Van Nieuwenhove I, Tytgat L, Ryx M, Blondeel P, Stillaert F, Thienpont H et al (2017) Soft tissue fillers for adipose tissue regeneration: from hydrogel development toward clinical applications. Acta Biomater 63:37–49
Wang W, Zhang Y, Ye R, Ni Y (2015) Physical crosslinkings of edible collagen casing. Int J Biol Macromol 81:920–925
Woodruff MA, Hutmacher DW (2010) The return of a forgotten polymer—polycaprolactone in the 21st century. Prog Polym Sci 35(10):1217–1256
Wu EC, Zhang S, Hauser CA (2012) Self-assembling peptides as cell-interactive scaffolds. Adv Funct Mater 22(3):456–468
Xiao H, Chen X, Liu X, Wen G, Yu Y (2023) Recent advances in decellularized biomaterials for wound healing. Mater Today Bio 19:100589
Xu C, Wang Q, Chen J, Wu Y, Zhao L (2019) Collagen matrix for periodontal plastic surgery procedures: a meta-analysis update. Int J Periodontics Restorative Dent 39(4)
Yang P, Ju Y, Hu Y, Xie X, Fang B, Lei L (2023) Emerging 3D bioprinting applications in plastic surgery. Biomater Res 27(1):1

Yeong WY, Chua CK (2014) Bioprinting: principles and applications, vol 1. World Scientific Publishing Co Inc.

Zhou G, Jiang H, Yin Z, Liu Y, Zhang Q, Zhang C et al (2018) In vitro regeneration of patient-specific ear-shaped cartilage and its first clinical application for auricular reconstruction. EBioMedicine 28:287–302

Zhu Z, Wang YM, Yang J, Luo XS (2017) Hyaluronic acid: a versatile biomaterial in tissue engineering. Plast Aesthet Res 4:219–227

Zielińska A, Nowak I (2017) Abundance of active ingredients in sea-buckthorn oil. In: Lipids in health and disease, vol 16, pp 1–11

Bioscaffold and Their Applications in Tissue Regeneration: Angiogenesis, Bone, Skin, and Nerve Tissue Regeneration

8

Hamed Manoochehri, Hanie Mahaki, Hamid Tanzadehpanah, Amirhossein Ahmadieh-Yazdi, Mohsen Sheykhhasan, Piao Yang, Naser Kalhor, and Reihaneh Seyedebrahimi

Hamed Manoochehri, Hanie Mahaki, Hamid Tanzadehpanah, Amirhossein Ahmadieh-Yazdi and Mohsen Sheykhhasan contributed equally with all other contributors.

H. Manoochehri
The Persian Gulf Marine Biotechnology Research Center, The Persian Gulf Biomedical Sciences Research Institute, Bushehr University of Medical Sciences, Bushehr, Iran

H. Mahaki
Vascular and Endovascular Surgery Research Center, Mashhad University of Medical Sciences, Mashhad, Iran

H. Tanzadehpanah
Antimicrobial Resistance Research Center, Mashhad University of Medical Sciences, Mashhad, Iran

A. Ahmadieh-Yazdi
Stem Cell Biology Research Center, Yazd Reproductive Sciences Institute, Shahid Sadoughi University of Medical Sciences, Yazd, Iran

M. Sheykhhasan (✉)
Cellular and Molecular Research Center, Qom University of Medical Sciences, Qom, Iran

P. Yang
Department of Molecular Genetics, College of Arts and Sciences, The Ohio State University, Columbus, OH, USA

N. Kalhor
Department of Mesenchymal Stem Cells, Academic Center for Education, Culture and Research, Qom, Iran

R. Seyedebrahimi
Cellular and Molecular Research Center, Qom University of Medical Sciences, Qom, Iran

Anatomy Department, Faculty of Medicine, Qom University of Medical Sciences, Qom, Iran

M. Kumar et al. (eds.), *Natural Product Inspired Scaffolds*,
https://doi.org/10.1007/978-981-97-3111-4_8

Abstract

Tissue engineering aims to restore, maintain, or improve tissue function through the use of biomaterials, cells, and bioactive molecules. In recent years, natural products have emerged as promising candidates for incorporation into tissue engineering scaffolds due to their bioactive properties and biocompatibility. This chapter summarizes the recent advances in the development of natural product-loaded scaffolds and their applications in tissue engineering. We discuss various types of natural products, such as chitin and chitosan, cellulose, starch, hyaluronic acid, collagen, fibroin, pullulan (PUL), keratin, elastin, alginate, fibrin, albumin, gelatin, and proteins, and their potential benefits in tissue regeneration. The applications of natural product-loaded scaffolds in various tissue engineering fields, such as vascular, bone, skin, and nerve tissue regeneration, are also discussed. Finally, we provide perspectives on future research directions in this field. Overall, this chapter highlights the potential of natural product-loaded scaffolds in advancing tissue engineering and regenerative medicine in the context of vascular, bone, skin, and nerve tissue regeneration.

Keywords

Tissue engineering · Chitosan · Starch · Hyaluronic acid · Collagen

8.1 Introduction

In order to introduce genes, medications, and cells into the body, scaffolds can be administered as injections or implants. In order to create a suitable scaffold with the necessary capabilities, the scaffolding components must be biodegradable, biocompatible, tissue compatible, and capable of being designed (Zielińska et al. 2023). Natural biomaterials have emerged as the primary bioactive ingredients utilized in medicinal materials throughout the last several decades. Natural biopolymers can be broadly categorized into two classes based on their structure and monomer units: (1) polysaccharide-based, comprising chitosan, chitin, hyaluronic acid, alginate, cellulose, glycosaminoglycans, and agarose, and (2) polypeptide-based, comprising protein-based components such as collagen, fibrin, cellulose, hyaluronic acid, agarose, glycosaminoglycans, and decellularized extracellular matrix (dECM) (Krishani et al. 2023; Kellaway et al. 2023).

When coupled with cells, natural biopolymers exhibit greater bioactivity, reduced or absent toxicity, superior biocompatibility, increased hydrophilic properties, and useful biological function (Zielińska et al. 2023; Krishani et al. 2023).

Many natural product-based scaffolds for tissue regeneration have been created for biomedical uses during the past few decades, and some of them are now offered for sale (Zielińska et al. 2023; Krishani et al. 2023).

An overview of natural product-based scaffolds for angiogenesis, bone, skin, and nerve tissue regeneration applications is given in this paper. We provided a number of natural product-based scaffold characteristics that are important for tissue regeneration and repair. Additionally, a discussion of several scaffolds based on natural products has been presented, along with the latest developments in herbal structures.

8.2 Application of Natural Product-Based Scaffold for Angiogenesis Tissue Regeneration

8.2.1 Angiogenesis

The circulatory system, which forms throughout embryonic development and persists into maturity, is responsible for supplying nutrients and oxygen to biological systems continuously. Two processes are defined: vasculogenesis, which involves angiogenesis, or the creation of newly formed blood vessels. Angiogenesis is a process where blood vessels naturally recede in response to functional demands, such as exercise-stimulating angiogenesis in muscles and weight loss causing capillary regression in the adipose tissue, which is a response to specific pro-angiogenic signals (Min et al. 2019). The introduction of prevascularization procedures, both in vivo and in vitro; the addition of growth factors; the addition of adhesion peptides; and the co-culturing of various cell types are the strategies that may be used to induce angiogenesis (Fig. 8.1). A number of diseases, including cancer, myocardial infarction, and stroke, can be brought on by abnormal angiogenesis. New treatment techniques can be developed with the aid of comprehending and managing the angiogenic process. Age, the tissue or cell type, and physiological or pathological conditions all influence the angiogenic response (Adair and Montani 2010).

There are four primary phases in the process, which may involve diverse cell types, cytokines, growth factors, and extracellular matrix constituents. First,

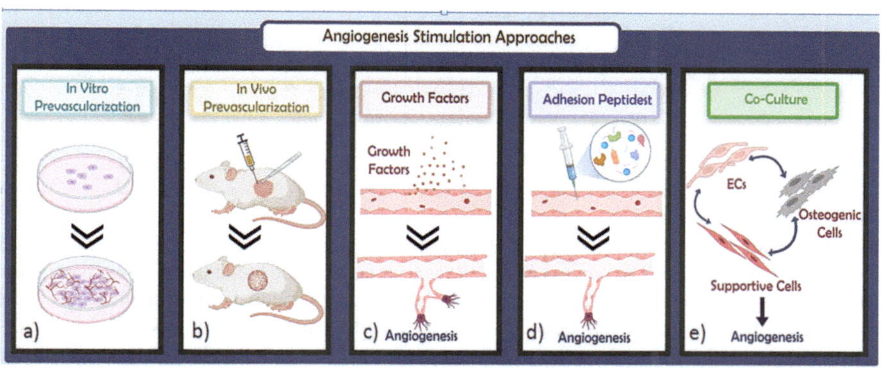

Fig. 8.1 There are five methods to promote angiogenesis: introducing prevascularization techniques both in vivo (**a**) and in vitro (**b**) adding growth factors (**c**), adding adhesion peptides (**d**), and co-culturing diverse cell types (**e**)

vascular sprouting occurs when endothelial cells (ECs) proliferate in response to cues that promote angiogenesis, like vascular endothelial growth factor (VEGF), fibroblast growth factors (FGF), angiopoietins, and hypoxia-inducible factors (Fig. 8.2). These signals may be generated by wounded cells or tissue that is low in nutrients and oxygen. Secondly, tubule morphogenesis creates a lumen by means of fibroblasts secreting endothelial cell matrix (ECM) molecules and encouraging tubulogenesis. Thirdly, freshly generated vascular sprouts either regress or stable as they respond to the demands of the tissue (Carmeliet and Jain 2011). The angiogenic process involves the formation of new endothelial tubes via a combination of factors such as platelet-derived growth factor (PDGF), VEGF, ECM, angiopoietin 1 (Ang1), Ang-2, and blood flow. The tubes regress and go through apoptosis if these components are abruptly eliminated and blood is not passing through the developing vessels. However, a mixture of these factors stabilizes the new endothelial tubes, allowing for vessel stabilization. As the last stage of this process, quiescent endothelial cells are substituted for proliferative ones, blood flow is adjusted, and adhesion between neighboring cells and the extracellular matrix is restored. Tissue engineering (TE), which replaces damaged biological tissues with replacements that mirror

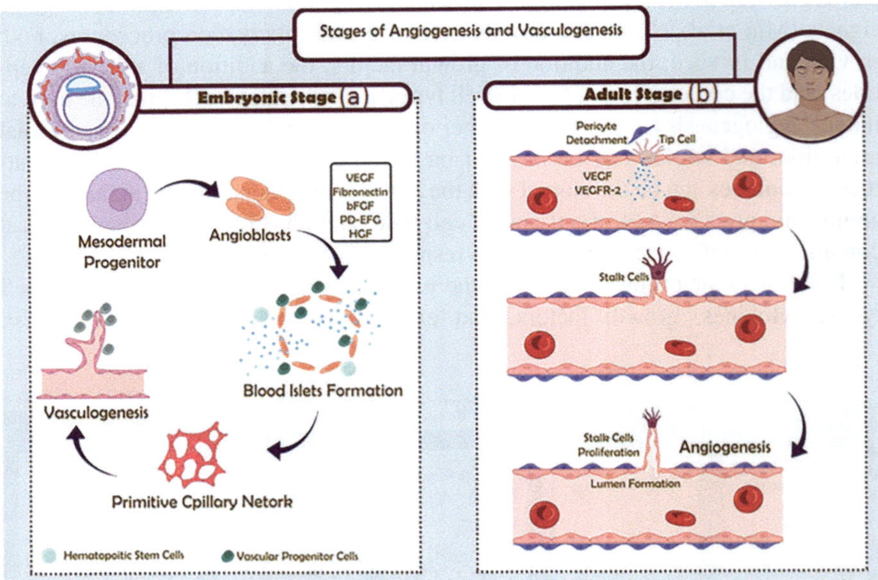

Fig. 8.2 Angiogenesis, the first step of the embryonic process, is when blood vessels start to develop and remodel (**a**). The process of vasculogenesis starts in the blood islets during pregnancy, where vascular progenitor cells differentiate into angioblasts. In contrast, during adult stages, these progenitor cells differentiate into pericytes, which support the physiological process of angiogenesis, which forms or renews blood vessels from pre-existing vessels (**b**). This diagram illustrates the key proteins, cells, and growth factors that stimulate the processes of vasculogenesis and angiogenesis

their appearance and function, depends on factors such as PDGF, sphingosine 1, Ang1, transforming growth factor-beta (TGF-β), ECM, and blood flow for stabilization (Irvin et al. 2014). Proper vascularization is crucial for large-size grafts, while small implants can maintain cell viability through oxygen and nutrient diffusion from pre-existing vasculature. Larger implants may cause cell death in the core due to diffusion only reaching outer cell layers. Living cells and blood supplies should be separated by around 200 μm for optimal health.

The vascularization of transplanted tissues facilitates the interchange of metabolites, the provision of oxygen and nutrients, and the existence of specialized cells for tissue remodeling and regeneration (Banfi et al. 2018). Various methods are employed to comprehend and monitor the angiogenic process, enhancing blood supply integration in larger structures, including growth factors, tumors, wound models, functionalized biomaterials, and the impact of ECM on blood vessel development as well as its function in TE strategies (Marchand et al. 2019). Advances in biomaterials have brought about novel approaches to the development of vasculature inside tissue scaffolds.

8.2.2 Natural Biomaterial-Derived Scaffolds

Because of their superior microstructure, high interconnectivity, and innate bioactivity, which resemble the extracellular matrix of real tissue, natural biomaterials are commonly employed in tissue engineering. The extracellular matrix is essential for supporting cells in repairing damaged tissue, facilitating cell adhesion, and delivering nutrients and oxygen. These materials have many biological advantages, but they frequently lack the precise degradation pattern and insufficient mechanical strength needed to match the pace of tissue regeneration. To make up for this, synthetic materials are needed. Many strategies for fast vascularization have been investigated by researchers; these include changing chemical compositions, scaffold architecture, employing growth factors, mixing pro-angiogenic ECM components (e.g., hyaluronic acid, collagen, or chitosan) with solid scaffold biomaterials, and using stem cells to form blood vessels. These techniques are intended to promote cell division particularly that of stem cells and enhance the vascularization process. The finest scaffold construction and cell seeding strategies are covered along with other approaches that may be utilized separately or in combinations (Fig. 8.3). A number of new polymers have also been modified in the field of biomaterials to promote angiogenesis.

For example, Chen et al. suggested that the production of a porous gelatin hydrogel modified with polyhedral oligomeric silsesquioxane (POSS) at varying dosages can sustain the release of VEGF and BMP-2 to induce angiogenesis. POSS is added to gelatin to increase cell integrin expression and protein adsorption, which improves cell adhesion to scaffolds and the formation of vascular tubes (Fig. 8.4) (Chen et al. 2020). This section will discuss several scaffolds derived from natural materials (as well as other materials) that facilitate angiogenesis (Fig. 8.5) (Shokrani et al. 2022).

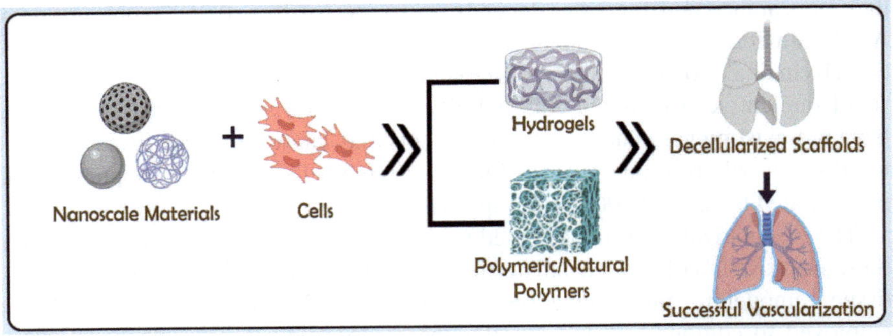

Fig. 8.3 Angiogenesis can be accelerated by combining scaffolds that have been seeded with cells with microcarriers. By adjusting their porosity and interconnectivity, microcarriers can be used with hydrogels or scaffolds to promote angiogenesis. This makes it possible to comprehend angiogenesis more thoroughly. It is possible to seed several cell types, including mesenchymal stem cells, and promote their proliferation and development with nanoscale materials. Microcarrier-containing polymeric scaffolds offer an effective angiogenesis template. While microcarriers can be coupled to carry drugs or biomolecules continuously, hydrogels can be utilized for encapsulating or seeding cells. Tissue angiogenesis follows, resulting in the creation of a whole designed tissue with a vast vascular network

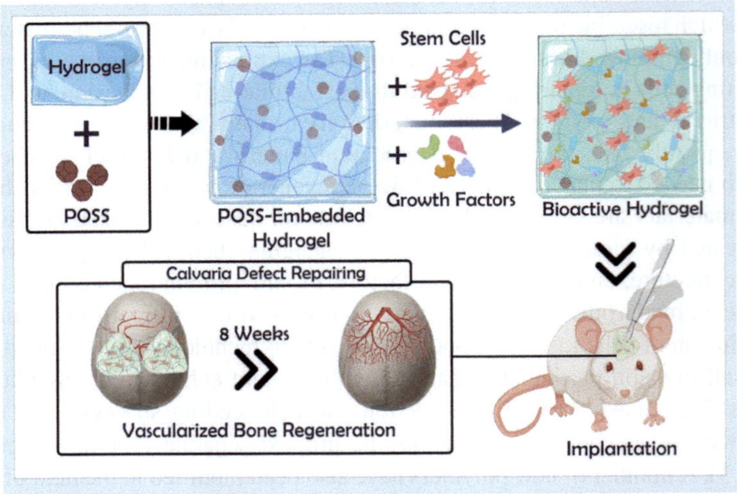

Fig. 8.4 Applications of an octafunctional POSS cage-primed Gel-POSS hybrid hydrogel in vivo. Among the applications of this platform are cell adhesion, regulated growth factor release, tissue defect healing, and angiogenesis acceleration. Gelatin hydrogel containing polyhedral oligomeric silsesquioxane stimulates angiogenesis during vascularized bone regeneration

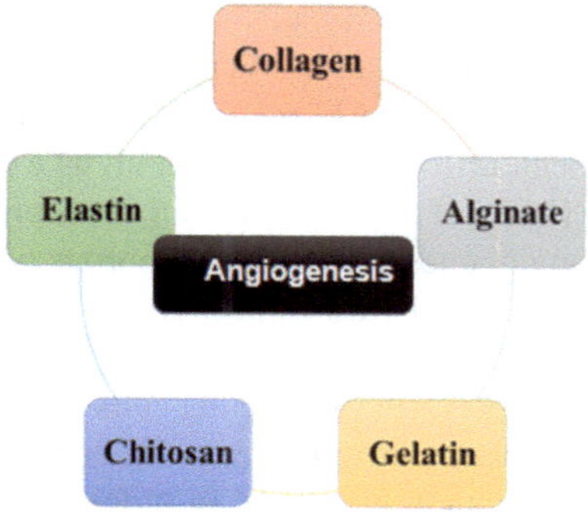

Fig. 8.5 Several scaffolds derived from natural materials that facilitate angiogenesis

8.2.2.1 Pro-angiogenic Biomaterials and Anti-angiogenic Biomaterials

Natural biomaterials called hyaluronic acid (HA) and fibrin stimulate fibroblast activity and increase angiogenesis through breakdown products like Ca^{2+}, which aid in the wound healing process. It was discovered that the effects of type IV collagen on vascular elongation were dose-dependent, with lower and larger dosages having opposing effects. Vessel advancement was restricted by 300 lg/mL, but new vessel creation was encouraged by a moderate concentration of 30 lg/mL (Bahramsoltani et al. 2014). After being implanted in diabetic rat skin wound models, collagen-based scaffolds have been demonstrated to improve angiogenesis and lower inflammation. Increased angiogenesis was seen in wounds treated with alginate/vitamin D3, according to the histological assessment (Ehterami et al. 2020). It has been determined that chitosan, a depolymerized byproduct of chitooligosaccharides and low-molecular-weight chitosan (LMWC/COs), inhibits angiogenesis. Angiogenesis has been reported to be accelerated by some chitosan composites. It has been demonstrated that chitosan/gelatin and chitosan/polycaprolactone modified by heparin can improve vascularization. These chitosan composites are being used by other groups to create novel angiogenesis biomaterials (Mohebbi et al. 2019). A possible natural biomaterial for angiogenesis, skin defect healing, and wound healing is silk fibroin (SF). Individuals in a randomized clinical experiment showed that silk fibroin film, as opposed to other commercial dressing materials, considerably shortens the time it takes for skin defects to regenerate (Zhang et al. 2017a). In owing to their microstructure, which encourages neo-tissue development, cell adhesion, migration, differentiation, or necrosis and facilitates cell activities, collagen, gelatin, and alginate are common natural biomaterials.

8.2.2.1.1 Collagen-Based Scaffolds

Wong et al. developed a template using PLGA microfibers integrated into a type I collagen scaffold and seeded with HUVECs/VEGF in order to encourage angiogenesis even after fiber breakdown (Wong et al. 2016). Chan et al. seeded primary human microvascular endothelial cells onto type I collagen-based scaffolds, which led to the creation of capillaries with CD31 positivity (Chan et al. 2016). Heparin-modified type I collagen-based scaffolds implanted with bone marrow cells have been utilized by Gérard et al. as a possible model for characterizing angiogenesis (Haifei et al. 2014). Haifei et al. proposed a more complex paradigm for boosting angiogenesis: heparinized porous type I collagen-chitosan scaffolds in combination with *N*-hydroxysuccinimide (NHS) and *N*-(3-dimethylaminopropyl)-*N*′-ethylcarbodiimide (EDC). Chiu et al. used 100 type I collagen hydrogels to explore how pericytes, branching cells that have a major influence on blood flow regulators and vascular reinforcement, contribute to endothelial cell migration and stability (Morgan et al. 2013). In order to mimic angiogenesis in the urethra's nanofibrous structure, a three-dimensional scaffold composed of bladder acellular matrix, bacterial cellulose, glycosaminoglycans, and type I collagen expressed endothelial proteins and created capillary-like tubes. A research team created type I collagen hydrogels with sulfhydryl groups and HUVEC/epidermal growth factor, similar to previous collagen-based scaffolds, for porous type I and type III collagen scaffolds. In order to evaluate the impact of cell seeding pattern, Schechner et al. suggested a novel framework for vascularization expansion: cultivating HUVECs in type I collagen/fibronectin gels (Schechner et al. 2000). During bone reformation, sulfated chitosan in gelatin hydrogels is an innovative synergistic angiogenesis formulation that effectively upregulates the angiogenesis genes of BMP-2 and VEGF (Zhang et al. 2019).

Physical and Chemical Modifications on Collagen to Facilitate Angiogenesis

Because of its mechanical characteristics and abundance in the ECM, type I collagen is a useful biomaterial for researching cell-matrix interactions and developing scaffolds for tissue engineering. Creating functional platforms for the vasculature through tissue engineering is a task that involves modifying collagen scaffolds to enhance the mechanical structure or biological activity of the tissue (Irawan et al. 2018). Understanding the molecular processes behind collagen remodeling may improve endothelial cell migration and vascular expansion by altering its physical and chemical properties. The migration and adherence of cells as well as the flow of nutrients through the collagen scaffold are significantly influenced by the size of the scaffold's pores. Certain cell types exhibit preferential adhesion to scaffolds possesing holes that are either larger or smaller than their typical cell size, demonstrating a selective attachemnet behaviour influenced by the structural properties of scaffold. Fibroblasts and endothelial cells adhere to scaffolds that have apertures larger than 90 μm and smaller than 80 μm, respectively. Collagen scaffold pore sizes may be changed by using degradable particles, gas foaming, rapid prototyping, and freeze-drying (Lu et al. 2011). Ice particles are mixed with an aqueous type I collagen solution to create porous collagen scaffolds (Zhang et al. 2014). As the ice crystals melt during the freeze-drying process, pores are left behind. Porous scaffolds were made with pore sizes close to the four sets of particles (diameters ranging from 150 to 500 μm). Efficient cell dispersion was possible in both in vitro and in vivo environments due to comparable connectivity among the holes. While comparable techniques can be applied in other tissue engineering domains, the purpose of this study was to investigate these scaffolds in order to facilitate cartilage regeneration. Furthermore, porous collagen scaffolds may be made using collagen alone or in conjunction with additional ECM proteins including heparin, chitosan, or GAGs (Shi et al. 2008). EDC/NHS is a safe and successful chemical cross-linking method that enhances mean pore size and scaffold stability by further cross-linking the material (Davidenko et al. 2015). Increased collagen molecule cross-linking slows the process of scaffold breakdown and may influence the in vivo inflammatory response. According to Yao et al. (2008), collagen scaffolds with greater amounts of cross-linking have a noticeably larger capability for angiogenic development than unaltered structures. By using this technique, it was possible to produce hole diameters between 30 and 50 μm and to improve migration in vitro. Further evidence of the influence of cross-linked collagen on endothelial cell morphogenesis was provided by spoke-like vascular structures from chorioallantoic membranes (CAM) migrating toward modified scaffolds after 7 days in culture. By covalently connecting the polypeptide chains of collagen fibers, dehydrothermal (DHT) treatment may enhance the stiffness of collagen scaffolds (Madaghiele et al. 2016). In this case, the degree of denaturation is an important differentiator between the two cross-linking procedures, regardless of the response of inflammation to a stiff scaffold need to be addressed. Although the form of the triple-helical collagen is lost during denaturement, DHT is physically superior than EDC/NHS and could present more cell-binding sites than the attached collagen (Madaghiele et al. 2016). To get the right stiffness, porosity, and cell attachment, collagen scaffolds' cross-linking density must be carefully controlled, and the triple-helical structure of the collagen must be preserved. Scaffolds may be functionalized by signaling molecules to control the migration of cells and the development of blood vessel networks in tissues. On the other hand, growth factors can be immobilized to delay enzyme internalization or degradation, extending their biological activity in vivo and preventing degradation of cells or media (Ito 2008). Endothelial cell migration, proliferation, and tube formation are stimulated by the growth factors VEGF and Ang1. Chiu and Radisic (2010) covalently immobilized Ang1 and VEGF in collagen scaffolds using EDC chemistry and then evaluated the combined and individual effects on the in vitro morphogenesis of endothelial cells (ECs). Higher EC proliferation and tube formation were produced by growth factors co-immobilized in comparison to control scaffolds containing soluble growth factors. The stimulation of EC infiltration from chorioallantoic membranes by collagen scaffolds containing immobilized growth factors suggests that this response may occur in vivo. Collagen may consequently operate as a transporter of cells and signaling molecules to stimulate angiogenesis in cells that have been transplanted as well as in the host. By adding heparin-binding motifs to biomaterials, it is possible to further regulate the kinetics of protein release. Comparable techniques that might be applied in a range of tissue engineering fields are NHS regeneration and EDC. Additional ECM proteins, including heparin, chitosan, or GAGs, can also be added to the solution to create porous collagen scaffolds (Shi et al. 2008). The material is cross-linked with NHS and EDC to enhance scaffold stability and mean pore size, using the secure chemical cross-linking technique EDC/NHS (Davidenko et al.

2015). The rate of scaffold degradation is slowed down by increased collagen molecule cross-linking that might have an impact on the inflammatory response in vivo. Yao et al. showed that collagen scaffolds with greater degrees of cross-linking had a much larger potential for angiogenic development than unmodified structures (Yao et al. 2008). By using this technique, it was possible to produce hole diameters between 30 and 50 μm and to improve migration in vitro. Further demonstrating the effect of conjugated collagen on endothelial cell morphogenesis, spoke-like vascular structures from CAM were orientated toward modified scaffolds after 7 days in culture. On the other hand, by covalently tying together the polypeptide chains of collagen fibers, dehydrothermal (DHT) therapy may make collagen scaffolds more rigid (Madaghiele et al. 2016). One important distinction between different cross-linking techniques is the degree of denaturation that results; nevertheless, in this instance, it's also important to take into account the inflammatory response that occurs to a stiff scaffold. Collagen is not degraded by DHT, a physical mechanism that may be advantageous over EDC/NHS. Denatured collagen may contain more cell attachment sites than cross-linked collagen, despite losing its triple-helical structure and becoming less ordered (Madaghiele et al. 2016). To get the right porosity, stiffness, and cell attachment, collagen scaffolds need control over their cross-linking strength and maintenance of their triple-helical structure. Functionalizing scaffolds with different signaling molecules may modify collagen's chemical structure while also regulating cell migration and the creation of vascular networks inside the tissue. Growth factors' bioactivity is prolonged in vivo by immobilization on scaffolds, which stops them from being quickly absorbed by cells or broken down via local microenvironmental enzymes (Ito 2008). Potent growth factors including Ang1 and VEGF aid in proliferation, endothelial cell migration, and tube formation (Chiu and Radisic 2010). Chiu et al. investigated the effects of Ang1 and VEGF covalently bonded in collagen scaffolds utilizing EDC chemistry, both alone and in combination, on endothelial cell morphogenesis in vitro (Chiu and Radisic 2010). EC proliferation and tube formation were higher in co-immobilized growth factor-treated scaffolds than in soluble growth factor-containing control scaffolds. The EC penetration from chorioallantoic membranes was similarly enhanced by collagen scaffolds containing immobilized growth factors, indicating that a similar reaction may take place in vivo. Collagen could therefore be employed as a medium for cells and signaling molecules to stimulate angiogenesis in the host and transplanted cells. By including heparin-binding sites, biomaterials may be modified, and the rate of protein secretion may be further regulated (Minor and Coulombe 2020).

8.2.2.2 Alginate/Gelatin-Based Scaffolds

Extensive study has also been conducted on the ability of alginate and gelatin, or denatured collagen, to stimulate angiogenesis. Campbell et al. alginate lyase-injected alginate hydrogels, an enzyme that cleaves alginate polymer chains, without including outgrowth endothelial cells (OECs), in an effort to enhance angiogenesis (Campbell et al. 2018). Andrade et al. introduced injectable gelatin-based sponges as an additional template for new vascularization. Rat pneumocytes at the embryonic stage were injected into these sponges (Andrade et al. 2007).

Perets et al. utilized poly microspheres embedded in alginate scaffolds to regulate angiogenic factor release, demonstrating the development of capillaries (Jin et al. 2016). Remarkably, important angiogenic growth factors which include TGF-β1, PDGF, and VEGF that encourage vascularization were linked to the alginate-sulfate/alginate scaffolds. To facilitate the prolonged release of VEGF, Wang et al. loaded VEGF onto heparin-containing modified 3D porous bacterial cellulose/gelatin (B/G) scaffolds. This permitted up to 2 weeks of growth factor release for angiogenesis (Wang et al. 2018). It has been demonstrated that endothelial cell viability is successfully increased, while keratinocyte survival is maintained when gelatin is combined with a porcine cholecyst-derived scaffold (CDS). Combining these two characteristics may help heal diabetic wounds and angiogenesis. It was clear from a comparison of bare CDS and gelatin-coupled CDS that the gelatin addition resulted in the enhanced features (Mony et al. 2021).

8.2.2.3 Chitosan-Based Scaffold

Chitosan-based hydrogels also possess antibacterial and anti-inflammatory qualities. The inflammatory response involves a wide range of chemicals and cytokines, among which TNF-α and members of the IL-1 family are only two (Guo et al. 2015). Chitosan derivatives can interact with molecules linked to the inflammatory response of the immune system to skin injury as well as cellular signaling pathways. For instance, Hao et al.'s research discovered that a hydrogel comprising velvet antler blood peptides, sodium alginate, and chitosan primarily inhibited the acute inflammatory response by activating the phosphatidylinositol 3-kinase (PI3K)/AKT/mTOR/HIF-1α/VEGFA pathway and reducing the synthesis of IL-1β and TNF-α (Hao et al. 2022). For the purpose of triggering inflammation and boosting the immune system, these two components are essential pharmaceutics. For example, during an immunological response, TNF-α and IL-1β both help trigger the inflammasome's development (Guo et al. 2015). Furthermore, chitosan may have an effect on the expression of a number of cell surface markers. During the inflammatory phase of the immune response, endothelial cell surface markers like CD31, which indicate increased angiogenesis, can be expressed more frequently when the chitosan hydrogels are active. Conversely, they can significantly reduce the expression of markers like CD68, which are primarily expressed by macrophages and monocytes (Hao et al. 2022; Yang et al. 2022). Self-healing hydrogels have properties that include promoting skin proliferation, limiting bacterial formation, and reducing inflammation (Guo et al. 2022). These properties may be further increased by including bioactive materials into hydrogels, like exosomes that carry additional chemicals and antibiotics. For instance, exosome-loaded self-healing chitosan hydrogels has the capacity to encourage the deposition of collagen, fibroblast growth, and angiogenesis (Qin et al. 2022a). Another study by Liu et al. similarly found that the use of chitosan hydrogel, which prevented pyroptosis in vascular endothelial cells and reduced inflammatory response, increased bone marrow-derived MSCs' survival and healing of cardiac tissue (Liu et al. 2020a). Additionally, hMSC-containing nanodots encased in chitosan/collagen hydrogel matrix boosted the heart's angiogenesis, which improved the left ventricle's ejection fraction (Kim et al. 2023). The angiogenic response is enhanced by bioactive signals on CS/PEGDA scaffolds, which influences the expression of the marker (CD-31) and the development of endothelial tissue (tube formation). This implies that novel implants might be utilized to promote the growth of tissue-engineered constructs in regenerative medicine. Due to its innate antimicrobial and coagulation-enhancing properties, chitosan is a commonly used biomaterial for wound dressings (Li et al. 2021a). Biomedicines are showing a great deal of interest in chitosan hydrogel due to its gradual drug release profile. However, under some situations, its biological procoagulant activity may be harmful for uses in vascular and tissue engineering. One significant disadvantage of using chitosan hydrogels as a scaffold for tissue engineering is their poor mechanical characteristics. They also show no pro-therapeutic advantages in vascular regeneration, a mismatching disintegration rate, and restricted cell adhesion (Pistone et al. 2020).

8.2.2.4 Elastin-Based Scaffold

Reports on a few less common natural biomaterials are included in this section. Some biomaterials are not as well-known because of their excessive cost or insufficient mechanical and biological characteristics. For instance, Anderson et al. suggested that fibrin hydrogels with nanoparticles bound to VEGF and seeded with HUVECs could accelerate capillary development (Anderson et al. 2011). The cyclic peptide LXW7, introduced by He and colleagues, binds to integrin $\alpha v\beta 3$ on endothelial progenitor cells and EC receptors to stimulate angiogenesis. They then developed the pro-angiogenic chemical LXW7-DS-SILY, which functionalizes the extracellular matrix scaffolds to facilitate neovascularization (Zhang et al. 2017b). The next step involved seeding HUVECs onto nanohydroxyapatite to promote vascular tube development, enhanced cell survival, and vascularization. The researchers showed that inhibiting the production of nitric oxide (NO) in HUVECs led to the suppression of PI3K (Shi et al. 2017). Zarrintaj et al. developed 3D agarose-based cryogels for ischemic cardiomyopathy, showing improved cell viability, migration, and proliferation due to increased cytokine production (Zarrintaj et al. 2018). It was obvious that plumbagin altered the signaling pathways of nuclear factor (NF), phospholipase C (PLC), hypoxia-inducible factor (HIF)-1, extracellular signal-regulated kinase (ERK), and extracellular signal. Plumbagin also dramatically reduced the creation of micro-vessels (Lee et al. 2019a). *Momordica charantia* is a bitter melon type that has piqued the interest of researchers because of its enhanced ERK phosphorylation. According to researchers, it promoted angiogenesis by encouraging the development of tubes from bovine aortic endothelial cells (BAEC) (Morgan et al. 2013). Subsequently, after approximately a week, it was found that injectable platelet-rich fibrin (PRF) in conjunction with a co-culture of primary osteoblasts (pOBs) and prouting endothelial cells (OECs) constituted yet another unique template for the formation of angiogenesis networks (Morgan et al. 2013). The formation of a neurovascular unit (NVU) is facilitated by a unique framework, consisting of blood vessels, glial cells, neurons, and extracellular matrix proteins (Shokrani et al. 2022; Abebayehu et al. 2019).

8.3 Application of Natural Product-Based Scaffold for Bone Tissue Regeneration

Bone performs numerous roles in body as a multifunctional organ such as assisting locomotion, providing supporting skeleton for body tissues and organs, and producing or storing some hormones, growth factors, and minerals. Also, a generation of blood cells (hematopoiesis) is an essential role of bone marrow (Ansari and Eshghanmalek 2019). Numerous cell types, including hematopoietic, progenitor, endothelial, and inflammatory cells, are involved in bone fracture regeneration. Bone repair has three intersecting phases: inflammation, bone production, and bone remodeling. Bone production begins once the coagulated blood is replaced with the fibrous tissue and cartilage which named soft callus. As regeneration growths, the soft callus is switched with hard callus. The penultimate stage of bone healing is remodeling, which takes more than a few, which newly formed bone is reshaped into its mature and mechanically suitable structure. But, delayed healing can occur in some conditions including infection, bone tumor removal, critical size defects, disrupted vascularization, and osteoporosis (Ansari 2019).

Novel approaches have been provided by TE for enhanced regeneration of bone defects. The "tissue engineering triad" refers to three crucial elements of tissue engineering: biomaterial scaffolds, cells, and growth-stimulating factors. Scaffolds are crucial in TE, because they act as a transient platform or template that offers directional and structural support during the development of new tissues. Scaffold is a three-dimensional (3D) porous substance that used to create an environment that is conducive to cell regeneration and tissue and organ repair. It promotes cell attachment, proliferation, extracellular matrix regeneration, and the repair of bones, muscles, and nerves. In addition, scaffolds have the ability to convey bioactive substances (Krishani et al. 2023). Scaffolds are typically originated from biological sources or have synthetic origins. A naturally occurring scaffold resembles the extracellular matrix (ECM) of the natural bone and also has superior bioactivity and biocompatibility (Wong et al. 2023). A different type of natural-based scaffolds has been described in Fig. 8.6. The current section aims to provide some information regarding the application of natural product-based scaffolds in regeneration of the bone tissue.

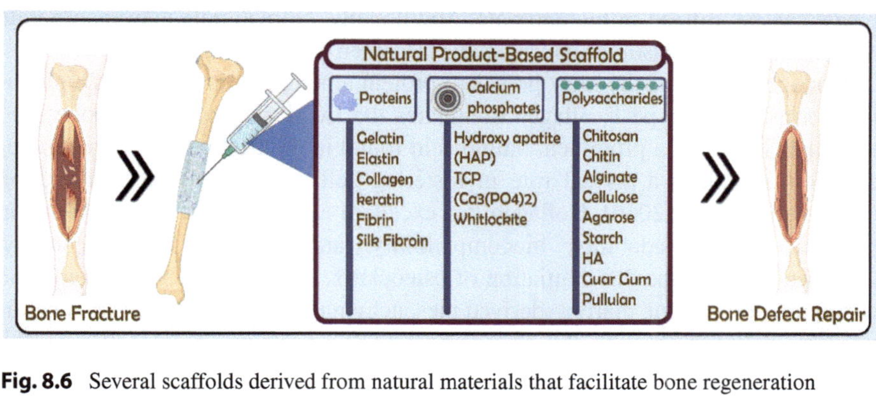

Fig. 8.6 Several scaffolds derived from natural materials that facilitate bone regeneration

8.3.1 Proteins

8.3.1.1 Gelatin

Gelatin is abundant in the Arg-Gly-Asp (RGD) peptide sequence, facilitating cell adhesion, proliferation, and differentiation. Gelatins comprise type I collagen, which makes a predominant (90%) organic constituent of the ECM. Furthermore, gelatin preserves a sequence that is recognized by matrix metalloproteinase (MMP), which permits enzymatic breakdown. The amino acid composition of gelatin mirrors that of collagen, encompassing 19 amino acids, with hydroxyproline proline, and glycine being particularly abundant (Filippi et al. 2020).

Mammals are the main origin of gelatin, particularly from porcine skin and bovine hides. It has been found that gelatin-coated scaffolds inhibit the opsonization and complement system, thereby reducing their immunogenicity. In vitro investigations demonstrate that gelatin-based scaffolds can regulate gene expression and cell differentiation (Krishani et al. 2023). Yao et al. have verified the capacity of gelatin-based scaffolds to stimulate osteogenesis both in vivo and in vitro (Yao et al. 2022). Promising outcomes have been achieved in bone tissue engineering through the utilization of gelatin-based scaffolds in combination with various biomaterials and polymers such as chitosan, hydroxyapatite, and polycaprolactone (PCL) (Kartikasari et al. 2016).

8.3.1.2 Elastin, Collagen, and Keratin

The ECM consists of fibers and protein fibrils intricately arranged in a hydrated framework of glycosaminoglycan chains, imparting a fibrillar and viscoelastic nature to it. The formation of extracellular fibrous networks, a result of the interplay between collagen, elastin, and keratin, enables tissues and organs to tolerate various strains and stresses (Chen and Liu 2016).

Fibrous proteins, notably collagens and elastin, constitute the primary and most abundant structural element of the ECM. In some tissues such as blood vessels, skin, lungs, and elastic ligaments, elastin ranks as the most prevalent protein. Elastin matrices with outstanding cytocompatibility are recommended for application in guided bone regeneration therapy. Various forms of elastin, including insoluble types present in decellularized ECM, autografts, allografts, xenografts, and purified types, can be utilized in biomaterials. Additionally, biomaterials may incorporate recombinant tropoelastin or its fragments (Daamen et al. 2007).

Collagen stands as the most prevalent protein among mammals, with 28 identified forms, where type I collagen dominates the ECM, particularly in bones and tendons. Glycine is a prominent amino acid found in the triple-helical structure of collagen and plays a pivotal role in fostering cell attachment and proliferation (Ramanathan et al. 2016). Collagen has excellent adhesion, osteogenic induction properties, biodegradability, biocompatibility, and minimal immunogenicity. Collagen prevents the differentiation of osteoclasts. Also, activation of osteogenic differentiation in bone marrow-derived mesenchymal stem cells (BMSCs) is facilitated by collagen through the engagement of the protein kinase B (PKB) and extracellular-regulated protein kinases (ERK) signaling pathways. Unfortunately, its application is constrained by suboptimal physicochemical and mechanical properties, particularly its limited mechanical strength. Consequently, in order to enhance collagen's performance, it must be altered or combined with a scaffold to be used as collagen-based scaffolds or hydrogels (Li et al. 2021b).

Various combinations of collagen and elastin have been utilized in the establishment of diverse scaffolds. The inclusion of elastin in collagen scaffolds produced through non-electrospinning methods has been observed to reduce scaffold stiffness, influence the contraction and degradation of collagen, and enhance some processes such as the formation of elastic fibers and angiogenesis. Mohammadzadeh Rad and colleagues recently developed a novel freeze-dried bionanocomposite, incorporating hydroxyapatite and polymethyl methacrylate (PMMA), reinforced with elastin polymer for orthopedic applications (Mohammadzadeh Rad et al. 2021).

Keratin, a plentiful source of protein, serves as the primary constituent in wool, hooves, horns, hair, feathers, and nails. It is a fiber protein with sulfur-containing amino acids which can establish robust disulfide bonds and provide a proper mechanical strength. Additionally, keratin offers benefits such natural abundance, antibacterial effects, biocompatibility, and biodegradability. This is attributed to keratin-unique signal motifs including glutamate-aspartic acid-serine (EDS) and leucine-aspartic acid-valine (LDV) domains facilitating strong cell adhesion. Moreover, keratin is rich in disulfide bonds and functional groups, including carboxyl and amino groups, offering reactive sites for modification as required (Sun et al. 2022a).

8.3.1.3 Silk Fibroin

Silk is an old protein and a biopolymer that encompasses various types, with mulberry silk being the most extensively researched in the biomedical field. Silk comprises two primary protein components: silk fibroin (SF) and sericin. Sericin, a water-soluble gelatinous protein, is eliminated during the SF purification process. SF emerges as a highly suitable biomaterial for bone regeneration due to its outstanding mechanical properties and biocompatibility, coupled with its minimal immunogenicity and tunable biodegradability (Wei et al. 2023a).

SF, characterized by a Gly-Ser-Gly-Ala-Gly-Ala amino acid sequence in its primary structure, is a structural protein that adopts an antiparallel β-sheet arrangement, endowing it with outstanding mechanical strength. SF-based scaffolds exhibit superior properties compared to other natural biodegradable polymers including collagen, hyaluronic acid, and chitosan. Moreover, the mechanical characteristics of SF-based scaffolds are customizable and predominantly influenced by factors such as scaffold preparation technique, SF concentration in solution, porosity, and pore size (Kim et al. 2005).

Despite having pure biologically active SF regions, they facilitate cell attachment, proliferation, and differentiation but are weak in stimulating angiogenesis and osteogenesis. To address this limitation, Wei et al. fabricated a 3D porous composite scaffold using of fibrin silk and collagen. This approach aims to circumvent the drawbacks of each component while capitalizing on their respective advantages through a targeted structural design (Wei et al. 2023a).

8.3.2 Polysaccharides

8.3.2.1 Chitosan

Chitosan (CS), obtained from partial deacetylation of chitin, is a copolymer of (1 → 4)-2-acetamido-2-deoxy-β-D-glucan (*N*-acetyl-D-glucosamine) and (1 → 4)-2-amino-2-deoxy-β-D-glucan (D-glucosamine) (Filippi et al. 2020). CS has a linear structure and is a natural polymer that serves as a structural element in fungi, insects, crustaceans (such as shrimp and crabs), and algae. Chemical hydrolysis is the primary method used to extract it from marine crustacean shells followed by demineralization and deproteinization steps and subsequently two final processes of discoloration and deacetylation. However, it may also be separated through enzymatic digestion of some fungi's cell walls (Cai et al. 2006). It resembles the extracellular matrix of mineralized tissue; for this reason it is used as a scaffold by researchers. CS has a variety of useful biological activities including antibacterial properties, provoking cell proliferation, biocompatibility, biodegradability, and a good hydrophilic surface. Also, it has been shown that CS can improve cell adhesion, osteoblast differentiation, and mineralization (Di Martino et al. 2005). The physiological role of CS is facilitated by its cationic characteristics, which allow it to interact with negatively charged molecules such as proteoglycans, glucose aminoglycans, nutrients, and other anions that control growth factors and cell activity. CS can be shaped into 3D scaffolds with diverse porosity architectures via different preparation processes such as 3D printing, and it can be combined with different materials for generating enhanced composite scaffolds. The CS scaffolds with porous structure could be made readily by lyophilizing and freezing CS solved in an acetic acid-containing solution (Filippi et al. 2020). Osteogenicity can be imparted by CS in combination with cells, polymers, and substances that induce bone formation. For bone TE applications, CS can be readily come in a variety of shapes, including fibers, films, sponges, beads, etc. (Martins et al. 2010).

8.3.2.2 Chitin

After cellulose, chitin is the second most frequent polysaccharide in the world. It can be found in the fungal cell walls and exoskeleton of arthropods such as shrimps, crabs, lobsters, prawns, and insects but usually extracted from shellfish sources. Chitin, a special kind of biopolymer, is mostly based on the *N*-acetyl-glucosamine monomer and is derived from shellfish sources. Depending on the manner of processing employed to produce the biopolymer, it is a copolymer of *N*-acetyl-glucosamine and *N*-glucosamine units randomly or block distributed throughout the biopolymer chain. The biopolymer is referred to be chitin when the proportion of *N*-acetyl-glucosamine units is greater than 50% (Jayakumar et al. 2011). One way to distinguish chitosan from chitin is by the level of deacetylation. Chitin offers a variety of applications in tissue engineering because of its exceptional qualities, which include biocompatibility, biodegradability, non-toxicity, and metal ion chelation. Moreover, it promotes migration, differentiation, and adherence of cells. Since chitin and *N*-glycosaminoglycans share a structural resemblance, it is an excellent choice for cartilage, skin, teeth, and bone tissue engineering (Chocholata et al. 2019). Chitin can be easily processed into different gels, nanofibers, membranes, and scaffold forms, the last two items are appropriate for use in bone engineering (Jayakumar et al. 2011).

8.3.2.3 Alginate

Alginate as the most prevalent marine biopolymer is mostly present in the cell walls of brown seaweeds, algae, and some bacteria. It is a naturally occurring polysaccharide that is linear and unbranched, containing D-mannuronic acid and α-l-glucuronic acid subunits that are linked via β-(1 → 4) linkage. Through ionotropic cross-linking, alginate can establish stable three-dimensional gel networks or “egg box” structures (Shi et al. 2019).

Alginate is a suitable material to be employed as bone engineering scaffolds because of high viscosity and gelling properties, which also give it more strength and flexibility (Lee and Mooney 2012). Unfortunately, because of its low osteoconductivity and hydrophilicity, alginate has demonstrated poor cell attachment in vitro. Alginate’s cell adhesion behavior toward bone tissue engineering can be enhanced by combining alginate with additional polymers, for example, collagen, chitosan, gelatin, and hyaluronate (Rajesh et al. 2017).

8.3.2.4 Cellulose

Cellulose is a linear polysaccharide widely distributed in natural sources from some organisms such as algae, fungi, and bacteria and several plants including cotton, wood, and bamboo. Cellulose is one of the most widely used polysaccharides for construction of bone substitutes, either in its pure or chemically modified form, because of its exceptional benefits, which include low production costs, source abundance, non-toxicity, non-immunogenicity, high specific mechanical properties, biodegradability, and hydrophilicity (Seddiqi et al. 2021). These properties are strongly influenced by the source of cellulose extraction. In these polysaccharides, D-glucopyranose monomers are connected by (1 → 4) glycosidic bonding. Numerous intra- and intermolecular hydrogen bonds within the cellulose biopolymer network result in high stability, hydrophilicity, and a high number of accessible sites for chemical alterations with various functional groups (Janmohammadi et al. 2023). From an application standpoint, cellulose etherification and esterification are the most important modification processes. The well-known and attractive cellulose derivatives in biomedicine and pharmaceutics are ethyl cellulose, methylcellulose, carboxymethyl cellulose, and hydroxyethyl cellulose as cellulose ethers and cellulose acetate as an ester (Marques-Marinho and Vianna-Soares 2013). Cellulose-based composites are developed by mixing two or more compounds to produce cellulose-based composites with enhanced properties. They can be loaded by varieties of therapeutic compounds to improve its anti-inflammatory, osteoinductivity, and osteoconductivity features. Many compounds are available to increase the bioactivity of cellulose scaffolds, including hydroxyapatite, alpha-tricalcium phosphate (α-TCP), and collagen (Pramanik et al. 2023).

8.3.2.5 Agarose

Agarose, a natural marine polysaccharide extracted from red algae and seaweed (Barua et al. 2018), is a biodegradable and biocompatible biopolymer made of a repeating unit of agarobiose. It can build a gel network to facilitate the transmission of nutrients and oxygen within the scaffold. Agarose is a suitable alternative organic matrix in which the stiffness of its hydrogels is adjustable, which makes it easy to manage the scaffold mechanical properties (Kazimierczak et al. 2019). Some advantages of agarose for biomedical applications are no immunogenicity, water solubility and contourable self-gelling, mechanical properties, and water adsorption capacity; all these provide an appropriate microenvironment for cellular activity (Zarrintaj et al. 2017). Despite this, agarose is known to hinder cell adherence, which is why it's frequently mixed with other polymers to increase its biocompatibility (Kazimierczak et al. 2019). Composite scaffolds containing agarose along with chitosan, hydroxyapatite, gelatin, fibrin, and graphene oxide have been made in studies in order to enhance agarose-based scaffolds (Khosalim et al. 2022).

8.3.2.6 Starch

Starch is a polysaccharide made of linear amylose, consisting of alpha 1 to alpha 4-linking glucose monomers, and branched amylopectin, consisting alpha 1 to alpha 6-linking glucose monomers. Starch is an easily available, inexpensive, bridgeable, biocompatible, osteoproductive, osteoconductive, and versatile polysaccharide with a significant promise as a biomaterial for bone tissue engineering (Krishani et al. 2023). In general, starch-based scaffolds are not bioactive or mechanically strong enough for bone regeneration. Instead, bioactive, biocompatible, and rigid fillers like polyvinyl alcohol (PVA), hydroxyapatite, citric acid cellulose, etc. should be added to the scaffold matrix to improve its mechanical and biological properties. In addition, numerous chemical modifications, such as esterification, etherification, oxidation, grafting, cationization, and cross-linking, can be used to improve the mechanical stability and bioactivity of starch scaffolds (Krishani et al. 2023; Ochubiojo and Rodrigues 2012). Anyway, chemical modifications may slow down the degradation rate of the starch-based biomaterials and also produce toxic byproducts. Other limitations for starch are very high water uptake and difficult processing. The best strategy to overcome these limitations is to use several types of starch isolated from various sources (Roslan et al. 2016).

8.3.2.7 Hyaluronic Acid

Hyaluronic acid (HA), also named as hyaluronate or hyaluronan, is a naturally occurring polysaccharide in the ECM of many tissues. It has acceptable biodegradability, biocompatibility, and ability to interact with crucial growth factors, for example, TGF-β, bone morphogenetic proteins (BMPs), and cell surface receptors (CD44) of cells which are involved in bone formation (osteoblasts, MSCs, osteoclasts). *N*-Acetylglucosamine and glucuronic acid repeating disaccharide units combine to form the linear molecule known as HA (Xing et al. 2020). HA-based bone substitutes can be modified into any sizes or shapes such as films, nanofibers, porous scaffolds, microspheres, and nanoparticles for bone repair. This is achieved by using various tissue engineering techniques like freeze-drying, electrospinning, and 3D printing. However, HA alone may exhibit a restricted osteogenic impact, so it can be exploited in combination with other biomaterials. For instance, in a research study, a freeze-drying method was employed to fabricate a tissue engineering scaffold comprised of HA, collagen, and chitosan enhanced with nanohydroxyapatite (Kaczmarek et al. 2018).

8.3.2.8 Guar Gum

Natural gums are polysaccharides which can increase the viscosity of a solution even at low concentrations. Gums are easily available, odorless, abundant, nontoxic, water-soluble, and chemically inert. They have various biomedical and industrial applications. Gums are considered sources of structural entities, water-binding components, and reserve nutrients and used as food ingredients (Mohammadinejad et al. 2020). Gum-based nanofibers can be produced under electrostatic forces using an electrospinning technique. A composite scaffold is fabricated by Anandan et al., using guar gum (GG), hydroxyapatite, and gellan, via the freeze-drying method for bone tissue regeneration (Anandan et al. 2019). Guar gum (GG) is a hydrated biopolymer that is extracted from bean *Cyamopsis* seeds of the *tetragonoloba.* GG is made up of a single galactose unit connected to a straight chain of β-1-4-mannose with α-1-6 connections to alternate mannose units (Mitra et al. 2018).

8.3.2.9 Pullulan

PUL is another natural polysaccharide produced by the fungus *Aureobasidium pullulans.* It is considered by excellent biocompatibility, biodegradable, non-mutagenic, non-toxic, non-immunogenic, and mechanical properties that make it a suitable hydrogel for tissue engineering purposes. PUL powder is white and non-hygroscopic; it dissolves readily in both hot and cold water. In addition to its thermal stability, it possesses antithrombotic, anti-inflammatory, and anticoagulant properties (Tabasum et al. 2018). PUL is linear, uncharged biopolymer composed of repeating maltotriose units connected through α-1,6-glycosidic linkages (Singh et al. 2008). Moreover, it has an FDA approval and is hemocompatible, making it appropriate for several food and pharmaceutical applications (Amrita et al. 2015). PUL is an excellent electroconductive polymer in order to make it thick by electrospinning approach. To make them last long enough to sustain the scaffold, cross-linking is required (Tabasum et al. 2018). They cannot provide a cell adhesion and proliferation supporting surface. In order to get around this issue, a number of compounds can be incorporated to a hydrophilic polymer such as gelatin, RGD ligands, hydroxyapatite, and small functional groups (Amrita et al. 2015).

8.3.2.10 Calcium Phosphates

About 60% of human bones are thought to be primarily composed of calcium phosphates. Calcium phosphate can promote bone regeneration with ion (calcium and phosphorus) release control, and also they can be encapsulated with different growth factors and drugs in order to enhance osteoinductivity (Jeong et al. 2019).

8.3.2.11 Hydroxyapatite (HAP)

Hydroxyapatite possesses a chemical formula of $Ca_{10}(PO_4)_6(OH)_2$, with a Ca/P ratio of 1.67. The majority of inorganic materials found in human bones are HAPs, which are a natural type of calcium phosphate. Many investigations have been done into the development of synthetic scaffolds to facilitate easy utilization and control over bioactivity, porosity, and mechanical strength (Thorpe et al. 2016).

8.3.2.12 Tricalcium Phosphate (TCP)

Tricalcium phosphate with a chemical formula of $Ca_3(PO_4)_2$) is an extensively researched calcium phosphate. Its Ca/P ratio is 1.5, and it contains two phases of α (α-TCP) and β (β-TCP). The α-TCP crystal structure resembles that of a monoclinic space group, while that of β-TCP is reminiscent of a rhombohedral space group. Compared to α-phase, β-phase has a higher biodegradation rate and higher stability; consequently, β-TCP is more suitable for use in bone regeneration. Although HAP is more stable than β-TCP, β-TCP degrades more quickly and is more soluble. β-Phase of tricalcium phosphate has a high rate of resorption and biocompatibility and can increase proliferation of stromal cells and osteoblast residence in bone marrow (Jeong et al. 2019).

8.3.2.13 Whitlockite (WH)

Whitlockite (WH) has a higher mechanical strength, when compared with HAP. Also, it has higher solubility when encountering physiological condition and could release more ions continuously (Kim et al. 2017). Techniques were used to create β-TCP/HAP scaffolds, which are anticipated to have improved compressive strength for bone tissue engineering (Jeong et al. 2019).

8.3.3 Directly-Derived Scaffolds from Whole Organisms

Researchers have employed scaffolds derived from various sources, including animals (fish, cuttlefish, cockle, and bovine) and humans, to investigate their impact on bone regeneration. The trabecular region of both animals and human bones serves as a basis for the development of 3D scaffolds for bone engineering (Lee et al. 2016).

Fish scales, considered as a biowaste resource, can be employed to make powdered hydroxyapatite. Fish scale scaffolds resemble the phase, content, shape, and crystallographic structure found in the natural bone (Bardhan et al. 2011). Mondal et al. utilized tilapia fish scales to make a naturally occurring hydroxyapatite scaffold, demonstrating notable bio-affinity and osteoconductivity (Mondal et al. 2016). Cockle shells, considered as biowaste from mollusks or cockles like scallops or mussels, comprise the pure aragonite form of calcium carbonate polymorph. This aragonite form can serve as a replacement for bone material, rendering cockle shells that may be substituted with bones (Islam et al. 2012). Bharatham et al. conducted a study where cockle shell powder was blended with alginate to produce a nanobio-composite bone scaffolds (Bharatham et al. 2014). Furthermore, a biomimetic composite scaffold can be made by utilizing an eggshell membrane as a template, supplemented with tricalcium phosphate from bovine bone ash, chitosan, and gelatin (Neacsu et al. 2020).

Scaffolds derived from plant tissue present a compelling alternative to those sourced from animals and humans, effectively addressing concerns related to immunogenicity and ethical considerations surrounding the use of human and animal tissues/organs. A wide range of 3D-decellularized cellulose scaffolds have been obtained from plants, including sweet, broccoli, apple, pepper, jujube, carrot, and persimmon. Notably, the apple scaffold has demonstrated efficacy in improving proliferation of pluripotent stem cells. In a rat model of calvarial defect treated with an apple-origin scaffold, bone regeneration was evident, accompanied by collagen deposition and the neoangiogenesis (Lee et al. 2019b).

Coral, as a marine invertebrate, has an exoskeleton made from $CaCO_3$ with several linked pores and resembles a cancellous bone pore structure. Also, hydroxyapatite obtained from corals can be used for fabricating porous scaffolds. Mineralized green and red algae also have been identified as a valuable calcium source. Composite bone scaffolds have been developed using HAP derived from *Corallina officinalis* (red algae) and *Phymatolithon calcareum* (green algae) (Walsh et al. 2008).

Marine sponges are promising candidates to be employed as natural scaffolds in bone regeneration, because of their remarkable porous structure, biosilica inorganic content, and spongin collagen-like organic content (Santos et al. 2023).

8.4 Application of Natural Product-Based Scaffold for Skin Tissue Regeneration

The application of natural product-based scaffolds in skin tissue regeneration is an area of significant interest in tissue engineering (Uppuluri et al. 2022). A scaffold in this context serves as a basic substance that is combined with other components during the development process to allow for the incorporation of growth factors and cells to build an artificial tissue (Krishani et al. 2023). Bio-based natural materials are especially valued due to their biocompatibility, safety, and non-toxic biodegradation properties (Krishani et al. 2023). There is a consensus that combining different materials can enhance scaffold characteristics, necessitating the selection of appropriate fabrication techniques (Krishani et al. 2023).

Skin tissue regeneration primarily relies on suitable scaffold matrices. For example, wound healing has made use of a variety of natural polymers, including alginate, cellulose, collagen, elastin, hyaluronic acid, silk, chitosan, and gelatin (Azimi et al. 2023; Fletes-Vargas et al. 2023). Different types of scaffolds, such as porous, fibrous, microsphere, hydrogel, composite, and acellular, each have distinct advantages and disadvantages (Suamte et al. 2023). These scaffolds must meet several fundamental requirements, including biocompatibility, biodegradability, processability, sterilizability, mechanical properties, and porosity, to be effective in tissue engineering applications (Chaudhari et al. 2016; Bonferoni et al. 2021).

Collagen, a key natural resource in the body and a component of skin tissue, plays a crucial role in all stages of the wound healing process (Mathew-Steiner et al. 2021). An ideal skin tissue regeneration scaffold should promote fibroblast activity for collagen secretion (Uppuluri et al. 2022). Additionally, the combination of aloe vera and silk fibroin, leveraging their wound healing and elasticity properties, respectively, has shown promising results in tissue engineering (Phimnuan 2022; Liu et al. 2021). These materials, when combined with and used in electrospinning techniques, have demonstrated enhanced hydrophilic properties, favorable tensile strain, and improved fibroblast proliferation, making them potent candidates for skin tissue regeneration scaffolds (Suganya et al. 2014).

The development of skin regeneration and wound healing is largely dependent on bioengineered scaffolds (Cummings et al. 2023). Both standard bioabsorbable and pro-regenerative scaffolds are used in current therapies for deep cutaneous injuries; the latter is more successful in producing whole skin structures as opposed to the scarred skin (Qin et al. 2022b). By imitating the three-dimensional porosity structure of natural extracellular matrices, tissue-engineered skin replacements hope to create an environment that is favorable for cell migration and proliferation—a process that is necessary for wound healing (Cummings et al. 2023). These scaffolds' structure and bioactive components are vital for their functionality (Xu et al. 2023). Making skin scaffolds by combining natural materials and synthetic polymers has become a standard procedure. The goal now is to regenerate the entire skin instead of only skin equivalents because of the recent advances in tissue engineering and skin biology, particularly with pro-regenerative scaffolds that trigger regenerative immune responses (Qin et al. 2022b).

The advancement in scaffold development not only depends on biomaterials but also on emerging fabrication technologies like electrospinning, stem cell technology, microfluidic technology, and 3D bioprinting (Qin et al. 2022b). These technologies play a significant role in producing scaffolds that effectively mimic natural extracellular matrices, crucial for tissue engineering and regenerative medicine (TERM) (Qin et al. 2022b). The goal is to develop scaffolds that can completely regenerate skin, addressing a critical unmet need in wound care and potentially leading to breakthroughs in clinical therapy for perfect skin regeneration (Qin et al. 2022b). The most important natural product-based scaffolds for skin regeneration are summarized in Fig. 8.7.

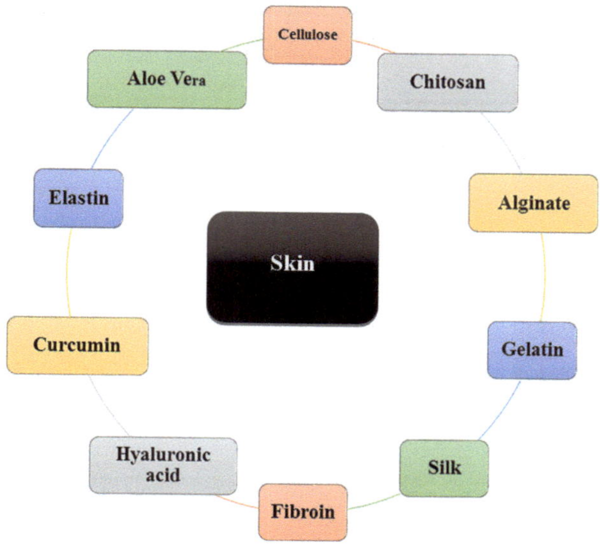

Fig. 8.7 Several scaffolds derived from natural materials that facilitate skin regeneration

8.5 Application of Natural Product-Based Scaffold for Nerve Tissue Regeneration

The nervous tissue includes the central (CNS) and peripheral nervous system (PNS). The nervous injuries and neurodegenerative diseases especially in CNS often cause permanent damages and inabilities of function (Varadarajan et al. 2022). In fact, CNS has a limited grade of spontaneously regeneration after injury, and it is an important challenge in neurodegenerative diseases (Nagappan et al. 2020). Moreover, the repair capacity of PNS depends on the patient's age, intensity of damage, and gap distance between injury point and neuron cell body (Faroni et al. 2015). Although small injuries can have self-repair, spontaneous treatment of large gaps of injury is impossible, and it needs medical interventions (Seyedebrahimi et al. 2021). So, a suitable choice of therapeutic technique can help improve injuries in CNS and PNS (Patel et al. 2018). Neural tissue engineering (NTE) can act as a new alternative medical approach (Papadimitriou et al. 2020).

Biomaterials can replace the injured soft tissue or act as a scaffold or conduit to facilitate tissue regeneration (Gaharwar et al. 2020). Though autografts are a gold standard in treatment of damaged tissues, there are still limitations of application (Rodrigues and Raz 2022).

Biomaterials have unique properties such as excellent biocompatibility and appropriate biodegradability, and they can be used in scaffold synthesis (Taghipour et al. 2020). Nerve tissue-engineered scaffolds should provide a supportive environment for neuritis growth and prevent fibrous scar formation in injured sites; also,

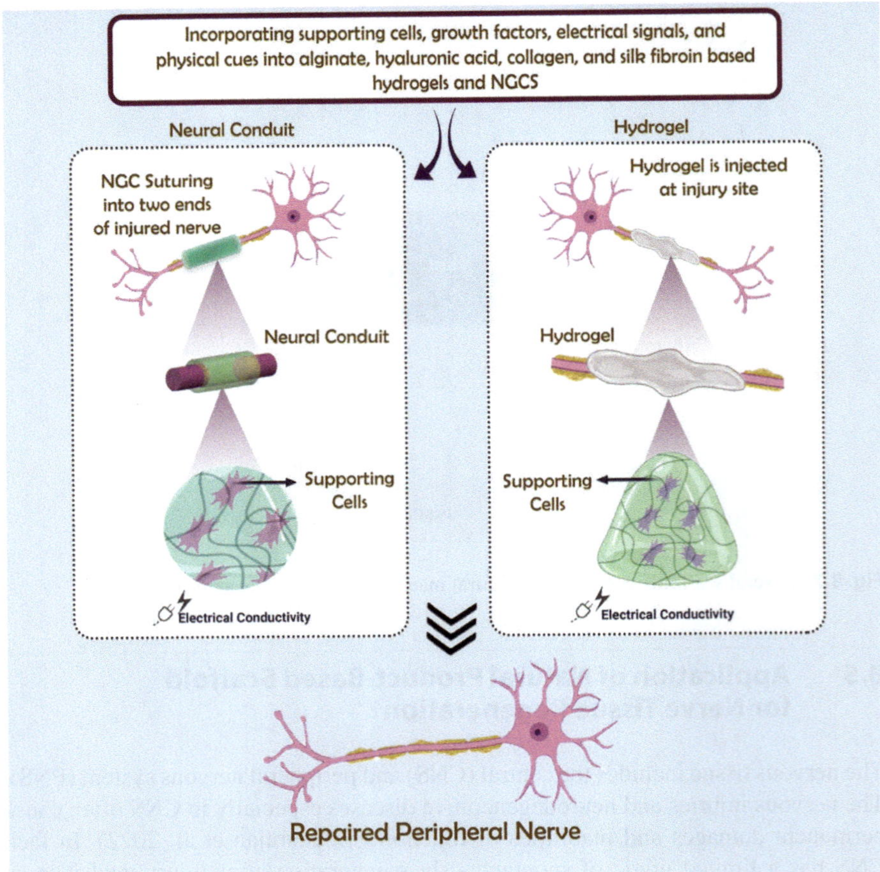

Fig. 8.8 For the purpose of healing PNS injuries, neural guidance channels (NGCs) have been the subject of much research. In addition, a lot of research have been done on the use of hydrogel-based medication delivery devices to treat PNS injuries. For the creation of NGCs and hydrogels, a variety of biomaterial alternatives are available, including as alginate, hyaluronic acid, collagen, and silk fibroin (Abdelbasset et al. 2022)

they can send biological factors to guide the axonal growth cone to the distal stump (Ghasemi-Mobarakeh et al. 2011).

In fact, nerve guide conduits (NGCs) can provide a bridge in the injury region and help in neural growth and extension (Fig. 8.8) (Zhu et al. 2020).

Natural polymers are plentiful, and they mimic like extracellular matrices, so it is rarely rejected by the host cells. Its structure is similar to macromolecular substances existing in the human body (Vasita and Katti 2006). Moreover, natural polymers offer an excellent environment for cell adhesion and proliferation (Doblado et al. 2021).

Collagen is an access natural biomaterial that is highly found in several body structures such as the bone, skin, tendon, and vessel wall. Collagen can be applied as a perfect biomaterial to prepare NGCs in nerve regeneration due to its low immune response and high absorbability (Liu et al. 2019). Much evidence shows that collagen acts like Schwann cells in PNS. Furthermore, collagen can play an important role in peripheral nerves regeneration. A study showed that the lack of type VI collagen in mice causes alterations in signaling pathways related to myelination (P-FAK, P-AKT, P-ERK1, P-ERK2, and P-p38) and reduction of vimentin (Chen et al. 2014). In fact, the expression of various collagen types such as VI and IV was upregulated in peripheral nerve injuries that type VI collagen encourages macrophage migration and polarization via AKT and PKA pathways (Chen et al. 2015). Also, the presence of type VI collagen increases macrophage recruitment, changes macrophages to M2 phenotype, and helps in nerve regeneration (Lv et al. 2017).

By directly attaching to the FNIII domain of neural cell adhesion molecule 1 (NCAM1) in the extracellular environment, type VI collagen also controls the development of nerve bundles (Sun et al. 2022b). Also, type XIII collagen can be found in the region of neuromuscular junction, connects to coding for the collagenous tail of acetylcholinesterase (ColQ), and plays a critical role in the synaptic process (Härönen et al. 2017). Generally, different types of collagens are necessary in the formation of nerve bundles and regeneration of nervous synapses after injury and can destroy the mistargeted axons (Li et al. 2022).

Laminin is a type of extracellular matrix protein including α, β, and γ chains that adjust cell viability and function. This protein produces Schwann cells and mediates the myelination in PNS (Chen and Strickland 2003). Moreover, laminin induces the β1-integrin receptor that is an important component of nerves' basal lamina and helps in Schwann cell migration and axonal elongation (Chang et al. 2020). A study showed that 2 mg/mL rate of laminin is an optimal concentration for PC-12 neurite outgrowth (Chang et al. 2020). Laminin acts as a promoter in a new blood vessel formation, so the presence of laminin in the region of nervous gap increases new vascularization and improves growth of axons in the injured area (Grant et al. 1991). So, laminin alone or in combination with another polymer can as a good guide facilitate nerve regeneration.

Gelatin is found to be a natural polymer, and it has good biocompatibility, but the poor mechanical properties have limited its use (Zhuang et al. 2016). A study showed that gelatin and graphene-based 3D scaffolds improve the stem cells' differentiation into Schwann-like cells in vitro (Uz et al. 2019).

Another natural polymer is HA that is found in the brain ECM, and it is identified by high water content as harmless degraded products. So, HA is a suitable biomaterial in NTE due to its distinct biological and physicochemical characteristics (Wang et al. 2012). Neurons and astrocytes are responsible for HA production (Song and Dityatev 2018). Also, HA was placed around myelinated fibers and helped in neuroplasticity and brain maturation (Giamanco and Matthews 2012). So, it is considered for the fabrication of NGCs (Liang et al. 2013). A hyaluronic acid conduit regenerated a 10-mm sciatic nerve gap through an axonal extension (Yang et al. 2023). A study showed that nerve conduit filled by chitosan-HA hydrogels along with a controlled release of nerve growth factor may induce axonal extension in sciatic nerve defect in rats (Xu et al. 2022). Moreover, HA injection into the epineurium region helps repair an injured sciatic nerve (Altinkaya et al. 2023).

A naturally occurring, biocompatible, hydrophilic polysaccharide is sodium alginate (SA) (Hou et al. 2018; Manzari-Tavakoli et al. 2020). This polymer is immune-compatible and non-toxic, emulating the extracellular matrix found in the human body (Hou et al. 2018; Wei et al. 2023b). Brown algae are the source of SA, which are biodegradable linear carbohydrate biopolymers (Hou et al. 2018). Alginate-based scaffolds have been demonstrated to enhance cell adhesion and proliferation, leading to the repair of various organs such as nerves (Hou et al. 2018). Furthermore, in order to manage nerve growth factors, sustain neuronal growth, and regulate synaptic formation, sulfated sodium alginate can trigger the human mesenchymal stem cell differentiation toward neural-like cells and resemble sulfated glycosaminoglycan (Wei et al. 2023b). Additionally, conductive scaffolds based on pyrrole monomers were constructed utilizing alginate-based scaffolds as an acceptable substrate through the usage of electrostatic interactions between the positively charged PPy16 and the carboxylate moieties of alginate (Manzari-Tavakoli et al. 2020). Scaffolds or hydrogels with electrical characteristics based on alginate-pyrrole have been developed in several research for nerve tissue engineering (Manzari-Tavakoli et al. 2020). Alginate has been shown in numerous trials to be a very successful scaffold for the transfer of drugs or cells. It is also a far better scaffold for functional rehabilitation following spinal cord damage (Jahandideh et al. 2022). For example, the findings of Liu et al. showed that following spinal cord damage, a 3D bioprinting scaffold in conjunction with oligodendrocytes and neural stem cells significantly enhanced hindlimb motor performance and encouraged nerve regeneration (Zhou et al. 2022). Furthermore, the result study conducted by Liu et al. demonstrated that valproic acid loaded on chitosan nanoparticles inside an alginate/chitosan hydrogel hybrid with or without human endometrial stem cells may repair injured nerve fibers and through histology stop spinal cord damaged-stimulated vacuolization gaps via regeneration (Jafarimanesh et al. 2023).

Biosaccharide chitosan, as well as its refined derivatives, are widely employed in tissue engineering (Wang et al. 2017). It has been demonstrated that an alginate-pyrrole scaffold with chitosan integrated offers an interactive surface for seeded cells to interact with an external electric field (Wang et al. 2017). In a research investigation, nerve tissue regeneration was studied using a chitosan-alginate hydrogel having a porous structure. The investigation on cell proliferation demonstrated that neural stem cells and olfactory ensheathing cells multiplied effectively on the chitosan-alginate hydrogel (Wang et al. 2017). These promising findings suggested that a chitosan-alginate hydrogel may be used as a better substitute for nerve tissue engineering (Wang et al. 2017).

Because of its hemostasis, biocompatibility, and three-dimensional appearance, fibrin—which is derived from the blood clotting proteins fibrinogen and thrombin—is a biopolymer with a variety of uses in the medical field. It can be employed as scaffolds in the regeneration of tissues. It has been demonstrated that fibrin, either alone or in conjunction with other biomaterials, can be helpful as a biological support to promote the regeneration of peripheral nerves and other injured tissues (Reis et al. 2022).

One of the main structural components of the cell wall of plants is cellulose. It can also be found in certain fungi, the extracellular portion of bacteria, and brown, red, and green algae (Krishani et al. 2023; Diekjürgen and Grainger 2017). Cellulose comprises D-glucose units joined by β-(1 → 4) glycosidic linkages (Krishani et al. 2023). The perfect substance for tissue development is cellulose. Plant and bacterial cellulose-based scaffolds were produced and employed in the regeneration of the nerve tissue (Krishani et al. 2023; Liu et al. 2020b). Furthermore, an exciting way to treat neurological damage and conditions requiring sustained assistance could be bacterial cellulose (BC), given its distinctive features (strong retained water content, excellent mechanical characteristics, biological compatibility, significant crystallinity, large area of surface, extremely pure, extremely fine network, and failure to absorb in the body of humans due to cellulase absence) (Jabbari et al. 2022). Though BC doesn't have any electrical functions, when combined with conductive building components, it can greatly increase the pace of nerve repair (Jabbari et al. 2022). It has been demonstrated that electrical stimulation works well to speed up and improve neuron repair (Jabbari et al. 2022). Hou et al. examined the possibility of using bacterial cellulose to treat peripheral nerve damage (Hou et al. 2018). This work demonstrates the possible use of bacterial cellulose as a strong contender for the healing of peripheral nerves (Hou et al. 2018).

Some larvae, including flies, spiders, silkworms, mites, and scorpions, create fibrin, which is protein silk. Commercial usage of silk from spiders like *Araneus diadematus* and *Nephila clavipes* as well as the silkworm *Bombyx mori* is common (Yonesi et al. 2021).

Although natural polymers commonly are present in the nervous system and play important roles in myelination, neuronal proliferation, and differentiation, they have unsuitable mechanical properties so they are better used in combination with other polymers in NTE (Li et al. 2022). The most important natural product-based scaffolds for nerve tissue regeneration are summarized in Fig. 8.9.

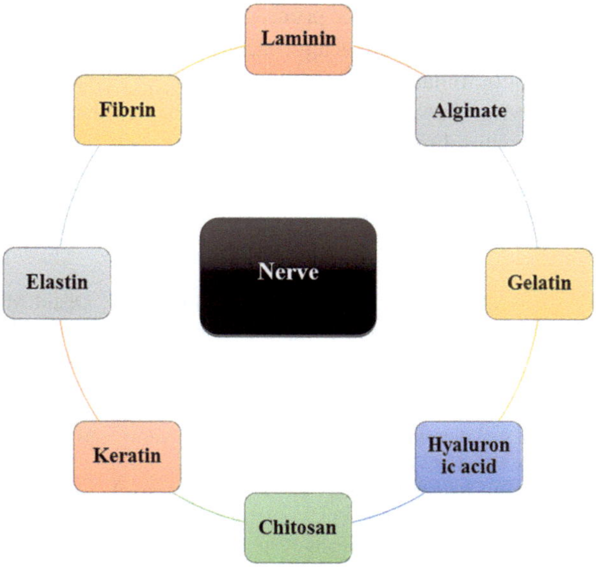

Fig. 8.9 Several scaffolds derived from natural materials that facilitate nerve regeneration

8.6 Concluding Remarks

In conclusion, the development and application of natural product-based scaffolds in tissue regeneration, particularly for vascular, bone, skin, and nerve tissue regeneration, represent a significant leap forward in the field of tissue engineering and regenerative medicine. These scaffolds, crafted from biodegradable, biocompatible, and tissue-compatible natural materials like chitin, chitosan, cellulose, starch, hyaluronic acid, collagen, and others, have shown an immense potential in facilitating the repair and regeneration of damaged tissues.

The unique properties of natural biopolymers, such as enhanced bioactivity, reduced toxicity, superior biocompatibility, and increased hydrophilicity, make them ideal candidates for creating scaffolds that effectively support tissue regeneration. The combination of different natural materials, leveraging their distinct advantages, is a strategic approach to enhance the functionality and effectiveness of these scaffolds. For instance, the synergistic properties of aloe vera and silk fibroin in skin tissue regeneration scaffolds have demonstrated promising results in promoting fibroblast proliferation and collagen secretion, crucial for the wound healing process.

Furthermore, the advancement in scaffold development is not only reliant on the choice of biomaterials but also significantly influenced by the evolution of fabrication technologies like electrospinning, 3D bioprinting, and microfluidic technology. These technologies have been instrumental in creating scaffolds that closely mimic the natural extracellular matrices, thereby providing an optimal environment for cell proliferation and migration essential for effective tissue regeneration.

The ongoing research and development in this field are driven by the goal to create scaffolds capable of completely regenerating skin and other tissues, thus addressing the unmet needs in wound care and offering new avenues for clinical therapies. The progress in this area holds a great promise for the future of tissue engineering, potentially revolutionizing the way we approach the treatment of injuries and degenerative diseases. The combination of innovative biomaterials and advanced fabrication techniques is paving the way for the creation of scaffolds that not only support but also enhance the natural regenerative processes of the human body.

References

Abdelbasset WK, Jasim SA, Sharma SK, Margiana R, Bokov DO, Obaid MA, Hussein BA, Lafta HA, Jasim SF, Mustafa YF (2022) Alginate-based hydrogels and tubes, as biological macromolecule-based platforms for peripheral nerve tissue engineering: a review. Ann Biomed Eng 50(6):628–653

Abebayehu D, Spence AJ, McClure MJ, Haque TT, Rivera KO, Ryan JJ (2019) Polymer scaffold architecture is a key determinant in mast cell inflammatory and angiogenic responses. J Biomed Mater Res A 107(4):884–892

Adair T, Montani J (2010) Chapter 1—Overview of angiogenesis. In: Angiogenesis. Morgan & Claypool Life Sciences, San Rafael (CA)

Altinkaya A, Cebi G, Çetinkale O (2023) Effects of subepineural hyaluronic acid injection on nerve recovery in a rat sciatic nerve defect model. Ulus Travma Acil Cerrahi Derg 29(3):277–283

Amrita, Arora A, Sharma P, Katti DS (2015) Pullulan-based composite scaffolds for bone tissue engineering: improved osteoconductivity by pore wall mineralization. Carbohydr Polym 123:180–189

Anandan D, Madhumathi G, Nambiraj NA, Jaiswal AK (2019) Gum based 3D composite scaffolds for bone tissue engineering applications. Carbohydr Polym 214:62–70

Anderson SM, Siegman SN, Segura T (2011) The effect of vascular endothelial growth factor (VEGF) presentation within fibrin matrices on endothelial cell branching. Biomaterials 32(30):7432–7443

Andrade CF, Wong AP, Waddell TK, Keshavjee S, Liu M (2007) Cell-based tissue engineering for lung regeneration. Am J Phys Lung Cell Mol Phys 292(2):L510–L518

Ansari M (2019) Bone tissue regeneration: biology, strategies and interface studies. Prog Biomater 8(4):223–237

Ansari M, Eshghanmalek M (2019) Biomaterials for repair and regeneration of the cartilage tissue. Bio-Des Manuf 2(1):41–49

Azimi B, Rasti A, Fusco A, Macchi T, Ricci C, Hosseinifard M, Guazzelli L, Donnarumma G, Bagherzadeh R, Latifi M, Roy I, Danti S, Lazzeri A (2023) Bacterial cellulose electrospun fiber mesh coated with chitin nanofibrils for eardrum repair. Tissue Eng A 30(7–8)

Bahramsoltani M, Slosarek I, De Spiegelaere W, Plendl J (2014) Angiogenesis and collagen type IV expression in different endothelial cell culture systems. Anat Histol Embryol 43(2):103–115

Banfi A, Holnthoner W, Martino MM, Ylä-Herttuala S (2018) Vascularization for regenerative medicine. Frontiers in bioengineering and biotechnology 6:175

Bardhan R, Mahata S, Mondal B (2011) Processing of natural resourced hydroxyapatite from eggshell waste by wet precipitation method. Adv Appl Ceram 110(2):80–86

Barua E, Deoghare AB, Deb P, Lala SD (2018) Naturally derived biomaterials for development of composite bone scaffold: a review. In: IOP conference series: materials science and engineering. IOP Publishing 377 p 012013

Bharatham BH, Bakar A, Zuki M, Perimal EK, Yusof LM, Hamid M (2014) Development and characterization of novel porous 3D alginate-cockle shell powder nanobiocomposite bone scaffold. Biomed Res Int 2014:146723

Bonferoni MC, Caramella C, Catenacci L, Conti B, Dorati R, Ferrari F, Genta I, Modena T, Perteghella S, Rossi S, Sandri G, Sorrenti M, Torre ML, Tripodo G (2021) Biomaterials for soft tissue repair and regeneration: a focus on Italian research in the field. Pharmaceutics 13(9):1341

Cai J, Yang J, Du Y, Fan L, Qiu Y, Li J, Kennedy JF (2006) Enzymatic preparation of chitosan from the waste Aspergillus niger mycelium of citric acid production plant. Carbohydr Polym 64(2):151–157

Campbell KT, Stilhano RS, Silva EA (2018) Enzymatically degradable alginate hydrogel systems to deliver endothelial progenitor cells for potential revasculature applications. Biomaterials 179:109–121

Carmeliet P, Jain RK (2011) Molecular mechanisms and clinical applications of angiogenesis. Nature 473(7347):298–307

Chan EC, Kuo S-M, Kong AM, Morrison WA, Dusting GJ, Mitchell GM, Lim SY, Liu G-S (2016) Three dimensional collagen scaffold promotes intrinsic vascularisation for tissue engineering applications. PLoS One 11(2):e0149799

Chang W, Shah MB, Zhou G, Walsh K, Rudraiah S, Kumbar SG, Yu X (2020) Polymeric nanofibrous nerve conduits coupled with laminin for peripheral nerve regeneration. Biomed Mater 15(3):035003

Chaudhari AA, Vig K, Baganizi DR, Sahu R, Dixit S, Dennis V, Singh SR, Pillai SR (2016) Future prospects for scaffolding methods and biomaterials in skin tissue engineering: a review. Int J Mol Sci 17(12):1974

Chen F-M, Liu X (2016) Advancing biomaterials of human origin for tissue engineering. Prog Polym Sci 53:86–168

Chen Z-L, Strickland S (2003) Laminin γ1 is critical for Schwann cell differentiation, axon myelination, and regeneration in the peripheral nerve. J Cell Biol 163(4):889–899

Chen P, Cescon M, Megighian A, Ronaldo P (2014) Collagen VI regulates peripheral nerve myelination and function. FASEB J 28(3):1145–1156

Chen P, Cescon M, Zuccolotto G, Nobbio L, Colombelli C, Filaferro M, Vitale G, Feltri ML, Bonaldo P (2015) Collagen VI regulates peripheral nerve regeneration by modulating macrophage recruitment and polarization. Acta Neuropathol 129:97–113

Chen M, Zhang Y, Zhang W, Li J (2020) Polyhedral oligomeric silsesquioxane-incorporated gelatin hydrogel promotes angiogenesis during vascularized bone regeneration. ACS Appl Mater Interfaces 12(20):22410–22425

Chiu LL, Radisic M (2010) Scaffolds with covalently immobilized VEGF and Angiopoietin-1 for vascularization of engineered tissues. Biomaterials 31(2):226–241

Chocholata P, Kulda V, Babuska V (2019) Fabrication of scaffolds for bone-tissue regeneration. Materials 12(4):568

Cummings J, Zhou Y, Lee G, Zhong K, Fonseca J, Cheng F (2023) Alzheimer's disease drug development pipeline: 2023. Alzheimers Dement (N Y) 9(2):e12385

Daamen WF, Veerkamp J, Van Hest J, Van Kuppevelt T (2007) Elastin as a biomaterial for tissue engineering. Biomaterials 28(30):4378–4398

Davidenko N, Schuster C, Bax D, Raynal N, Farndale R, Best S, Cameron R (2015) Control of crosslinking for tailoring collagen-based scaffolds stability and mechanics. Acta Biomater 25:131–142

Di Martino A, Sittinger M, Risbud MV (2005) Chitosan: a versatile biopolymer for orthopaedic tissue-engineering. Biomaterials 26(30):5983–5990

Diekjürgen D, Grainger DW (2017) Polysaccharide matrices used in 3D in vitro cell culture systems. Biomaterials 141:96–115

Doblado LR, Martínez-Ramos C, Pradas MM (2021) Biomaterials for neural tissue engineering. Front Nanotechnol 3:643507

Ehterami A, Salehi M, Farzamfar S, Samadian H, Vaez A, Sahrapeyma H, Ghorbani S (2020) A promising wound dressing based on alginate hydrogels containing vitamin D3 cross-linked by calcium carbonate/d-glucono-δ-lactone. Biomed Eng Lett 10:309–319

Faroni A, Mobasseri SA, Kingham PJ, Reid AJ (2015) Peripheral nerve regeneration: experimental strategies and future perspectives. Adv Drug Deliv Rev 82:160–167

Filippi M, Born G, Chaaban M, Scherberich A (2020) Natural polymeric scaffolds in bone regeneration. Front Bioeng Biotechnol 8:474

Fletes-Vargas G, Rodríguez-Preciado SY, Díaz-Zaragoza M, Rodríguez-Rodríguez R (2023) Natural hydrogels as wound dressing for skin wound-healing applications. In: Sheikh FA, Majeed S, Beigh MA (eds) Interaction of nanomaterials with living cells. Springer, Singapore, pp 439–469

Gaharwar AK, Singh I, Khademhosseini A (2020) Engineered biomaterials for in situ tissue regeneration. Nat Rev Mater 5(9):686–705

Ghasemi-Mobarakeh L, Prabhakaran MP, Morshed M, Nasr-Esfahani MH, Baharvand H, Kiani S, Al-Deyab SS, Ramakrishna S (2011) Application of conductive polymers, scaffolds and electrical stimulation for nerve tissue engineering. J Tissue Eng Regen Med 5(4):e17–e35

Giamanco K, Matthews R (2012) Deconstructing the perineuronal net: cellular contributions and molecular composition of the neuronal extracellular matrix. Neuroscience 218:367–384

Grant DS, Lelkes PI, Fukuda K, Kleinman HK (1991) Intracellular mechanisms involved in basement membrane induced blood vessel differentiation in vitro. In Vitro Cell Dev Biol Anim 27:327–336

Guo H, Callaway JB, Ting JP (2015) Inflammasomes: mechanism of action, role in disease, and therapeutics. Nat Med 21(7):677–687

Guo S, Ren Y, Chang R, He Y, Zhang D, Guan F, Yao M (2022) Injectable self-healing adhesive chitosan hydrogel with antioxidative, antibacterial, and hemostatic activities for rapid hemostasis and skin wound healing. ACS Appl Mater Interfaces 14(30):34455–34469

Haifei S, Xingang W, Shoucheng W, Zhengwei M, Chuangang Y, Chunmao H (2014) The effect of collagen–chitosan porous scaffold thickness on dermal regeneration in a one-stage grafting procedure. J Mech Behav Biomed Mater 29:114–125

Hao M, Ding C, Sun S, Peng X, Liu W (2022) Chitosan/sodium alginate/velvet antler blood peptides hydrogel promotes diabetic wound healing via regulating angiogenesis, inflammatory response and skin flora. J Inflamm Res 15:4921–4938

Härönen H, Zainul Z, Tu H, Naumenko N, Sormunen R, Miinalainen I, Shakirzyanova A, Oikarainen T, Abdullin A, Martin P (2017) Collagen XIII secures pre-and postsynaptic integrity of the neuromuscular synapse. Hum Mol Genet 26(11):2076–2090

Hou Y, Wang X, Yang J, Zhu R, Zhang Z, Li Y (2018) Development and biocompatibility evaluation of biodegradable bacterial cellulose as a novel peripheral nerve scaffold. J Biomed Mater Res A 106(5):1288–1298

Irawan V, Sung T-C, Higuchi A, Ikoma T (2018) Collagen scaffolds in cartilage tissue engineering and relevant approaches for future development. Tissue Eng Regen Med 15:673–697

Irvin MW, Zijlstra A, Wikswo JP, Pozzi A (2014) Techniques and assays for the study of angiogenesis. Exp Biol Med 239(11):1476–1488

Islam KN, Zuki A, Ali M, Hussein MZB, Noordin M, Loqman M, Wahid H, Hakim M, Hamid SBA (2012) Facile synthesis of calcium carbonate nanoparticles from cockle shells. J Nanomater 2012:2

Ito Y (2008) Covalently immobilized biosignal molecule materials for tissue engineering. Soft Matter 4(1):46–56

Jabbari F, Babaeipour V, Bakhtiari S (2022) Bacterial cellulose-based composites for nerve tissue engineering. Int J Biol Macromol 217:120–130

Jafarimanesh MA, Ai J, Shojaei S, Khonakdar HA, Darbemamieh G, Shirian S (2023) Sustained release of valproic acid loaded on chitosan nanoparticles within hybrid of alginate/chitosan hydrogel with/without stem cells in regeneration of spinal cord injury. Prog Biomater 12(2):75–86

Jahandideh A, Noori H, Rahimi B, Hamblin MR, Behroozi Z, Ramezani M, Ramezani F (2022) Alginate scaffolds improve functional recovery after spinal cord injury. Eur J Trauma Emerg Surg 48(3):1711–1721

Janmohammadi M, Nazemi Z, Salehi AOM, Seyfoori A, John JV, Nourbakhsh MS, Akbari M (2023) Cellulose-based composite scaffolds for bone tissue engineering and localized drug delivery. Bioact Mater 20:137–163

Jayakumar R, Chennazhi KP, Srinivasan S, Nair SV, Furuike T, Tamura H (2011) Chitin scaffolds in tissue engineering. Int J Mol Sci 12(3):1876–1887

Jeong J, Kim JH, Shim JH, Hwang NS, Heo CY (2019) Bioactive calcium phosphate materials and applications in bone regeneration. Biomater Res 23(1):1–11

Jin H, Pi J, Yang F, Wu C, Cheng X, Bai H, Huang D, Jiang J, Cai J, Chen ZW (2016) Ursolic acid-loaded chitosan nanoparticles induce potent anti-angiogenesis in tumor. Appl Microbiol Biotechnol 100:6643–6652

Kaczmarek B, Sionkowska A, Gołyńska M, Polkowska I, Szponder T, Nehrbass D, Osyczka A (2018) In vivo study on scaffolds based on chitosan, collagen, and hyaluronic acid with hydroxyapatite. Int J Biol Macromol 118:938–944

Kartikasari N, Yuliati A, Listiana I, Setijanto D, Suardita K, Ariani MD, Sosiawan A (2016) Characteristic of bovine hydroxyapatite-gelatin-chitosan scaffolds as biomaterial candidate for bone tissue engineering. In: 2016 IEEE EMBS conference on biomedical engineering and sciences (IECBES). IEEE, pp 623–626

Kazimierczak P, Benko A, Nocun M, Przekora A (2019) Novel chitosan/agarose/hydroxyapatite nanocomposite scaffold for bone tissue engineering applications: comprehensive evaluation of biocompatibility and osteoinductivity with the use of osteoblasts and mesenchymal stem cells. Int J Nanomedicine 14:6615–6630

Kellaway SC, Roberton V, Jones JN, Loczenski R, Phillips JB, White LJ (2023) Engineered neural tissue made using hydrogels derived from decellularised tissues for the regeneration of peripheral nerves. Acta Biomater 157:124–136

Khosalim IP, Zhang YY, Yiu CKY, Wong HM (2022) Synthesis of a graphene oxide/agarose/hydroxyapatite biomaterial with the evaluation of antibacterial activity and initial cell attachment. Sci Rep 12(1):1971

Kim U-J, Park J, Kim HJ, Wada M, Kaplan DL (2005) Three-dimensional aqueous-derived biomaterial scaffolds from silk fibroin. Biomaterials 26(15):2775–2785
Kim HD, Jang HL, Ahn H-Y, Lee HK, Park J, Lee E-S, Lee EA, Jeong Y-H, Kim D-G, Nam KT (2017) Biomimetic whitlockite inorganic nanoparticles-mediated in situ remodeling and rapid bone regeneration. Biomaterials 112:31–43
Kim Y, Zharkinbekov Z, Raziyeva K, Tabyldiyeva L, Berikova K, Zhumagul D, Temirkhanova K, Saparov A (2023) Chitosan-based biomaterials for tissue regeneration. Pharmaceutics 15(3):807
Krishani M, Shin WY, Suhaimi H, Sambudi NS (2023) Development of scaffolds from bio-based natural materials for tissue regeneration applications: a review. Gels 9(2):100
Lee KY, Mooney DJ (2012) Alginate: properties and biomedical applications. Prog Polym Sci 37(1):106–126
Lee DJ, Diachina S, Lee YT, Zhao L, Zou R, Tang N, Han H, Chen X, Ko C-C (2016) Decellularized bone matrix grafts for calvaria regeneration. J Tissue Eng 7:2041731416680306
Lee H-P, Chen P-C, Wang S-W, Fong Y-C, Tsai C-H, Tsai F-J, Chung J-G, Huang C-Y, Yang J-S, Hsu Y-M (2019a) Plumbagin suppresses endothelial progenitor cell-related angiogenesis in vitro and in vivo. J Funct Foods 52:537–544
Lee J, Jung H, Park N, Park S-H, Ju JH (2019b) Induced osteogenesis in plants decellularized scaffolds. Sci Rep 9(1):20194
Li J, Tian X, Hua T, Fu J, Koo M, Chan W, Poon T (2021a) Chitosan natural polymer material for improving antibacterial properties of textiles. ACS Appl Bio Mater 4(5):4014–4038
Li Y, Liu Y, Li R, Bai H, Zhu Z, Zhu L, Zhu C, Che Z, Liu H, Wang J (2021b) Collagen-based biomaterials for bone tissue engineering. Mater Des 210:110049
Li X, Zhang X, Hao M, Wang D, Jiang Z, Sun L, Gao Y, Jin Y, Lei P, Zhuo Y (2022) The application of collagen in the repair of peripheral nerve defect. Front Bioeng Biotechnol 10:973301
Liang Y, Walczak P, Bulte JW (2013) The survival of engrafted neural stem cells within hyaluronic acid hydrogels. Biomaterials 34(22):5521–5529
Liu X, Zheng C, Luo X, Wang X, Jiang H (2019) Recent advances of collagen-based biomaterials: multi-hierarchical structure, modification and biomedical applications. Mater Sci Eng C 99:1509–1522
Liu Y, Li P, Qiao C, Wu T, Sun X, Wen M, Zhang W (2020a) Chitosan hydrogel enhances the therapeutic efficacy of bone marrow–derived mesenchymal stem cells for myocardial infarction by alleviating vascular endothelial cell pyroptosis. J Cardiovasc Pharmacol 75(1):75
Liu W, Du H, Zhang M, Liu K, Liu H, Xie H, Zhang X, Si C (2020b) Bacterial cellulose-based composite scaffolds for biomedical applications: a review. ACS Sustain Chem Eng 8(20):7536–7562
Liu Y, Fan J, Lv M, She K, Sun J, Lu Q, Han C, Ding S, Zhao S, Wang G, Zhang Y, Zang G (2021) Photocrosslinking silver nanoparticles–aloe vera–silk fibroin composite hydrogel for treatment of full-thickness cutaneous wounds. Regen Biomater 8(6):rbab048
Lu P, Takai K, Weaver VM, Werb Z (2011) Extracellular matrix degradation and remodeling in development and disease. Cold Spring Harb Perspect Biol 3(12):a005058
Lv D, Zhou L, Zheng X, Hu Y (2017) Sustained release of collagen VI potentiates sciatic nerve regeneration by modulating macrophage phenotype. Eur J Neurosci 45(10):1258–1267
Madaghiele M, Calò E, Salvatore L, Bonfrate V, Pedone D, Frigione M, Sannino A (2016) Assessment of collagen crosslinking and denaturation for the design of regenerative scaffolds. J Biomed Mater Res A 104(1):186–194
Manzari-Tavakoli A, Tarasi R, Sedghi R, Moghimi A, Niknejad H (2020) Fabrication of nanochitosan incorporated polypyrrole/alginate conducting scaffold for neural tissue engineering. Sci Rep 10(1):22012
Marchand M, Monnot C, Muller L, Germain S (2019) Extracellular matrix scaffolding in angiogenesis and capillary homeostasis. In: Seminars in cell & developmental biology. Elsevier 89:147–156.
Marques-Marinho FD, Vianna-Soares CD (2013) Cellulose and its derivatives use in the pharmaceutical compounding practice. In: Cellulose-medical, pharmaceutical and electronic applications. IntechOpen, London

Martins AM, Alves CM, Kasper FK, Mikos AG, Reis RL (2010) Responsive and in situ-forming chitosan scaffolds for bone tissue engineering applications: an overview of the last decade. J Mater Chem 20(9):1638–1645

Mathew-Steiner SS, Roy S, Sen CK (2021) Collagen in wound healing. Bioengineering 8(5):63

Min S, Ko IK, Yoo JJ (2019) State-of-the-art strategies for the vascularization of three-dimensional engineered organs. Vasc Specialist Int 35(2):77

Minor AJ, Coulombe KL (2020) Engineering a collagen matrix for cell-instructive regenerative angiogenesis. J Biomed Mater Res B Appl Biomater 108(6):2407–2416

Mitra S, Ghosh N, Banerjee E (2018) Carboxymethyl guar gum nanoscaffold as matrix for cell growth in vitro. J Lung Pulm Respir Res 5(1):00156

Mohammadinejad R, Kumar A, Ranjbar-Mohammadi M, Ashrafizadeh M, Han SS, Khang G, Roveimiab Z (2020) Recent advances in natural gum-based biomaterials for tissue engineering and regenerative medicine: a review. Polymers 12(1):176

Mohammadzadeh Rad M, Saber-Samandari S, Sadighi M, Tayebi L, Mohammadi Aghdam M, Khandan A (2021) Macro-and micromechanical modelling of HA-elastin scaffold fabricated using freeze drying technique. J Nanoanal 8(1):17–31

Mohebbi S, Nezhad MN, Zarrintaj P, Jafari SH, Gholizadeh SS, Saeb MR, Mozafari M (2019) Chitosan in biomedical engineering: a critical review. Curr Stem Cell Res Ther 14(2):93–116

Mondal S, Pal U, Dey A (2016) Natural origin hydroxyapatite scaffold as potential bone tissue engineering substitute. Ceram Int 42(16):18338–18346

Mony MP, Shenoy SJ, Raj R, Geetha CS, Pratheesh KV, Nair RS, Purnima C, Anilkumar TV (2021) Gelatin-modified cholecyst-derived scaffold promotes angiogenesis and faster healing of diabetic wounds. ACS Appl Bio Mater 4(4):3320–3331

Morgan JP, Delnero PF, Zheng Y, Verbridge SS, Chen J, Craven M, Choi NW, Diaz-Santana A, Kermani P, Hempstead B (2013) Formation of microvascular networks in vitro. Nat Protoc 8(9):1820–1836

Nagappan PG, Chen H, Wang D-Y (2020) Neuroregeneration and plasticity: a review of the physiological mechanisms for achieving functional recovery postinjury. Mil Med Res 7(1):1–16

Neacsu IA, Serban AP, Nicoara AI, Trusca R, Ene VL, Iordache F (2020) Biomimetic composite scaffold based on naturally derived biomaterials. Polymers 12(5):1161

Ochubiojo EM, Rodrigues A (2012) Starch: from food to medicine. In: Scientific, health and social aspects of the food industry. IntechOpen, London, pp 355–380

Papadimitriou L, Manganas P, Ranella A, Stratakis E (2020) Biofabrication for neural tissue engineering applications. Mater Today Bio 6:100043

Patel NP, Lyon KA, Huang JH (2018) An update–tissue engineered nerve grafts for the repair of peripheral nerve injuries. Neural Regen Res 13(5):764

Phimnuan P (2022) Investigation ofblended fibroin/aloe gel extract film on the biomolecular mechanism (s) relating to fibroblast wound healing activity. Université Bourgogne Franche-Comté, Université de Naresuan

Pistone A, Iannazzo D, Celesti C, Scolaro C, Giofré SV, Romeo R, Visco A (2020) Chitosan/PAMAM/hydroxyapatite engineered drug release hydrogels with tunable rheological properties. Polymers 12(4):754

Pramanik S, Kharche S, More N, Ranglani D, Singh G, Kapusetti G (2023) Natural biopolymers for bone tissue engineering: a brief review. Eng Regen 4(2):193–204

Qin D, Zhang A, Wang N, Yao Y, Chen X, Liu Y (2022a) Hydroxybutyl chitosan/oxidized glucomannan self-healing hydrogels as BMSCs-derived exosomes carriers for advanced stretchable wounds. Appl Mater Today 26:101342

Qin J, Chen F, Wu P, Sun G (2022b) Recent advances in bioengineered scaffolds for cutaneous wound healing. Front Bioeng Biotechnol 10:841583

Rajesh R, Ravichandran YD, Kuo Y-C (2017) Alginate in bone tissue engineering. In: Seaweed polysaccharides. Elsevier pp 349–368

Ramanathan G, Singaravelu S, Raja M, Nagiah N, Padmapriya P, Ruban K, Kaveri K, Natarajan TS, Sivagnanam UT, Perumal PT (2016) Fabrication and characterization of a collagen coated electrospun poly (3-hydroxybutyric acid)–gelatin nanofibrous scaffold as a soft bio-mimetic material for skin tissue engineering applications. RSC Adv 6(10):7914–7922

Reis CHB, Buchaim DV, Ortiz AC, Fideles SOM, Dias JA, Miglino MA, Teixeira DB, Pereira ESBM, da Cunha MR, Buchaim RL (2022) Application of fibrin associated with photobiomodulation as a promising strategy to improve regeneration in tissue engineering: a systematic review. Polymers (Basel) 14(15):3150

Rodrigues P, Raz S (2022) The burden of reoperations and timeline of problems in 1,530 cases of mesh-related complications. Urol Int 106(3):235–242

Roslan MR, Nasir NM, Cheng E, Amin N (2016) Tissue engineering scaffold based on starch: a review. In: 2016 International conference on electrical, electronics, and optimization techniques (ICEEOT). IEEE, pp 1857–1860

Santos CP, Prado JP, Fernandes KR, Kido HW, Dorileo BP, Parisi JR, Silva JA, Cruz MA, Custódio MR, Rennó AC (2023) Different species of marine sponges diverge in osteogenic potential when therapeutically applied as natural scaffolds for bone regeneration in rats. J Funct Biomater 14(3):122

Schechner JS, Nath AK, Zheng L, Kluger MS, Hughes CC, Sierra-Honigmann MR, Lorber MI, Tellides G, Kashgarian M, Bothwell AL (2000) In vivo formation of complex microvessels lined by human endothelial cells in an immunodeficient mouse. Proc Natl Acad Sci U S A 97(16):9191–9196

Seddiqi H, Oliaei E, Honarkar H, Jin J, Geonzon LC, Bacabac RG, Klein-Nulend J (2021) Cellulose and its derivatives: towards biomedical applications. Cellulose 28(4):1893–1931

Seyedebrahimi R, Razavi S, Varshosaz J, Vatankhah E, Kazemi M (2021) Beneficial effects of biodelivery of brain-derived neurotrophic factor and gold nanoparticles from functionalized electrospun PLGA scaffold for nerve tissue engineering. J Clust Sci 32:631–642

Shi H, Han C, Mao Z, Ma L, Gao C (2008) Enhanced angiogenesis in porous collagen–chitosan scaffolds loaded with angiogenin. Tissue Eng A 14(11):1775–1785

Shi X, Zhou K, Huang F, Wang C (2017) Interaction of hydroxyapatite nanoparticles with endothelial cells: internalization and inhibition of angiogenesis in vitro through the PI3K/Akt pathway. Int J Nanomedicine 12:5781–5795

Shi D, Shen J, Zhang Z, Shi C, Chen M, Gu Y, Liu Y (2019) Preparation and properties of dopamine-modified alginate/chitosan–hydroxyapatite scaffolds with gradient structure for bone tissue engineering. J Biomed Mater Res A 107(8):1615–1627

Shokrani H, Shokrani A, Sajadi SM, Seidi F, Mashhadzadeh AH, Rabiee N, Saeb MR, Aminabhavi T, Webster TJ (2022) Cell-seeded biomaterial scaffolds: the urgent need for unanswered accelerated angiogenesis. Int J Nanomedicine 17:1035–1068

Singh RS, Saini GK, Kennedy JF (2008) Pullulan: microbial sources, production and applications. Carbohydr Polym 73(4):515–531

Song I, Dityatev A (2018) Crosstalk between glia, extracellular matrix and neurons. Brain Res Bull 136:101–108

Suamte L, Tirkey A, Barman J, Jayasekhar Babu P (2023) Various manufacturing methods and ideal properties of scaffolds for tissue engineering applications. Smart Mater Manuf 1:100011

Suganya S, Venugopal J, Ramakrishna S, Lakshmi BS, Dev VR (2014) Naturally derived biofunctional nanofibrous scaffold for skin tissue regeneration. Int J Biol Macromol 68:135–143

Sun L, Li S, Yang K, Wang J, Li Z, Dan N (2022a) Polycaprolactone strengthening keratin/bioactive glass composite scaffolds with double cross-linking networks for potential application in bone repair. J Leather Sci Eng 4(1):1–13

Sun J-H, Huang M, Fang Z, Li T-X, Wu T-T, Chen Y, Quan D-P, Xu Y-Y, Wang Y-M, Yang Y (2022b) Nerve bundle formation during the promotion of peripheral nerve regeneration: collagen VI-neural cell adhesion molecule 1 interaction. Neural Regen Res 17(5):1023

Tabasum S, Noreen A, Maqsood MF, Umar H, Akram N, Nazli Z-I-H, Chatha SAS, Zia KM (2018) A review on versatile applications of blends and composites of pullulan with natural and synthetic polymers. Int J Biol Macromol 120:603–632

Taghipour YD, Hokmabad VR, Bakhshayesh D, Rahmani A, Asadi N, Salehi R, Nasrabadi HT (2020) The application of hydrogels based on natural polymers for tissue engineering. Curr Med Chem 27(16):2658–2680

Thorpe AA, Creasey S, Sammon C, Le Maitre CL (2016) Hydroxyapatite nanoparticle injectable hydrogel scaffold to support osteogenic differentiation of human mesenchymal stem cells. Eur Cell Mater 32:1–23

Uppuluri V, Thukani Sathananthan S, Bhimavarapu SK, Elumalai L (2022) Polymeric hydrogel scaffolds: skin tissue engineering and regeneration. Adv Pharm Bull 12(3):437–448

Uz M, Donta M, Mededovic M, Sakaguchi DS, Mallapragada SK (2019) Development of gelatin and graphene-based nerve regeneration conduits using three-dimensional (3D) printing strategies for electrical transdifferentiation of mesenchymal stem cells. Ind Eng Chem Res 58(18):7421–7427

Varadarajan SG, Hunyara JL, Hamilton NR, Kolodkin AL, Huberman AD (2022) Central nervous system regeneration. Cell 185(1):77–94

Vasita R, Katti DS (2006) Nanofibers and their applications in tissue engineering. Int J Nanomedicine 1(1):15–30

Walsh P, Buchanan F, Dring M, Maggs C, Bell S, Walker G (2008) Low-pressure synthesis and characterisation of hydroxyapatite derived from mineralise red algae. Chem Eng J 137(1):173–179

Wang X, He J, Wang Y, Cui F-Z (2012) Hyaluronic acid-based scaffold for central neural tissue engineering. Interface Focus 2(3):278–291

Wang G, Wang X, Huang L (2017) Feasibility of chitosan-alginate (Chi-Alg) hydrogel used as scaffold for neural tissue engineering: a pilot study in vitro. Biotechnol Biotechnol Equip 31(4):766–773

Wang B, Lv X, Chen S, Li Z, Yao J, Peng X, Feng C, Xu Y, Wang H (2018) Use of heparinized bacterial cellulose based scaffold for improving angiogenesis in tissue regeneration. Carbohydr Polym 181:948–956

Wei S, Wang Y, Sun Y, Gong L, Dai X, Meng H, Xu W, Ma J, Hu Q, Ma X (2023a) Biodegradable silk fibroin scaffold doped with mineralized collagen induces bone regeneration in rat cranial defects. Int J Biol Macromol 235:123861

Wei Q, Zhou J, An Y, Li M, Zhang J, Yang S (2023b) Modification, 3D printing process and application of sodium alginate based hydrogels in soft tissue engineering: a review. Int J Biol Macromol 232:123450

Wong HK, Lam CRI, Wen F, Chong SKM, Tan NS, Jerry C, Pal M, Tan LP (2016) Novel method to improve vascularization of tissue engineered constructs with biodegradable fibers. Biofabrication 8(1):015004

Wong SK, Yee MMF, Chin K-Y, Ima-Nirwana S (2023) A review of the application of natural and synthetic scaffolds in bone regeneration. J Funct Biomater 14(5):286

Xing F, Zhou C, Hui D, Du C, Wu L, Wang L, Wang W, Pu X, Gu L, Liu L (2020) Hyaluronic acid as a bioactive component for bone tissue regeneration: fabrication, modification, properties, and biological functions. Nanotechnol Rev 9(1):1059–1079

Xu H, Yu Y, Zhang L, Zheng F, Yin Y, Gao Y, Li K, Xu J, Wen J, Chen H, Lei Z, Huang Z, Yu D, Jiang N, Xu P (2022) Sustainable release of nerve growth factor for peripheral nerve regeneration using nerve conduits laden with bioconjugated hyaluronic acid-chitosan hydrogel. Compos Part B 230:109509

Xu Y, Wu X, Zhang Y, Yu Y, Gan J, Tan Q (2023) Engineered artificial skins: current construction strategies and applications. Eng Regen 4(4):438–450

Yang Y, Liang Y, Chen J, Duan X, Guo B (2022) Mussel-inspired adhesive antioxidant antibacterial hemostatic composite hydrogel wound dressing via photo-polymerization for infected skin wound healing. Bioact Mater 8:341–354

Yang J, Hsu C-C, Cao T-T, Ye H, Chen J, Li Y-Q (2023) A hyaluronic acid granular hydrogel nerve guidance conduit promotes regeneration and functional recovery of injured sciatic nerves in rats. Neural Regen Res 18(3):657–663

Yao C, Markowicz M, Pallua N, Noah EM, Steffens G (2008) The effect of cross-linking of collagen matrices on their angiogenic capability. Biomaterials 29(1):66–74

Yao C-H, Yang B-Y, Li Y-CE (2022) Remodeling effects of the combination of GGT scaffolds, percutaneous electrical stimulation, and acupuncture on large bone defects in rats. Front Bioeng Biotechnol 10:832808

Yonesi M, Garcia-Nieto M, Guinea GV, Panetsos F, Pérez-Rigueiro J, González-Nieto D (2021) Silk fibroin: an ancient material for repairing the injured nervous system. Pharmaceutics 13(3):429

Zarrintaj P, Bakhshandeh B, Rezaeian I, Heshmatian B, Ganjali MR (2017) A novel electroactive agarose-aniline pentamer platform as a potential candidate for neural tissue engineering. Sci Rep 7(1):17187

Zarrintaj P, Manouchehri S, Ahmadi Z, Saeb MR, Urbanska AM, Kaplan DL, Mozafari M (2018) Agarose-based biomaterials for tissue engineering. Carbohydr Polym 187:66–84

Zhang Q, Lu H, Kawazoe N, Chen G (2014) Pore size effect of collagen scaffolds on cartilage regeneration. Acta Biomater 10(5):2005–2013

Zhang W, Chen L, Chen J, Wang L, Gui X, Ran J, Xu G, Zhao H, Zeng M, Ji J (2017a) Silk fibroin biomaterial shows safe and effective wound healing in animal models and a randomized controlled clinical trial. Adv Healthc Mater 6(10):1700121

Zhang K, Song L, Wang J, Yan S, Li G, Cui L, Yin J (2017b) Strategy for constructing vascularized adipose units in poly (l-glutamic acid) hydrogel porous scaffold through inducing in-situ formation of ASCs spheroids. Acta Biomater 51:246–257

Zhang S, Chen J, Yu Y, Dai K, Wang J, Liu C (2019) Accelerated bone regenerative efficiency by regulating sequential release of BMP-2 and VEGF and synergism with sulfated chitosan. ACS Biomater Sci Eng 5(4):1944–1955

Zhou J, Wu Y, Tang Z, Zou K, Chen J, Lei Z, Wan X, Liu Y, Zhang H, Wang Y, Blesch A, Lei T, Liu S (2022) Alginate hydrogel cross-linked by Ca^{2+} to promote spinal cord neural stem/progenitor cell differentiation and functional recovery after a spinal cord injury. Regen Biomater 9:rbac057

Zhu T, Cui Y, Zhang M, Zhao D, Liu G, Ding J (2020) Engineered three-dimensional scaffolds for enhanced bone regeneration in osteonecrosis. Bioact Mater 5(3):584–601

Zhuang H, Bu S, Hua L, Darabi MA, Cao X, Xing M (2016) Gelatin-methacrylamide gel loaded with microspheres to deliver GDNF in bilayer collagen conduit promoting sciatic nerve growth. Int J Nanomedicine 11:1383–1394

Zielińska A, Karczewski J, Eder P, Kolanowski T, Szalata M, Wielgus K, Szalata M, Kim D, Shin SR, Słomski R, Souto EB (2023) Scaffolds for drug delivery and tissue engineering: the role of genetics. J Control Release 359:207–223

Applications of Natural Product-Loaded Scaffold for Wound Healing

9

Navjot Kaur, Baljeet Singh, Rahul Kumar Sharma, Shailesh Sharma, and Simrandeep Singh

Abstract

The goal of the intricate physiological process of wound healing is to restore tissue integrity and functionality through a variety of cellular and molecular processes. Chronic and non-healing wounds continue to present major issues to healthcare systems globally, despite breakthroughs in wound care. Biocompatible matrices known as scaffolds can replicate the extracellular matrix found in tissues, creating an environment that is favorable for cell migration, proliferation, and tissue regeneration. The healing capacity of these scaffolds can be increased by adding natural bioactive chemicals to them. Numerous naturally occurring substances, including peptides, polyphenols, alkaloids, and polysaccharides, have been studied for their ability to heal wounds. These bioactive substances have a variety of biological properties that are essential for wound healing and regeneration, such as anti-inflammatory, antibacterial, angiogenic, and antioxidant properties.

Utilizing methods like electrospinning, freeze-drying, and 3D bioprinting, scaffolds containing naturally occurring bioactive compounds have been created with controlled release, guaranteeing long-lasting therapeutic benefits at the wound site. Furthermore discussed are the biocompatibility and safety of these scaffolds loaded with natural products, since patient safety is a vital factor in wound healing therapies. Comprehending the possible unfavorable consequences and biodegradability of these scaffolds is crucial for their effective transition from laboratory to clinical use.

N. Kaur (✉) · B. Singh · R. K. Sharma · S. Sharma
Amar Shaheed Baba Ajit Singh Jujhar Singh Memorial College of Pharmacy, Bela, Ropar, Punjab, India

S. Singh
UCRD, Chandigarh University, Gharuan, Mohali, Punjab, India

M. Kumar et al. (eds.), *Natural Product Inspired Scaffolds*,
https://doi.org/10.1007/978-981-97-3111-4_9

The scaffolds have the potential to improve overall wound healing outcomes, decrease inflammation, encourage tissue regeneration, and hasten wound closure. Optimizing the wound healing process can be achieved by combining the structural advantages of scaffolds with the therapeutic potential of natural substances. However, more investigation and clinical testing are needed to completely comprehend.

Keywords
Wound healing · Scaffolds · Growth factor · Electrospinning · 3D bioprinting

9.1 Introduction

Wound is a disruption of normal skin structure and its integrity (Pugliese et al. 2018). Wound healing signifies a living organism's natural reaction to chemical, physical, mechanical, or thermal damage, involving a coordinated interplay of cells, matrix components, and various biological mechanisms. This biologically self-regulating process encompasses vasoconstriction and coagulation, inflammation, cellular propagation, and wound remodeling, all working in harmony to expedite recovery and restore tissue integrity (Rodrigues et al. 2019). There are diverse phases which work together to facilitate the complete healing and restoration of a damaged tissue (Negut et al. 2020).

9.1.1 Homeostasis and Inflammation

The primary stages, hemostasis and inflammation, begin very soon after the emergence of a wound damage. To attract cells to the wound site, cytokines, growth factors, and reactive oxygen species are created (Sekhon and Sen Gupta 2018). Along with the release of the IL-1 and tumor necrosis factor (TNF) from the keratinocytes, the epidermal barrier is disrupted. These elements alert neighboring cells to blockade the injury. The clot that has formed causes hemostasis and provides a surface for inflammatory cells to invade. Platelet growth factors including PDGF, EGF, and TGF-α encourage the employment of inflammatory leukocytes to the injured area (Mumford et al. 2015). Following the removal of the contaminated bacteria by neutrophils, activated macrophages phagocytose them (Yeaman 2014). Monocytes undergo development into macrophages due to the interactions at the site of the lesion, leading to the inflammatory response (Ishihara et al. 2019).

9.1.2 Proliferative Phase

It starts around day 3 after the damage occurs and lasts for 2–4 weeks. Endothelial, fibroblast, and keratinocyte cells generate a new tissue during this stage, which is also marked by the development of a granulation tissue. At around four days following injury, granulation tissue begins to grow at the site of the wound, because the fibroblasts and macrophages supplying a continuous supply of VEGF and FGF necessary to promote angiogenesis as well as fibroplasias (Ishihara et al. 2019). After then, fibroblasts undergo phenotypic change to become myofibroblasts, which themselves have an impact on the edges of the ECM (Trinh et al. 2022) to create a constrictive force that will allow wound closure (Quaresma 2019). Re-epithelialization is the process by which epithelial cells conceal the surface of the wound after the progression of the granulation tissue (Piipponen et al. 2020). Keratinocytes, one of the primary cell types in the epidermis, are typically involved in this stage. Additionally, newly created blood vessels pass through the granulation tissue to permote blood stream and supply of substances important for wound healing.

9.1.3 Remodeling Phase

It is the final stage and often starts on day 21 and lasts for a year or longer following the injury. The fibroblasts undergo apoptosis and differentiate into myofibroblasts, and the creation of granulation tissue comes to an end. This stage mostly refers to the collagen refining and related ECM. Maturity of the scar tissue is indicated by the resorption of the vascular capillary connection (Turksen 2018). The growing scar exhibits balanced anabolic and catabolic phenomenon leading to scar maturation in a connective tissue under normal physiological circumstances (Negut et al. 2020). The skin wounds are categorized as either acute or chronic based on their pathophysiology and outcomes.

9.2 Types of Wound Healing

The wounds can be considered into two types on the basis of duration of wound healing.

9.2.1 Acute Wound Healing

In response to an acute wound, the initial step is hemostasis, aimed at stemming bleeding and minimizing blood loss. Subsequently, the inflammatory phase is triggered, initiating a complex immune response to clear potential pathogens and prepare the tissue for restoration. After that comes the proliferative phase, which includes neovascularization, re-epithelialization, and the development of the granulation tissue. After that, during the remodeling phase, immune cells are eliminated from the epidermis either by migration to the dermis or apoptosis, replacing the granulation tissue with a scar tissue. Immune cells that orchestrate the acute wound healing process include neutrophils, basophils, macrophages, mast cells, Langerhans cells, B cells, and T cells (Raziyeva et al. 2021).

9.2.2 Chronic Wound Healing

Chronic wound healing, seen in conditions like pressure ulcers and diabetic foot ulcers, involves a disrupted immune response leading to prolonged inflammation and delayed recovery. Diabetic ulcers, for instance, exhibit imbalanced expression of inflammatory cytokines with heightened pro-inflammatory mediators like TNF-α and reduced pro-healing factors such as IL-10 and TGF-α (see Table 9.1). This imbalance fosters an abundance of pro-inflammatory immune cells, particularly M1 macrophages and activated CD8+ T cells, which contribute to tissue damage. In chronic wound healing, certain myeloid cells persist while other immune cell types diminish. Non-healing wounds are characterized by the involvement of mast cells, T cells, and specific keratinocyte miRNAs that maintain a pro-inflammatory environment (Raziyeva et al. 2021).

9.3 Challenges in Wound Healing

Management of chronic wounds is still difficult, partly because of lingering inflammation that is frequently brought on by the formation of bacterial biofilms. Increased amounts of MMP and pro-inflammatory cytokines like TNF- and IL-6 are produced as a result of these biofilms stimulating neutrophils and pro-inflammatory macrophages. A damaging loop of continuing inflammation and biofilm development is thus perpetuated by the dysregulated immunological milieu that develops in chronic wounds. Due to a variety of microbial species, poor antimicrobial agent penetration, fast bacterial development of antibiotic resistance, and other variables, biofilms are extremely difficult to treat. To increase the duration of their effects and assure controlled release, these chemicals can be coupled with biomaterials. Additionally, biomaterials can be utilized on their own to shield the injured skin from physical harm. Numerous studies have shown that cellular therapy speeds up angiogenesis and re-epithelization, hence accelerating wound healing (Raziyeva et al. 2021).

9.4 Scaffolds in Wound Healing

Scaffolds are specialized materials or structures used in the arena of tissue engineering and regenerative medication to facilitate the repair and regeneration of damaged or injured tissues. These scaffolds provide support and a favorable environment for cells to proliferate, migrate, and differentiate, which is essential for wound healing

Table 9.1 The cellular aspects and growth factors involved in wound healing (Wilkinson and Hardman 2020)

Cells	Role in wound healing
Platelets	Creation of thrombosis-abundant stockpiles of inflammatory mediators (including cytokines), anti-inflammatory agents
Neutrophils	Phagocytosis, intracellular bacterial death, and first cells to reach the wound site
Macrophages	Removing waste and dead tissue, phagocytosing and eliminating invasive bacteria and rich inflammatory mediator pools (including cytokines), and promoting angiogenesis, collagen synthesis, and fibroblast proliferation
Fibroblasts	Produce a range of ECM substances, including collagen, fibronectin, hyaluronic acid, and other substances, and produce a granulation tissue
Growth factors	Function in wound healing
VEGF	Stimulates angiogenesis and endothelial cell development in the granulation tissue
FGFS	Proliferation of fibroblast as well as epithelial cells, matrix buildup, healing of wounds, and angiogenesis. Enhances the growth of the granulation tissue
KGFS	Migration and propagation of keratinocytes
PDGF	Fibroblasts and endothelial cells exhibit mitogenic properties. They act as chemoattractants for neutrophils and fibroblasts. They promote fibroblast propagation and influence collagen breakdown. It's worth noting that they are the sole growth agents currently permitted by the FDA
EGF	Differentiation, expansion, migration, and bond of keratinocytes. Development of the granulation tissuc
IL-1	Induces neutrophil chemotaxis and higher fibroblast activity
TGF-ß	Mitogenic in terms of fibroblasts. It chemoattracts macrophages to it. Subsidiary stimulus is provided for collagen metabolism and angiogenesis

and tissue regeneration. The extracellular matrix (ECM), which is the natural environment that surrounds cells in tissues and is crucial to tissue repair, is what they are intended to resemble (Negut et al. 2020). By a variety of means, they aid in the healing of wounds. Scaffolds for wound healing offer a structural foundation for preserving and supporting the damaged tissue. Their purpose is to facilitate the attachment and migration of various tissue-specific cells, such as fibroblasts, to the location of the wound (Raziyeva et al. 2021). They can be altered to release growth hormones and bioactive compounds that encourage cell division and tissue repair. The scaffold's structure, which is often porous to permit the passage of nutrients, oxygen, and waste products, creates a favorable milieu for cells. Because biocompatible scaffolds must not be poisonous or trigger an adverse immune response in the body, they are essential for the healing of wounds (Qin et al. 2022). Certain scaffolds are made to break down gradually as the tissue grows again, eliminating the need for removal techniques. They can be customized to meet the needs of the patient by being made to fit particular tissue types and wound sizes. These scaffolds are employed in numerous clinical settings, including surgical incisions, burn injuries, and chronic wounds. Certain scaffolds are made to break down gradually as the tissue grows again, eliminating the need for removal techniques. They can be customized to meet the needs of the patient by being made to fit particular tissue types and wound sizes. The features of wound dressings can be found in Table 9.2. These scaffolds are employed in numerous clinical settings, including surgical incisions, burn injuries, and chronic wounds.

Recent research has described that cell proliferation and differentiation are affected by the type and strength of intermolecular interactions present in scaffolds. It has been discovered that the stiffness of the matrix where cells are planted controls how cell lineage is specified. While stiffer matrices encourage myogenic lineage commitment, softer matrices promote neurogenic commitment. This is especially important when distributing stem cells through matrices since mesenchymal stem cells are affected by microenvironmental variables that can change their course.

Table 9.2 Features of wound dressings (Negut et al. 2020)

Physical	Chemical	Technological
Keep the wound bed's moisture level appropriate. Permeability to gas and water to absorb exudates and blood at the site of a wound mechanical safety capacity to adjust for body type and wound kind defending or combating infectious pathogens preserving the temperature quickly applied and withdrawn and to be used infrequently	Biocompatible appropriate deterioration rate, non-toxic, non-inflammatory, non-allergenic, antibacterial effect	Inexpensive method of preparation extremely reproducible, mechanical robustness, easily sterilizable, extended shelf life

9.4.1 Matrix Composition

It is essential for maintaining stem cells. The activity of stem cells is affected by integrin-mediated adhesion, primarily through β1 integrins. Different matrices, such as those containing laminin, can promote endothelial cells' ability to form tube-like structures. The right extracellular matrix (ECM) selection is crucial for controlling the behavior of particular cellular populations. The three scaffolding proteins that are most frequently employed are collagen, laminin, and fibrin.

9.4.2 Topographic Cues

By altering cell shape and biochemistry, the cues on material surfaces can affect stem cell activity. Particular integrin receptors and focal adhesions used by cells to adhere to nano-topographical characteristics influence cell fate and differentiation.

9.4.3 Hypoxia

Particularly in the case of MSCs produced from bone marrow, physiological hypoxia, defined as a range of 1–9% oxygen, can affect cell proliferation and differentiation. To preserve the pluripotency and quiescence of stem cells, a hypoxic environment is necessary.

9.4.4 Growth Factors

Delivery devices for growth factors and other biomolecules essential for wound healing include scaffolds that are depicted in Fig. 9.1. Growth factors play a critical function in coordinating cellular signals for tissue regeneration. Growth factors that are frequently employed include VEGF, SDF-1, and FGF. The secretion of growth factors from scaffolds can be maintained by variables like scaffold porosity and surface area to generate a biomimetic atmosphere and long-lasting therapeutic effects. Another strategy involves attaching growth agents to the scaffold chemically or electrostatically. This makes it possible for a more steady and regulated release.

9.4.5 Tissue Architecture

The physical characteristics of tissues and, in turn, cell behavior are influenced by tissue architecture. In order to avoid scarring and maintain flexibility during wound healing, natural tissue architecture must be recreated. Scaffold signals can guide matrix deposition and cell alignment. Furthermore, it is now well acknowledged that the very structure of the tissue can affect the behavior and ultimate fate of cells.

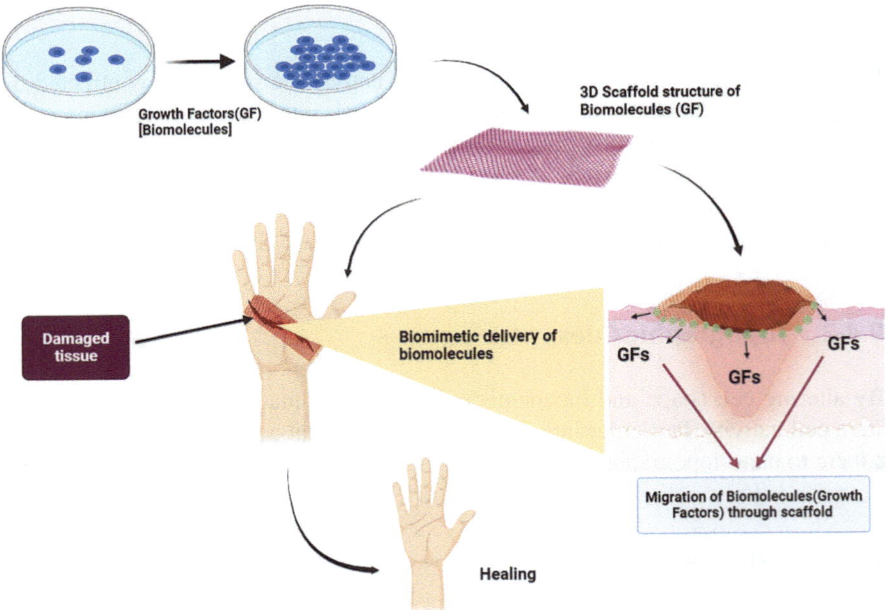

Fig. 9.1 The role of scaffolds in wound healing

For instance, the primary axis of strain is a principal alignment along which cells in aligned tissues tend to align and deposit matrix.

Numerous components of the microenvironment, such as scaffold material, topography, hypoxia, growth factors, and tissue architecture, play a critical role in guiding stem cell behavior and tissue regeneration. The efficiency of cell-based therapies can be improved by comprehending and modifying these variables (Qin et al. 2022).

9.5 Current Polymeric Materials in Wound Healing and Their Properties

Polymers are valuable in wound dressing due to their chemical adaptability. They serve as excellent drug delivery systems, offering various sizes and chemical combinations. In wound treatment, polymers can act as skin substitutes, creating 3D scaffolds that shield wounds from contamination and nourish tissue growth. These scaffolds should interact with cells, facilitating processes like adhesion, proliferation, and differentiation. They must also be porous and biocompatible. When designing wound healing scaffolds, it's crucial to consider these attributes while maintaining cost-effectiveness. An ideal scaffold should prevent infections, provide the right mechanical and physical properties, and support cell adhesion, proliferation, and differentiation for an effective wound healing environment. The main features about the physical and mechanical characteristics of wound healing scaffolds are as follows:

9.5.1 Fluid Absorption and Moist Environment

Excellent fluid retention/swelling, excellent absorbent qualities, and an adequate moisture vapor transmission rate are requirements for scaffolds (MVTR). It is possible to increase the water-uptake capacity of scaffolds with channeled pores. For example, cinnamon increased the PCL/gel nanofibrous scaffolds' ability to absorb water. Appropriate pore diameters are necessary for the delivery of nutrients and oxygen to the wound site (e.g., 17.25 and 72.80 μm) and porosity (60–90%) are crucial (Shyna et al. 2020).

9.5.2 Mechanical Properties

To accommodate various wound types, scaffolds should have mechanical qualities that correspond to the range of the skin's Young's modulus (10 kPa to 50 MPa). In order to fend off quick deterioration and preserve stability while in operation, mechanical strength is essential. To avoid contaminating the wound bed, scaffolds must remain intact. Soft elasticity has a crucial role in controlling cell migration and responses. Faster cell migration is supported by low fiber stiffness (Shu et al. 2019).

9.5.3 Healing Properties

Wound dressings can be combined with therapeutic agents, such as cells, antimicrobials, anti-inflammatory chemicals, and antibiotics, to aid in the healing process and fight infections. The bioactive properties of natural polymers, which are frequently utilized as scaffold materials, can promote fibroblast migration and quicken wound healing. Local therapeutic agent administration can speed up the healing of wounds, and scaffolds may help with sustained release (Ajmal et al. 2019; Shefa et al. 2019).

9.5.4 Anti-infection Properties

By encasing or attaching anti-infectious elements like medicines, nanoparticles, and antimicrobial agents, scaffolds can be created to prevent infections. Natural herbal extracts incorporated into scaffolds can exhibit strong pathogen-fighting properties. Controlling inflammatory responses and fostering tissue regeneration address excessive inflammation and decreased angiogenesis. Properly constructed scaffolds can boost the release of anti-inflammatory cytokines and decrease the secretion of pro-inflammatory cytokines, which makes them ideal for wound healing (Bhullar et al. 2019; Wang et al. 2019b).

The scaffolds have to guarantee the following: the delivery and retention of cells and various biochemical factors; appropriate cell attachment and migration, enabling cells to interact and connect; the movement of essential cell nutrients and secreted products; the application of mechanical and biological stimuli to modify cell behavior; and the three-dimensional simulation of an environment similar to the extracellular matrix (Li et al. 2018; Negut et al. 2020).

9.5.5 Natural and Synthetic Polymers

Natural polymers from plants, animals, and microbes outshine synthetic counterparts in human health applications (see Table 9.3). These organic materials can replicate the skin's extracellular matrix, offering inherent biological activity and often biodegradability. Their chemical structure flexibility meets various tissue engineering needs. Though controlling their degradation is tricky, by-products from enzymatic breakdown are well-tolerated. Synthetic polymers, while stable and pure, lack the biological benefits of natural polymers and may be toxic. For wound healing, synthetic polymers are used but don't match the natural ones in terms of biological compatibility and utility (Li et al. 2018).

9.5.5.1 Natural Polymers

Collagen Collagen, whether derived from animals or created using recombinant technology, is crucial in both hard and soft tissues, composing a significant part of the extracellular matrix (ECM). It's highly compatible with various tissues, though processing and sterilization can be challenging. With 28 known forms, collagen is present in the bone, teeth, skin, and more, produced by different cells depending on the body's location. Collagen scaffolds are biocompatible, biodegradable, cost-effective, and low-toxic. They support wound healing by maintaining moisture, promoting granulation tissue formation, and enhancing cytokine secretion, but overstimulating fibroblasts can be an issue (Thomas and Mercuri 2023).

Glycosaminoglycans (GAGs) GAGs are crucial for tying growth factors to collagen structures and utilized in burn therapy products such as Integra. These are genuine ECM elements that serve a variety of purposes, including cell adhesion, cell signaling, homeostasis maintenance, and inflammation management (Malliappan et al. 2022).

Fibrin Fibrin is a system for naturally mending wounds, forms during injury, and can act as a biological support structure for different tissues. Advantages: Excellent biocompatibility, low immune system impact, and high mechanical strength. By causing clots, promoting cell migration, and serving as a substrate for platelets, endothelial cells, fibroblasts, and macrophages, they have these effects. However, they are prone to inflammation and deterioration (Rojas-Murillo et al. 2022).

Table 9.3 Various polymers and their scaffolds

Polymer	Plant	Fabrication technique	Activity	References
Alginate, PEG	*Satureja cuneifolia*	3D printing	Antibacterial activity	Ilhan et al. (2020)
Carbopol gel, chitosan nanoparticles	*Pterocarpus marsupium*	Loaded in chitosan nanoparticles	Potent antibacterial activity	Manne et al. (2021)
Chitosan, gelatin-PVA	*Lithospermi radix*	Electrospinning	Bilayer wound dressing and potent antibacterial activity	Yao et al. (2017)
Chitosan scaffold	*Hemigraphis alternate*	Electrospinning	Hemostatic and antibacterial	Annapoorna et al. (2013)
Chitosan, gelatin	*Cinnamomum zeylanicum*	Electrospinning	Potent antibacterial activity	Ahmadi et al. (2021)
Chitosan, PEO	*Lawsonia inermis*	Electrospinning	Potential antibacterial activity	Hsouna et al. (2011)
Chitosan, polycaprolactone, and keratin	*Aloe vera*	Coaxial electrospinning	Wound healing	Zahedi et al. (2019)
Chitosan/gelatin bilayer scaffolds	*Lithospermi radix*	Electrospinning	Enhance cell adhesion and proliferation	Yao et al. (2017)
Chitosan/ polyethylene oxide	*Calendula officinalis*	Electrospinning	Antibacterial and collagen synthesis, re-epithelization, and remodeling of the tissue	Kharat et al. (2021)
Decellularized G-SIS (goat small intestine submucosa)	Curcumin	Embedding	Free radical scavenging, biocompatibility, and antibacterial properties	Singh et al. (2022)
Gelatin, PVA	*Centella asiatica*	Electrospinning	Potential antibacterial activity	Yao et al. (2017)
Gelatin-chitosan-based nanofibers	Curcumin and *Lithospermi radix* extract	Electrospinning	Anti-inflammatory and wound healing effects	Yang et al. (2019)

(continued)

Table 9.3 (continued)

Polymer	Plant	Fabrication technique	Activity	References
Gelatin methacryloyl, and alginate biopolymers	*Panax notoginseng* saponins and microalgae *Chlorella pyrenoidosa*	Facile microfluidic bioprinting	Protective effect against hypoxia-induced cell death	Wang et al. (2023)
Guar gum/PVA	*Acalypha indica*, *Aristolochia bracteolata*, *Lawsonia inermis* and *Thespesia populnea*	Electrospinning	Increased immune modulation of gingival MSCs	Kalachaveedu et al. (2020)
Methacrylated gelatin/poly-L-lactic acid radially oriented nanofiber mats	*Salvia miltiorrhiza Bunge-Radix Puerariae*	Electrospun hydrogel	Effectively reducing the inflammation, facilitating the vascularization, regeneration of hair follicles	Wu et al. (2022)
PCL	*Terminalia bellirica*	Electrospinning	Potent antibacterial activity	Ramamoorthy et al. (2020)
PCL, chitosan	*Melilotus officinalis*	Electrospinning	Potent antibacterial activity	Shahrousvand et al. (2021)
PCL, gelatin	*Althea officinalis*	Direct blending	Antioxidant and antimicrobial	Ghaseminezhad et al. (2020)
PCL, PEG	*Nasturtium officinale*	Electrospinning	Potential antioxidant activity	Shahani et al. (2017)
Poly-ε-caprolactone/gum Arabic	*Calendula officinalis*	Electrospinning	Antimicrobial, promoted cell attachment and proliferation	Pedram Rad et al. (2019)
Poly-ε-caprolactone/gum Arabic	*Calendula officinalis*	Electrospinning	Antimicrobial, promoted cell attachment and proliferation	Pedram Rad et al. (2019)
Poly-ε-caprolactone/polylactic acid	*Nigella sativa*	Double-nozzle electrospinning	Antibacterial activity	Sharifi et al. (2020)

(continued)

Table 9.3 (continued)

Polymer	Plant	Fabrication technique	Activity	References
Polyvinyl alcohol/ poly-ε-caprolactone nanofibrous mats	*Calendula officinalis* and coffee extracts	Electrospinning	Antioxidant activity and wound healing	Doostan et al. (2022)
Polycaprolactone	*Aloe vera*	Electrospinning	Rapid proliferation	Agnes Mary and Giri Dev (2015)
Poly-L-lactic acid nanofiber	*Aloe vera* gel	Electrospinning	Wound healing	Jouybar et al. (2017)
Polyvinyl alcohol	*Glycyrrhiza glabra*	Electrospinning	Antibacterial efficacy against *Staphylococcus aureus*	Shahid et al. (2022)
Poly-ε-caprolactone/ gelatin	*Gymnema sylvestre*	Coaxial electrospinning	Potent antimicrobial and antibiofilm properties	Ramalingam et al. (2019)
PVA	*Cordia myxa*	Electrospinning	Potent antioxidant activity	Seyedian et al. (2021)
TOCNF	*Aloe barbadensis* Miller	3D bioprinting	Bioinks' efficient printability	Forno-Bell et al. (2019)
ß-Tricalcium phosphate	*Cissus quadrangularis*	3D printing	Before extract loading, polydopamine was applied to the constructed scaffolds	Robertson and Bose (2020)

Fibronectin It is used to heal wounds and regenerate the brain tissue. It is derived from either human or cow plasma. Fibronectin (Fn), a key component of the extracellular matrix (ECM), has important roles in processes including wound healing and vascularization. The Fn structure may change in response to cell traction, which impacts cell signaling and ultimately causes cancer (Asadishekari et al. 2022).

Gelatin Gelatin is a collagen-based organic polymer which offers structural diversity and flexibility, two qualities necessary for cell attachment and proliferation. Blending gelatin with other polymers or composites is one way to potentially improve its mechanical, chemical, biological, and physical qualities as a bone graft. Gelatin, a denatured version of collagen, functions by encouraging cell adhesion and proliferation and has excellent biodegradability, immunogenicity, and economy. But they break down and regularly cross-link with other polymers at or above 37 °C (Singh and Dasgupta 2022).

Elastin The extracellular matrix (ECM) of elastic tissues is where the structural protein elastin, which gives tissues their elasticity, is found. It is used in tissues where recovery of shape and energy is essential. Dermal tissue engineering has made use of collagen and tropoelastin composite scaffolds (Posniak et al. 2023).

Chitosan Chitosan, a versatile polysaccharide, aids in skin repair and bone scaffold applications. It enhances cell adhesion, proliferation, and bone matrix mineralization. Its amino groups offer mucoadhesion, in situ gelation, transfection, and controlled drug release. Chemically modifiable, chitosan is non-toxic, non-antigenic, biodegradable, and biocompatible with oxygen permeability. It boosts fibroblast and inflammatory cell functions and supports antibacterial activity, rapid bone regeneration, wound granulation, and ECM development, but it has stability challenges (Bharathi et al. 2022).

Keratin Keratin is a protein with a structural role in the skin, nails, and hair and has a wide range of uses, such as tissue engineering and wound healing. Given its great biological activity, biocompatibility, biodegradability, desirable material mechanical qualities, and abundance in nature, keratin, one of the natural biomolecules, has a significant biological and biomedical application potential (Yan et al. 2022).

Alginate Alginates are naturally occurring polysaccharides that can be extracted from bacteria and brown sea algae. They have a healthy biodegradability and biocompatibility for a natural polysaccharide and are utilized as tissue engineering scaffolds and for wound treatments. There are qualities like biocompatibility, biodegradability, affordability, low toxicity, ability to gel, and resilience to acidic media. Alginates absorb liquids, keep wounds wet, encourage the production of granulation tissue, and compel monocytes to secrete more cytokines than usual. (Sharma et al. 2023).

Hyaluronic Acid The repeating unit of *N*-acetyl-glucosamine and D-glucuronic acid disaccharides makes up the naturally occurring polysaccharide known as hyaluronic acid (HA). It can be found in the extracellular matrix (ECM) of several tissues, including the skin, cartilage, neuron, and muscle. It is a naturally occurring glycosaminoglycan involved in inflammation, angiogenesis, and wound healing. It is non-allergenic, soluble in water, and a component of the ECM. It promotes the growth of keratinocytes, collagen deposition, and fibroblasts. They degrade quickly by enzymatic means in physiological media (Djoudi et al. 2022).

9.5.5.2 Non-biological Polymers (Qin et al. 2022)

Poly(hydroxyortho Esters) Bioengineering uses polymers such as polyglycolic acid (PGA) and polylactic acid (PLA) because they provide specific mechanical properties and breakdown times. Hydrophobicity, biodegradability, biocompatibility, and structural stability are all provided by polylactic acid (PLA). Compared to PLA, polyglycolic acid (PGA) is more hydrophilic, biocompatible, and biodegradable by hydrolysis. It also has a high tensile strength.

Polyethylene Glycol (PEG) PEG is a widely used polyether that resists protein adsorption and cell adhesion, minimizing the immune response following implantation. Polyethylene glycol (PEG) is biocompatible, non-biodegradable, bioinert, and hydrophilic.

Poly-β-Hydroxybutyrate (PHB) PHB is a biodegradable and biocompatible polyester used for wound scaffolding and tissue engineering.

Polyvinyl Alcohol (PVA) PVA is known for its biocompatibility and biodegradability and used in bone implants and as artificial skin. PVA is non-hazardous, hydrophilic, water-soluble, and pH-sensitive.

Polycaprolactone (PCL) PCL is a cost-effective and elastic polyester with good mechanical properties and slow degradation time. It is used for cell-based tissue engineering and skin regeneration. Polycaprolactone (PCL) is biocompatible, is biodegradable with a slower rate, and has good elastic properties.

9.5.5.3 Sample Hybrid Scaffolds

Chitosan, Hyaluronic Acid, and Collagen They were assembled into cytocompatible scaffolds that facilitated cell growth and infiltration. These scaffolds serve as ideal matrices for tissue engineering and transdermal drug delivery applications (Attasgah et al. 2022).

Collagen/Tussah Silk Fibroin (Col/TSF) Hybrid of collagen and tussah silk fibroin laden with bone mesenchymal stem cells show promise in skin repair and tissue engineering. They offer excellent skin compatibility and air and water penetrability (Cui et al. 2020).

Collagen-Silk Fibroin The hybridization of collagen (Col) and silk fibroin (SF) accelerated cutaneous healing, achieving rapid wound closure and re-epithelialization without any adverse effects (Naomi et al. 2020).

Nanofibrous Scaffold-Growth Factor A nanofibrous scaffold loaded with two growth factors for complete full-thickness wound healing (Aberoumandi et al. 2017).

Chitosan/PLGA Utilizing chemical cross-linking, chitosan/PLGA materials exhibit low cytotoxicity and have been found to significantly enhance wound healing in diabetic rat models. This application not only reduces inflammation but also promotes neovascularization, showing promise for wound healing (Viezzer et al. 2020).

CMCS/SA Employing EDC/NHS cross-linking, CMCS/SA materials possess antibacterial activity, making them effective in inhibiting bacterial growth and promoting wound healing, particularly in burn-infection models. Rapid epithelialization and higher collagen deposition are additional benefits (Mai et al. 2020).

Alginate/PVA Created through a solvent casting method, alginate/PVA materials stand out due to their good mechanical properties. These hydrogel membranes may efficiently promote wound healing by offering sustained release and promoting granulation tissue formation (Abbasi et al. 2020).

Dextran/HA/PEI Utilizing UV light cross-linking, dextran/HA/PEI materials are biocompatible, inhibit inflammation, and encourage microvascular activity. They present a promising solution for wound dressing in the cure of various wound types (Wang et al. 2019a).

PVA/Dextran-Aldehyde PVA/dextran-aldehyde materials, which are created via the freeze-thaw process, have qualities including high exudate absorption, appropriate transmission rates, and biocompatibility. These hydrogel scaffolds have proven to have the capacity to quicken the healing of wounds, particularly in models with full-thickness skin defects (Zheng et al. 2019).

PEG/Fibrin Through co-polymerization, PEG/fibrin materials provide adjustable properties. Their macroporous and mechanically reinforced fibrin-based sequential IPN hydrogels are valuable for dermal tissue regeneration, showcasing their potential in wound healing applications (Gsib et al. 2020).

9.6 Commonly Employed Manufacturing Techniques

Different fabrication methods offer unique advantages and disadvantages, making them suitable for different applications in herbal scaffold development as mentioned in Table 9.4. Researchers select the most appropriate method based on the desired scaffold properties and intended applications in order to prepare the commercial scaffolds as depicted in Table 9.5 (Agarwal et al. 2021; Yusuf Aliyu and Adeleke 2023).

9.7 Tactics to Engineer Wound Healing Scaffolds

Bioengineered scaffolds have emerged as a promising artificial skin alternative for wound healing in recent years. The material qualities and architecture of these scaffolds are key factors in their success, and they have improved in their ability to understand how wounds heal. To improve their capacity for regeneration, these scaffolds can be designed to incorporate immunomodulatory substances, bioactive compounds, and stem cells and have wide applications (Fig. 9.2).

Immuno-engineered Pro-regenerative Scaffolds Immunomodulatory biomaterials can be used to modify scaffolds in order to alter the immunological milieu and direct endogenous cells toward tissue regeneration and repair. Macrophages are essential immune cells in the process of wound healing. They can develop into phenotypes that promote growth (M2) or are pro-inflammatory (M1). Chemical and physical scaffold alterations can be made to encourage M2 macrophage polarization and improve wound healing (Tylek et al. 2020).

Incorporating Growth Factors into Scaffolds Growth factors are essential because they promote cell migration and proliferation. To establish a regenerative microenvironment, scaffolds can be engineered to encapsulate and release growth factors in a regulated fashion. This encourages angiogenesis, the creation of granulation tissue, and the deposition of collagen—processes essential to the healing of deep wounds. Different scaffold designs enable spatiotemporally controlled release, increasing their effectiveness (Xiao et al. 2021).

Encapsulating Stem Cells into Scaffolds Stem cells, when encapsulated within scaffolds, can directly contribute to new tissue formation or release bioactive molecules that promote wound healing. These stem cell-embedded scaffolds improve cell viability, differentiation, and their ability to enhance the repair and regeneration of tissues. Alternatively, stem cells can release bioactive molecules without directly contributing to tissue formation. While encapsulating stem cells into scaffolds offers potential benefits, challenges include maintaining their viability and controlling their behavior in vivo (Lu et al. 2021). Cell-free technologies like small extracellu-

Table 9.4 Fabrication techniques used in scaffold making

Fabrication technique	Methodology	Advantages	Disadvantages	Application
Electrospinning	Electrospinning is a versatile method for creating nanofibrous scaffolds. Electrostatic forces are used to eject a polymer solution by the use of a syringe pump, collector, and high-voltage power source, resulting in the production of fibers	Allows for the creation of nanofibrous scaffolds, various parameters for fiber diameter control	Complex scaffold structures and homogeneous pore distribution can be challenging to achieve	Herbal extracts have been incorporated into electrospun fibers through blending, coaxial, and emulsion-based techniques for various applications, including wound healing
Solvent casting	Solvent casting is a straightforward and cost-effective technique that exploits the evaporation of volatile solvents. A polymer is first dissolved in a volatile solvent. After the mixture is put into a mold, the solvent evaporates, leaving behind a solid material membrane that takes on the shape of the mold	Simplicity, minimal equipment requirements	Use of potentially toxic solvents, requiring thorough drying to remove solvents	Solvent casting has been used to develop nanocomposite films with herbal extracts and polymers for applications like wound healing

(continued)

Table 9.4 (continued)

Fabrication technique	Methodology	Advantages	Disadvantages	Application
Lyophilization/ freeze-drying	Low-temperature dehydration is known as lyophilization or freeze-drying. In order to produce dry scaffolds, it entails freezing a solution at extremely low temperatures, followed by primary and secondary drying steps	Maintains the stability of bioactive components, eliminates rinsing steps	Long processing time, risk of heterogeneous freezing	Lyophilization has been applied to create biopolymeric dressings loaded with herbal extracts for wound healing
Hydrogel formation	A 3D network of hydrophilic polymer chains cross-linked through different interactions is called a hydrogel. In aquatic conditions, they swell as a result of their high absorption and retention of water. Herbal extracts as well as synthetic medications can be encapsulated in hydrogels	Tunable biodegradability, mechanical properties, injectability, self-healing capabilities	Low mechanical properties and fragility	Hydrogels have been used for encapsulating herbal extracts from *Theobroma cacao* and *Dimocarpus longan* seeds for wound healing and other applications
3D printing/ additive manufacturing	Additive manufacturing (AM) is a method for producing 3D scaffolds by adding a material layer by layer. It offers precise and customizable scaffold fabrication	Precision, customization, automated and reproducible fabrication	Limited nozzle resolution, challenges with multi-scale hierarchical structures	3D printing techniques, such as extrusion 3D printing and direct-ink-writing, have been used to create herbal scaffolds for applications like diabetic wound healing

(continued)

Table 9.4 (continued)

Fabrication technique	Methodology	Advantages	Disadvantages	Application
Phase separation	Phase separation takes advantage of differences in polymer solubilities. It can be thermally induced or diffusion-induced and provides control over scaffold thickness and porosity	Control over scaffold properties and features	Limited to specific polymers and laboratory-scale applications	Phase separation has been employed for creating vascular grafts for treating peripheral vascular disease and other tissue engineering applications
Self-assembly	Self-assembly involves creating nano-dimensional patterned structures by organizing molecular building blocks with appropriate interactions, resulting in fibrous matrices	Versatile, no advanced instruments required	Typically produces mechanically weak fibrous matrices and can be costly	Self-assembly has been used to create fibrous matrices for various applications
Melt blowing	Melt blowing is a unique technique that produces micro- and nanofibers by extruding molten polymer through orifices with high-velocity hot air	No reliance on solvents or high-voltage devices, easily industrialized	Limited to specific polymers and applications	Applications like sutures and bone tissue engineering have made use of melt blowing
Templating system	This approach creates nanofibers with different diameters by synthesizing biomaterials, usually polymers, inside the holes of a template membrane	Control over fiber diameter and porosity	Limited to specific materials and template configurations	Structural scaffolds for skin tissue engineering and other uses have been made using templating methods

Table 9.5 Commercial scaffolds used in wound healing (Ho et al. 2017; Hajialyani et al. 2018)

Acellular products	
Promogram	This product consists of oxidized regenerated cellulose and collagen and is produced by Acelity L.P., Inc.
PuraPly	PuraPly is derived from porcine type I collagen and is manufactured by Organogenesis, Inc.
MatriDerm	Manufactured by MedSkin Solutions Dr. Suwelack AG, MatriDerm is made from bovine collagen fibrils with elastin
Tisseel	Tisseel is a product from Baxter International, Inc. and contains fibrin
Beriplast P	Produced by CSL Behring, Beriplast P is a freeze-dried combination of fibrinogen, factor XIII, and thrombin
EVICEL	EVICEL contains fibrin and is produced by Ethicon
Hyaff	Hyaff is composed of hyaluronic acid and is manufactured by ATGmed-AT Technologies GmbH
HYCOAT	Made by The Hymed Group, HYCOAT contains sodium hyaluronate
Suprathel	PolyMedics Innovation produces Suprathel, which is a membrane made from D,L-lactide-co-trimethylene carbonate and Epsilon caprolactone
Terudermis	Terudermis is composed of bovine dermal cross-linked atelocollagen and may or may not include silicone. It is produced by Olympus Terumo Biomaterials Corp.
Pelnac	Pelnac is derived from porcine tendon and consists of atelocollagen type I, with or without a silicone film. It is produced by Gunze Ltd., Medical Materials Centre
Biologically processed matrix	
PriMatrix	PriMatrix, an acellular fetal bovine dermis, is produced by Integra Life Science Corp.
MatriStem	ACell, Inc. manufactures MatriStem, an acellular porcine urinary bladder
SurgiMend	SurgiMend is an acellular fetal or neonatal bovine dermis manufactured by Integra Life Science Corp.
AlloDerm	AlloDerm is an acellular human dermis produced by LifeCell Corp.
EZ Derm	EZ Derm is an acellular silver-impregnated aldehyde cross-linked porcine dermis produced by Mölnlycke Healthcare, LLC.
AmnioExcel	AmnioExcel is made from dehydrated amnion-derived tissue and is produced by Derma Sciences, Inc.
Biovance	Biovance is composed of dehydrated amnion-derived tissue and is produced by Alliqua BioMedical, Inc.
Grafix	Grafix is cryopreserved amnion-derived tissue manufactured by Osiris Therapeutics, Inc.
EpiFix/EpiBurn	EpiFix/EpiBurn consists of dehydrated and sterilized human amnion/chorion tissue and is produced by MiMedx Group, Inc.
Cellular products	
Epidermal products	
Epicel	Epicel is composed of cultured autologous keratinocytes and is produced by Genzyme Tissue Repair Corporation
Keratinozyten Sheets (DIZG)	Keratinozyten Sheets are cultured autologous keratinocytes
ReCell	ReCell contains autologous epidermal cells in liquid suspension and is produced by Avita Medical

(continued)

Table 9.5 (continued)

MySkin	MySkin consists of cultured autologous keratinocytes on a membrane and is produced by CellTran Ltd.
LaserSkin/ Vivoderm	LaserSkin/Vivoderm is made from cultured autologous keratinocytes in laser-perforated hyaluronic acid and is produced by Fidia Advanced Biopolymers
Dermal/epidermal (composite) products	
TheraSkin	TheraSkin is a human allogeneic splitskin graft that includes keratinocytes and fibroblasts. It is produced by Soluble Systems LLC
Dermagraft	Dermagraft contains neonatal foreskin-derived fibroblasts seeded in polyglycolic acid or polyglactin 910 mesh and is produced by Organogenesis, Inc.
Apligraf	Apligraf is a bilayered product consisting of human neonatal epidermal keratinocytes and neonatal foreskin-derived fibroblasts seeded in a bovine collagen matrix. It is produced by Organogenesis, Inc.
StrataGraft	StrataGraft is a bilayered product consisting of human dermal fibroblasts and human keratinocyte-derived fully stratified epidermis. It is produced by Stratatech Corp.
Growth factor products	
Regranex	Regranex contains human recombinant platelet-derived growth factor and is produced by Ortho-McNeil Pharmaceutical
AutoloGel	AutoloGel is made from platelet-rich plasma and is produced by Cytomedix, Inc.

lar vesicles (sEVs) obtained from stem cells are being explored as an alternative to overcome some of these challenges. Additionally, the development of efficient and easily managed scaffold systems remains a goal for improving stem cell-based wound healing therapies (Shen et al. 2021; Qin et al. 2022).

9.7.1 Cellular and Molecular Mechanisms of Wound Healing

Throughout history, traditional remedies have included plants and herbs from all around the world. They consist of a diverse array of phytochemicals, such as flavonoids, phenolic acids, terpenes, alkaloids, and functional phytohormones, all of which are known to offer a variety of therapeutic advantages, either directly or indirectly. However, we would restrict our discussion to effects such as pro-angiogenic, antioxidant, antimicrobial, and stem cell differentiation that are specifically relevant to tissue engineering applications in the following subsections (Fernandes et al. 2023).

Antimicrobial, Antioxidant, and Anti-inflammatory Properties Herbal extracts contain phytochemicals that can combat oxidative stress and inflammation, both of which can impede the wound healing process. These extracts offer a safer alternative to pharmacological agents and nanomaterials, with fewer associated concerns. The extracts exhibit antioxidant effects by interacting with free radicals and anti-inflammatory effects by inhibiting cytokine production and the nitric oxide path-

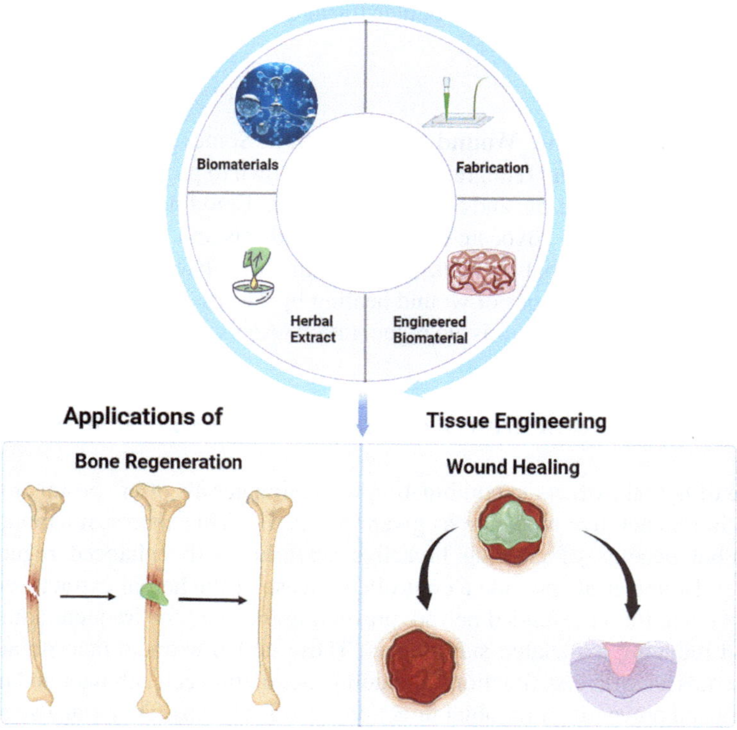

Fig. 9.2 Application of natural product-loaded scaffolds in wound healing

way. The antimicrobial actions of these extracts work by disrupting bacterial functions, including protein synthesis and DNA replication. Additionally, they can promote neovascularization, leading to the development of new blood vessels in the wound area.

Pro-angiogenic Properties The development of new blood vessels, or angiogenesis, is essential for tissue regeneration. Though they have their uses, pro-angiogenic growth factors such as VEGF are not without restrictions. An alternative that can be used to encourage angiogenesis is bioactive plant extracts. There have been reports of pro-angiogenic qualities in a variety of plant extracts.

Stem/Progenitor Cell Proliferation and Differentiation Promotion Properties Although stem cells like mesenchymal stem cells have an enormous potential for mending tissue, they nonetheless face difficulties in preserving their stemness and directing development. Herbal extracts provide a different strategy to deal with these issues. They prevent senescence and encourage the growth of stem cells. It has been discovered that a number of herbal extracts improve the abilities of stem cells to differentiate into particular lineages, such as osteogenesis, chondrogenesis, and neurogenesis.

Re-epithelialization and Wound Regeneration Some plant-based nanostructures, such as emodin and fenugreek, have been shown to promote collagen production, fibroblast propagation, and tissue regeneration. Emodin, when encapsulated in fibrous mats, acts as an advocate of collagen synthesis and fibroblast propagation. Overall, the use of herbal extracts and their nanosized forms shows promise in improving the various stages of wound healing by addressing inflammation, oxidative stress, angiogenesis, and tissue regeneration (Agarwal et al. 2021).

9.8 Tissue Engineering Applications of Herbal Scaffolds

The use of herbal extracts in combination with biomaterials offers promising applications in regenerative medicine as given in Fig. 9.2. This approach capitalizes on the double benefits of creating bioactive scaffolds with enhanced regenerative potential. Biomaterials provide a controlled release of the herbal extracts, ensuring their presence for an extended period, preventing the need for frequent administration and reducing associated side effects. These herbal extracts incorporated into biomaterials have diverse functions, including modulating cell behavior and immune response and preventing microbial infections (see Table 9.6). They can also alter the final structure's physicochemical properties, including its microstructure, tensile strength, wettability, swelling ratio, porosity, biodegradation, shape, and thermal properties (Agarwal et al. 2021).

9.9 Future Prospective

Prospects for scaffolds' use in tissue engineering and wound healing are bright, yet there are still many unanswered questions. The objective of scar-free skin regeneration is still unattainable. The development of effective scaffolds for wound therapy has been hampered by our limited understanding of the full process of cutaneous wound healing, despite the fact that several studies have been carried out to improve skin tissue engineering. The pursuit of flawless skin regeneration has yielded mixed results, especially when dealing with a variety of wound types and difficult-to-repair body sites including massive skin defects and deep traumas like third-degree burns. Neovascularization is necessary for the movement of vital components, but it can also cause excessive inflammation, which hinders the healing process. Clinical transplantation and the upkeep of sizable, functional tissue-engineered scaffolds provide further difficulties. Thus, there is a need for efficient scaffolds that are easy to handle and customize, with added antibacterial and anti-inflammatory properties (Qin et al. 2022). It is possible to develop customized wound scaffolds by utilizing the chemical and physical architectures of various components that are incorporated

Table 9.6 Natural product-loaded scaffolds

Plant	Family	Chemical component	Activity	References
A. absinthium	Asteraceae	Phenolic acids (caffeoylquinic, chlorogenic acid, dicaffeoylquinic) and flavonoids (apigenin, artemetin, luteolin, orientin)	Reducing local inflammation and facilitating tissue re-epithelization and granulation	Boudjelal et al. (2020)
Allium saralicum	Amaryllidaceae.	Linolenic acid, neofitadine, phytol	Remarkable improvement in antioxidant properties, non-cytotoxicity, antibacterial and antifungal activities, and cutaneous wound healing effects	Zangeneh et al. (2019)
Aloe vera	Asphodelaceae	Aloe emodin, aloesin, aloin, barbaloin, emodin	Enhancing natural healing and regenerative processes, promoting the production of new cells and collagen	Chelu et al. (2023)
Artemisia judaica	Compositae	Oxygenated monoterpenes (57.2%) in the plant essential oils	Elevated levels of pro-angiogenic TGF-β1 and anti-inflammatory IL-10	Mohammed et al. (2022)
Boerhavia diffusa	Nyctaginaceae	Caffeic acid, D-pinitol, ferulic acid	Improved survival and mobility of human keratinocyte cells	Juneja et al. (2019)
Bursera morelensis	Burseraceae	*m*-Menthane, *p*-menthane	Inducing fibroblast relocation to wound and enhancing collagen synthesis	Salas-Oropeza et al. (2020)
Carica papaya	Caricaceae	Anthraquinones, lycoside, β-carotene, caffeic acid, carpaine, chlorogenic acid, ferulic acid, dehydrocarpaine I and II, kaempferol, lycopene, myricetin, pseudocarpaine	Stimulated cell proliferation, migration, and collagen synthesis	Chabane et al. (2021)

(continued)

Table 9.6 (continued)

Plant	Family	Chemical component	Activity	References
Cinnamomum verum	Lauraceae	Cinnamaldehyde, cinnamyl acetate, eugenol, linalool	Modulating the expression of IGF-1, FGF-2, and VEGF, leading to enhanced cell proliferation, collagen synthesis, and re-epithelialization ratio	Chabane et al. (2020)
Curcuma longa	Zingiberaceae	Curcumin	Enhanced cell migration in normal fibroblasts	Maghimaa and Alharbi (2020)
Evolvulus alsinoides	Convolvulaceae	Anthocyanidin, betulin, glycerol, lupeol, viridiflorol	Effective antimicrobial action against *S. typhimurium*, *Micrococcus luteus*, and *Trichophyton rubrum* is crucial for wound healing	Ambika and Nair (2018)
Helichrysum italicum	Asteraceae	α-Curcumin and γ-curcumin, α-pinene, geranyl acetate, neryl acetate, nerol	Enhanced wound contraction rate and increased hydroxyproline content	Andjić et al. (2021)
Honey (*Apis mellifera*)	Apidae	Enzymes, flavonoids, organic acids, phenolic acid, vitamins	Exhibits antibacterial and antimicrobial effects, stimulates vascular endothelial growth factor production, and disinfects the wound site	Tashkandi (2021)
Ipomoea carnea	Convolvulaceae	Flavonoids, tannins, total phenols	Collagen synthesis, blood vessel constriction followed by dilation, fibroblast multiplication, and keratinocyte replication are key processes in wound healing	Ambika and Nair (2018)
Lavender essential oil	Lamiaceae	1,8-Cineole, β-ocimene, camphor, linalool, linalyl acetate, terpinen-4-ol	Elevated collagen expression and increased activity of proteins associated with tissue remodeling	Samuelson et al. (2020)

(continued)

Table 9.6 (continued)

Plant	Family	Chemical component	Activity	References
Lippia citrodora	Verbenaceae	Cineol, geranial, geraniol, limonene, neral	Demonstrated potent antifungal and antibacterial activity against *A. aureus* and *S. aureus*, respectively	Li et al. (2019)
Mentha piperita	Lamiaceae	α-Terpinene, β-caryophyllene, somenthone, menthone, trans-carveol	Enhanced fibroblast migration, collagen synthesis, and re-epithelization	Modarresi et al. (2019)
Mentha pulegium	Lamiaceae	Isomenthone, menthone, piperitenone, piperitone, pulegone	Expediting the proliferation phase and reducing inflammation during wound healing	Khezri et al. (2020)
Origanum vulgare	Lamiaceae	Carvacrol, linalool, *p*-cymene, thymol	Addressing inflammation and facilitating cell mobility in the wound healing process	Avola et al. (2020)
Rosmarinus officinalis	Lamiaceae	Betulinic, camphor, carnosic, caffeic, rosmarinic, and ursolic	Enhancement of angiogenesis, INO synthase expression, alpha 1 type 1 collagen, and antioxidant characteristics	Firdous and Sautya (2018)
Teucrium polium	Lamiaceae	α-Pinene, β-bourbonene, β-pinene, bicyclogermacrene, germacrene, limonene, myrcene, spathulenol, trans-β-guaiene	Positive impact on the healing of wounds due to antioxidant qualities	Chabane et al. (2021)

into scaffolds. For a synergistic wound healing effect, two or more scaffolds made of the same biomaterials can be combined. Scaffolds with several functions show potential; but, in order to keep costs down, over-engineering must be avoided. Although adding growth factors or stem cells to scaffolds increases the effectiveness of wound healing, doing so presents difficulties for clinical acceptance and regulatory approval (Ho et al. 2017). Scaffold structures are becoming more efficient thanks to innovative fabrication techniques. Asymmetric membranes, which closely resemble natural skin, are becoming more and more popular as remarkable wound dressings. Novel techniques, such as bioprinting, are investigated for the fabrication of these membranes and the integration of stem or skin cells obtained from patients. Acknowledging the role that the 3D microenvironment plays in supporting tissue regeneration is a major development in the research. The goal of biomimetic scaffolds is to replicate the natural extracellular matrix in cell niches. These scaffolds may be synthetic or biological. Collagen and hyaluronic acid are examples of matrix proteins, whereas organic molecular polymers are used to create biological scaffolds. In chronic wounds, decellularized dermal matrices are thought to hold promise for reducing inflammation. Biologically produced acellular matrices are widely used in commercial wound dressing solutions because they offer the required (Mousavi et al. 2020). The customizable and mass production benefits of synthetic scaffolds make them perfect for point-of-care tissue-engineered wound care devices. 3D bioprinting is a rapidly developing discipline that has the ability to create intricate skin structures and precisely position biological components and living cells. This method works especially well for building layered structures, such as the skin. Even while the scaffold attracts the interest of the business community, it is still in its infancy and needs more development before being made freely accessible (Ho et al. 2017).

To sum up, advancements in tissue engineering and wound healing will likely require the creation of scaffolds that are more effective, customized, and multipurpose. Working together, professionals from many fields such as chemists, engineers, scientists, and surgeons are essential to bringing these discoveries into the clinic and enhancing the results of skin regeneration and wound healing.

References

Abbasi AR, Sohail M, Minhas MU, Khaliq T, Kousar M, Khan S, Hussain Z, Munir A (2020) Bioinspired sodium alginate based thermosensitive hydrogel membranes for accelerated wound healing. Int J Biol Macromol 155:751–765

Aberoumandi SM, Mohammadhosseini M, Abasi E, Saghati S, Nikzamir N, Akbarzadeh A, Panahi Y, Davaran S (2017) An update on applications of nanostructured drug delivery systems in cancer therapy: a review. Artif Cells Nanomed Biotechnol 45(6):1–11

Agarwal T, Tan S, Onesto V, Law JX, Agrawal G, Pal S, Lim WL, Sharifi E, Moghaddam FD, Maiti TK (2021) Engineered herbal scaffolds for tissue repair and regeneration: recent trends and technologies. Biomed Eng Adv 2:100015

Agnes Mary S, Giri Dev VR (2015) Electrospun herbal nanofibrous wound dressings for skin tissue engineering. J Text Inst 106(8):886–895

Ahmadi S, Hivechi A, Bahrami SH, Milan PB, Ashraf SS (2021) Cinnamon extract loaded electrospun chitosan/gelatin membrane with antibacterial activity. Int J Biol Macromol 173:580–590

Ajmal G, Bonde GV, Mittal P, Khan G, Pandey VK, Bakade BV, Mishra B (2019) Biomimetic PCL-gelatin based nanofibers loaded with ciprofloxacin hydrochloride and quercetin: a potential antibacterial and anti-oxidant dressing material for accelerated healing of a full thickness wound. Int J Pharm 567:118480

Ambika AP, Nair SN (2018) Wound healing activity of plants from the Convolvulaceae family. Adv Wound Care 8(1):28–37

Andjić M, Božin B, Draginić N, Kočović A, Jeremić JN, Tomović M, MilojevićŠamanović A, Kladar N, Čapo I, Jakovljević V, Bradić JV (2021) Formulation and evaluation of Helichrysum italicum essential oil-based topical formulations for wound healing in diabetic rats. Pharmaceuticals 14(8):813

Annapoorna M, Sudheesh Kumar PT, Lakshman LR, Lakshmanan VK, Nair SV, Jayakumar R (2013) Biochemical properties of Hemigraphis alternata incorporated chitosan hydrogel scaffold. Carbohydr Polym 92(2):1561–1565

Asadishekari M, Mpoyi EN, Li Y, Eslami J, Walker M, Cantini M, Gourdon D (2022) Three-dimensional tunable fibronectin-collagen platforms for control of cell adhesion and matrix deposition. Front Phys 10:806554

Attasgah RB, Velasco-Rodríguez B, Pardo A, Fernández-Vega J, Arellano-Galindo L, Rosales-Rivera LC, Prieto G, Barbosa S, Soltero JFA, Mahmoudi M, Taboada P (2022) Development of functional hybrid scaffolds for wound healing applications. iScience 25(4):104019

Avola R, Granata G, Geraci C, Napoli E, Graziano ACE, Cardile V (2020) Oregano (Origanum vulgare L.) essential oil provides anti-inflammatory activity and facilitates wound healing in a human keratinocytes cell model. Food Chem Toxicol 144:111586

Bharathi R, Ganesh SS, Harini G, Vatsala K, Anushikaa R, Aravind S, Abinaya S, Selvamurugan N (2022) Chitosan-based scaffolds as drug delivery systems in bone tissue engineering. Int J Biol Macromol 222(Pt A):132–153

Bhullar SK, Ruzgar DG, Fortunato G, Aneja GK, Orhan M, Saber-Samandari S, Sadighi M, Ahadian S, Ramalingam M (2019) A facile method for controlled fabrication of hybrid silver nanoparticle-poly(-caprolactone) fibrous constructs with antimicrobial properties. J Nanosci Nanotechnol 19(11):6949–6955

Boudjelal A, Smeriglio A, Ginestra G, Denaro M, Trombetta D (2020) Phytochemical profile, safety assessment and wound healing activity of Artemisia absinthium L. Plan Theory 9(12):1744

Chabane S, Boudjelal A, Napoli E, Benkhaled A, Ruberto G (2020) Topical application of Cinnamon verum essential oil accelerates infected wound healing process by increasing tissue antioxidant capacity and keratin biosynthesis. Kaohsiung J Med Sci 33:143–151. https://doi.org/10.1080/10412905.2020.1842260

Chabane S, Boudjelal A, Napoli E, Benkhaled A, Ruberto G (2021) Phytochemical composition, antioxidant and wound healing activities of Teucrium polium subsp. capitatum (L.) Briq. essential oil. J Essent Oil Res 33:143–151. https://doi.org/10.1080/10412905.2020.1842260

Chelu M, Musuc AM, Popa M, Calderon Moreno J (2023) *Aloe vera*-based hydrogels for wound healing: properties and therapeutic effects. Gels 9(7):539. https://doi.org/10.3390/gels9070539

Cui B, Zhang C, Gan B, Liu W, Liang J, Fan Z, Wen Y, Yang Y, Peng X, Zhou Y (2020) Collagen-tussah silk fibroin hybrid scaffolds loaded with bone mesenchymal stem cells promote skin wound repair in rats. Mater Sci Eng C Mater Biol Appl 109:110611

Djoudi A, Molina-Peña R, Ferreira N, Ottonelli I, Tosi G, Garcion E, Boury F (2022) Hyaluronic acid scaffolds for loco-regional therapy in nervous system related disorders. Int J Mol Sci 23(20):12174

Doostan M, Doostan M, Maleki H, Faridi Majidi R, Bagheri F, Ghanbari H (2022) Co-electrospun poly (vinyl alcohol)/poly (ε-caprolactone) nanofiber scaffolds containing coffee and Calendula officinalis extracts for wound healing applications. J Bioact Compat Polym 37:437

Fernandes A, Rodrigues PM, Pintado M, Tavaria FK (2023) A systematic review of natural products for skin applications: targeting inflammation, wound healing, and photo-aging. Phytomedicine 115:154824

Firdous SM, Sautya D (2018) Medicinal plants with wound healing potential. Bangladesh J Pharmacol 13(1):41–52. https://doi.org/10.3329/bjp.v13i1.32646

Forno-Bell N, Bucarey SA, García D, Iragüen D, Chacón O, San Martín B (2019) Antimicrobial effects caused by Aloe barbadensis Miller on bacteria associated with mastitis in dairy cattle. Nat Prod Commun 14:1934578X1989667. https://doi.org/10.1177/1934578X19896670

Ghaseminezhad K, Zare M, Lashkarara S, Yousefzadeh M, Aghazadeh Mohandesi J (2020) Fabrication of Althea officinalis loaded electrospun nanofibrous scaffold for potential application of skin tissue engineering. J Appl Polym Sci 137:48587. https://doi.org/10.1002/APP.48587

Gsib O, Eggermont LJ, Egles C, Bencherif SA (2020) Engineering a macroporous fibrin-based sequential interpenetrating polymer network for dermal tissue engineering. Biomater Sci 8:7106–7116. https://doi.org/10.1039/D0BM01161D

Hajialyani M, Tewari D, Sobarzo-Sánchez E, Nabavi SM, Farzaei MH, Abdollahi M (2018) Natural product-based nanomedicines for wound healing purposes: therapeutic targets and drug delivery systems. Int J Nanomedicine 13:5023–5043. https://doi.org/10.2147/IJN.S174072

Ho J, Walsh C, Yue D, Dardik A, Cheema U (2017) Current advancements and strategies in tissue engineering for wound healing: a comprehensive review. Adv Wound Care 6(6):191–209

Hsouna AB, Trigui M, Culioli G, Blache Y, Jaoua S (2011) Antioxidant constituents from Lawsonia inermis leaves: isolation, structure elucidation and antioxidative capacity. Food Chem 125:193–200. https://doi.org/10.1016/J.FOODCHEM.2010.08.060

Ilhan E, Cesur S, Guler E, Topal F, Albayrak D, Guncu MM, Cam ME, Taskin T, Sasmazel HT, Aksu B, Oktar FN, Gunduz O (2020) Development of Satureja cuneifolia-loaded sodium alginate/polyethylene glycol scaffolds produced by 3D-printing technology as a diabetic wound dressing material. Int J Biol Macromol 161:1040–1054

Ishihara J, Ishihara A, Starke RD, Peghaire CR, Smith KE, McKinnon TAJ, Tabata Y, Sasaki K, White MJV, Fukunaga K, Laffan MA, Lutolf MP, Randi AM, Hubbell JA (2019) The heparin binding domain of von Willebrand factor binds to growth factors and promotes angiogenesis in wound healing. Blood 133(24):2559–2569

Jouybar A, Seyedjafari E, Ardeshirylajimi A, Zandi-Karimi A, Feizi N, Khani MM, Pousti I (2017) Enhanced skin regeneration by herbal extract-coated poly-L-lactic acid nanofibrous scaffold. Artif Organs 41(11):E296–E307

Juneja K, Mishra R, Chauhan S, Gupta S, Roy P, Sircar D (2019) Metabolite profiling and wound-healing activity of *Boerhavia diffusa* leaf extracts using *in vitro* and *in vivo* models. J Tradit Complement Med 10(1):52–59

Kalachaveedu M, Jenifer P, Pandian R, Arumugam G (2020) Fabrication and characterization of herbal drug enriched Guar galactomannan based nanofibrous mats seeded with GMSC's for wound healing applications. Int J Biol Macromol 148:737–749

Kharat Z, AmiriGoushki M, Sarvian N, Asad S, Dehghan MM, Kabiri M (2021) Chitosan/PEO nanofibers containing Calendula officinalis extract: preparation, characterization, in vitro and in vivo evaluation for wound healing applications. Int J Pharm 609:121132

Khezri K, Farahpour MR, Mounesi Rad S (2020) Efficacy of Mentha pulegium essential oil encapsulated into nanostructured lipid carriers as an in vitro antibacterial and infected wound healing agent. Colloids Surf A Physicochem Eng Asp 589:124414

Li H, Xue Y, Jia B, Bai Y, Zuo Y, Wang S, Zhao Y, Yang W, Tang H (2018) The preparation of hyaluronic acid grafted pullulan polymers and their use in the formation of novel biocompatible wound healing film. Carbohydr Polym 188:92–100

Li R, Chen Z, Ren N, Wang Y, Wang Y, Yu F (2019) Biosynthesis of silver oxide nanoparticles and their photocatalytic and antimicrobial activity evaluation for wound healing applications in nursing care. J Photochem Photobiol B 199:111593

Lu K, Li K, Zhang M, Fang Z, Wu P, Feng L et al (2021) Adipose-derived stem cells (ADSCs) and platelet-rich plasma (PRP) loaded gelatin/silk fibroin hydrogels for improving healing in a murine pressure ulcer model. Chem Eng J 424:130429

Maghimaa M, Alharbi SA (2020) Green synthesis of silver nanoparticles from Curcuma longa L. and coating on the cotton fabrics for antimicrobial applications and wound healing activity. J Photochem Photobiol B 204:111806. https://doi.org/10.1016/j.jphotobiol.2020.111806

Mai B, Jia M, Liu S, Sheng Z, Li M, Gao Y, Wang X, Liu Q, Wang P (2020) Smart hydrogel-based DVDMS/bFGF nanohybrids for antibacterial phototherapy with multiple damaging sites and accelerated wound healing. ACS Appl Mater Interfaces 12(9):10156–10169

Malliappan SP, Yetisgin AA, Sahin SB, Demir E, Cetinel S (2022) Bone tissue engineering: Anionic polysaccharides as promising scaffolds Carbohydrate Polymers 283:119142. https://doi.org/10.1016/J.CARBPOL.2022.119142

Manne AA, Arigela B, Giduturi AK, Komaravolu RK, Mangamuri U, Poda S (2021) Pterocarpus marsupium Roxburgh heartwood extract/chitosan nanoparticles loaded hydrogel as an innovative wound healing agent in the diabetic rat model. Materials Today Communications, 26. https://doi.org/10.1016/J.MTCOMM.2020.101916

Modarresi M, Farahpour MR, Baradaran B (2019) Topical application of Mentha piperita essential oil accelerates wound healing in infected mice model. Inflammopharmacology 27(3):531–537

Mohammed HA, Qureshi KA, Ali HM, Al-Omar MS, Khan O, Mohammed SAA (2022) Bio-evaluation of the wound healing activity of *Artemisia judaica* L. as part of the plant's use in traditional medicine; phytochemical, antioxidant, anti-inflammatory, and antibiofilm properties of the plant's essential oils. Antioxidants 11(2):332

Mousavi SM, Zarei M, Hashemi SA, Ramakrishna S, Chiang W, Lai CW, Gholami A, Omidifar N, Shokripour M (2020) Asymmetric membranes: a potential scaffold for wound healing applications. Symmetry 12(7):1100

Mumford AD, Frelinger AL III, Gachet C, Gresele P, Noris P, Harrison P, Mezzano D (2015) A review of platelet secretion assays for the diagnosis of inherited platelet secretion disorders. Thromb Haemost 114(1):14–25

Naomi R, Ratanavaraporn J, Fauzi MB (2020) Comprehensive review of hybrid collagen and silk fibroin for cutaneous wound healing. Materials 13(14):3097

Negut I, Dorcioman G, Grumezescu V (2020) Scaffolds for wound healing applications. Polymers 12(9):2010

Pedram Rad Z, Mokhtari J, Abbasi M (2019) Preparation and characterization of Calendula officinalis-loaded PCL/gum arabic nanocomposite scaffolds for wound healing applications. Iranian Polymer Journal (English Edition) 28(1):51–63. https://doi.org/10.1007/S13726-018-0674-X

Piipponen M, Li D, Landén NX (2020) The immune functions of keratinocytes in skin wound healing. Int J Mol Sci 21(22):8790. https://doi.org/10.3390/ijms21228790

Posniak S, Chung JHY, Liu X, Mukherjee P, Wallace GG (2023) The importance of elastin and its role in auricular cartilage tissue engineering. Bioprinting 32:e00276. https://doi.org/10.1016/J.BPRINT.2023.E00276

Pugliese E, Coentro JQ, Raghunath M, Zeugolis DI (2018) Wound healing and scar wars. Adv Drug Deliv Rev 129:1–3

Qin J, Chen F, Wu P, Sun G (2022) Recent advances in bioengineered scaffolds for cutaneous wound healing. Front Bioeng Biotechnol 10:841583

Quaresma JAS (2019) Organization of the skin immune system and compartmentalized immune responses in infectious diseases. Clin Microbiol Rev 32(4):e00034-18

Ramalingam R, Dhand C, Leung CM, Ong ST, Annamalai SK, Kamruddin M, Verma NK, Ramakrishna S, Lakshminarayanan R, Arunachalam KD (2019) Antimicrobial properties and biocompatibility of electrospun poly-ε-caprolactone fibrous mats containing Gymnema sylvestre leaf extract. Mater Sci Eng C Mater Biol Appl 98:503–514

Ramamoorthy R, Andiappan M, Muthalagu M (2020) Preparation and characterization of Terminalia bellerica loaded PCL nanofibrous mats for biomedical applications. Mater Today Proc 45:7247–7252

Raziyeva K, Kim Y, Zharkinbekov Z, Kassymbek K, Jimi S, Saparov A (2021) Immunology of acute and chronic wound healing. Biomol Ther 11(5):700

Robertson SF, Bose S (2020) Enhanced osteogenesis of 3D printed β-TCP scaffolds with Cissus quadrangularis extract-loaded polydopamine coatings. J Mech Behav Biomed Mater 111:103945

Rodrigues M, Kosaric N, Bonham CA, Gurtner GC (2019) Wound healing: a cellular perspective. Physiol Rev 99(1):665–706. https://doi.org/10.1152/physrev.00067.2017

Rojas-Murillo JA, Simental-Mendía MA, Moncada-Saucedo NK, Delgado-Gonzalez P, Islas JF, Roacho-Pérez JA, Garza-Treviño EN (2022) Physical, mechanical, and biological properties of fibrin scaffolds for cartilage repair. Int J Mol Sci 23(17):9879

Salas-Oropeza J, Jimenez-Estrada M, Perez-Torres A, Castell-Rodriguez AE, Becerril-Millan R, Rodriguez-Monroy MA, Canales-Martinez MM (2020) Wound healing activity of the essential oil of *Bursera morelensis*, in mice. Molecules 25(8):1795

Samuelson R, Lobl M, Higgins S, Clarey D, Wysong A (2020) The effects of lavender essential oil on wound healing: a review of the current evidence. J Altern Complement Med 26:680–690. https://doi.org/10.1089/ACM.2019.0286

Sekhon UDS, Sen Gupta A (2018) Platelets and platelet-inspired biomaterials technologies in wound healing applications. ACS Biomater Sci Eng 4(4):1176–1192

Seyedian R, Isavi F, Najafiasl M, Zaeri S (2021) Electrospun fibers loaded with Cordia myxa L. fruit extract: fabrication, characterization, biocompatibility and efficacy in wound healing. J Drug Deliv Sci Technol 63:102528

Shahani S, Behzadfar F, Jahani D, Ghasemi M, Shaki F (2017) Antioxidant and anti-inflammatory effects of Nasturtium officinale involved in attenuation of gentamicin-induced nephrotoxicity. Toxicol Mech methods 27(2):107–114. https://doi.org/10.1080/15376516.2016.1258748

Shahid MA, Khan MS, Hasan MM (2022) Licorice extract-infused electrospun nanofiber scaffold for wound healing. OpenNano 8:100075

Shahrousvand M, Haddadi-Asl V, Shahrousvand M (2021) Step-by-step design of poly (ε-caprolactone)/chitosan/Melilotus officinalis extract electrospun nanofibers for wound dressing applications. Int J Biol Macromol 180:36–50

Sharifi M, Bahrami H, Nejad N, Milan P (2020) Electrospun PCL and PLA hybrid nanofibrous scaffolds containing Nigella sativa herbal extract for effective wound healing. J Appl Polym Sci 137:49528

Sharma R, Malviya R, Singh S, Prajapati B (2023) A critical review on classified excipient sodium-alginate-based hydrogels: modification, characterization, and application in soft tissue engineering. Gels 9(5):430

Shefa AA, Taz M, Hossain M, Kim YS, Lee SY, Lee BT (2019) Investigation of efficiency of a novel, zinc oxide loaded TEMPO-oxidized cellulose nanofiber based hemostat for topical bleeding. Int J Biol Macromol 126:786–795

Shen Y, Xu G, Huang H, Wang K, Wang H, Lang M, Gao H, Zhao S (2021) Sequential release of small extracellular vesicles from bilayered thiolated alginate/polyethylene glycol diacrylate hydrogels for scarless wound healing. ACS Nano 15(4):6352–6368

Shu M, Long S, Huang Y, Li D, Li H, Li X (2019) High strength and antibacterial polyelectrolyte complex CS/HS hydrogel films for wound healing. Soft Matter 15:7686–7694. https://doi.org/10.1039/C9SM01380F

Shyna S, Shanti Krishna A, Nair PD, Thomas LV (2020) A nonadherent chitosan-polyvinyl alcohol absorbent wound dressing prepared via controlled freeze-dry technology. Int J Biol Macromol 150:129–140

Singh YP, Dasgupta S (2022) Gelatin-based electrospun and lyophilized scaffolds with nano scale feature for bone tissue engineering application: review. J Biomater Sci Polym Ed 33(13):1704–1758

Singh H, Purohit SD, Bhaskar R, Yadav I, Bhushan S, Gupta MK, Mishra NC (2022) Curcumin in decellularized goat small intestine submucosa for wound healing and skin tissue engineering. J Biomed Mater Res B Appl Biomater 110(1):210–219

Tashkandi H (2021) Honey in wound healing: an updated review. Open Life Sci 16(1):1091–1100

Thomas V, Mercuri J (2023) In vitro and in vivo efficacy of naturally derived scaffolds for cartilage repair and regeneration. Acta Biomater 171:1–18. https://doi.org/10.1016/j.actbio.2023.09.008

Trinh XT, Long NV, Van Anh LT, Nga PT, Giang NN, Chien PN, Nam SY, Heo CY (2022) A comprehensive review of natural compounds for wound healing: targeting bioactivity perspective. Int J Mol Sci 23(17):9573. https://doi.org/10.3390/ijms23179573

Turksen K (2018) Wound healing: stem cells repair and restorations, basic and clinical aspects. Wiley, Hoboken, NJ

Tylek T, Blum C, Hrynevich A, Schlegelmilch K, Schilling T, Dalton PD, Groll J (2020) Precisely defined fiber scaffolds with 40 μm porosity induce elongation driven M2-like polarization of human macrophages. Biofabrication 12(2):025007

Viezzer C, Mazzuca R, Machado DC, de Camargo Forte MM, Gómez Ribelles JL (2020) A new waterborne chitosan-based polyurethane hydrogel as a vehicle to transplant bone marrow mesenchymal cells improved wound healing of ulcers in a diabetic rat model. Carbohydr Polym 231:115734

Wang J, Cheng Y, Chen L, Zhu T, Ye K, Jia C, Wang H, Zhu M, Fan C, Mo X (2019a) In vitro and in vivo studies of electroactive reduced graphene oxide-modified nanofiber scaffolds for peripheral nerve regeneration. Acta Biomater 84:98–113

Wang P, Huang S, Hu Z, Yang W, Lan Y, Zhu J, Hancharou A, Guo R, Tang B (2019b) In situ formed anti-inflammatory hydrogel loading plasmid DNA encoding VEGF for burn wound healing. Acta Biomater 100:191–201

Wang X, Jia J, Niu M, Li W, Zhao Y (2023) Living Chinese herbal scaffolds from microfluidic bioprinting for wound healing. Research (Wash D C) 6:0138

Wilkinson HN, Hardman MJ (2020) Wound healing: cellular mechanisms and pathological outcomes. Open Biol 10(9):200223

Wu S, Zhao W, Sun M, He P, Lv H, Wang Q, Zhang S, Wu Q, Ling P, Chen S, Ma J (2022) Novel bi-layered dressing patches constructed with radially-oriented nanofibrous pattern and herbal compound-loaded hydrogel for accelerated diabetic wound healing. Appl Mater Today 28:101542

Xiao Z, Zheng X, An Y, Wang K, Zhang J, He H et al (2021) Zwitterionic hydrogel for sustained release of growth factors to enhance wound healing. Biomater Sci 9:882–891. https://doi.org/10.1039/D0BM01608J

Yan RR, Gong JS, Su C, Liu YL, Qian JY, Xu ZH, Shi JS (2022) Preparation and applications of keratin biomaterials from natural keratin wastes. Appl Microbiol Biotechnol 106(7):2349–2366

Yang BY, Hu CH, Huang WC, Ho CY, Yao CH, Huang CH (2019) Effects of bilayer nanofibrous scaffolds containing curcumin/Lithospermi radix extract on wound healing in streptozotocin-induced diabetic rats. Polymers 11(11):1745

Yao CH, Yeh JY, Chen YS, Li MH, Huang CH (2017) Wound-healing effect of electrospun gelatin nanofibres containing Centella asiatica extract in a rat model. J Tissue Eng Regen Med 11(3):905–915. https://doi.org/10.1002/term.1992

Yeaman MR (2014) Platelets: at the nexus of antimicrobial defence. Nat Rev Microbiol 12(6):426–437

Yusuf Aliyu A, Adeleke OA (2023) Nanofibrous scaffolds for diabetic wound healing. Pharmaceutics 15(3):986

Zahedi E, Esmaeili A, Eslahi N, Shokrgozar MA, Simchi A (2019) Fabrication and characterization of core-shell electrospun fibrous mats containing medicinal herbs for wound healing and skin tissue engineering. Mar Drugs 17(1):27

Zangeneh A, Zangeneh MM, Moradi R (2019) Ethnomedicinal plant-extract-assisted green synthesis of iron nanoparticles using Allium saralicum extract, and their antioxidant, cytotoxicity, antibacterial, antifungal and cutaneous wound-healing activities. Appl Organomet Chem 34(1):e5247

Zheng C, Liu C, Chen H, Wang N, Liu X, Sun G, Qiao W (2019) Effective wound dressing based on poly(vinyl alcohol)/dextran-aldehyde composite hydrogel. Int J Biol Macromol 132:1098–1105. https://doi.org/10.1016/j.ijbiomac.2019.04.038

Enhancing Breast Reconstruction with Bio-Scaffolds- A Supportive Approach for Optimal Outcomes

10

Sunil Sable, Mitesh Ikar, and Aaditi Pendse

Abstract

Breast reconstruction is vital for women who have undergone mastectomy or have breast deformities. Autologous fat grafting offers a natural and minimally invasive approach, but fat survival remains a challenge. This research explores bio-scaffolds in breast reconstruction to optimize outcomes. Bio-scaffolds promote tissue regeneration, creating a supportive environment for transplanted fat. They stimulate angiogenesis, enhance tissue growth, and provide mechanical support, improving fat graft survival and aesthetic results. Types of bio-scaffolds, mechanisms, clinical applications, and factors influencing selection are discussed. Evidence-based medicine is emphasized, along with clinical outcomes and potential complications. Future directions in bio-scaffold research are explored, advancing breast reconstruction. This research provides plastic surgeons with a comprehensive understanding of bio-scaffolds' role in breast reconstruction, enabling informed decision-making for optimal outcomes.

Keywords

Angiogenesis · Autologous fat grafting · Bio-scaffolds · Breast reconstruction · Tissue regeneration

S. Sable · M. Ikar (✉)
Department of Chemical Engineering, Vishwakarma Institute of Technology, Pune, Maharashtra, India
e-mail: mitesh.ikar21@vit.edu

A. Pendse
Department of Ayurveda, KLE, Belagavi, Karnataka, India

M. Kumar et al. (eds.), *Natural Product Inspired Scaffolds*,
https://doi.org/10.1007/978-981-97-3111-4_10

10.1 Introduction

Breast reconstruction is a transformative surgical procedure aimed at restoring the form and function of the breast for women who have undergone mastectomy or experienced breast deformities. In recent years, significant advancements in regenerative medicine and tissue engineering have revolutionized the field of breast reconstruction, offering innovative and promising techniques that provide more natural and minimally invasive approaches. One such groundbreaking technique is tissue-engineered breast reconstruction with Brava-assisted fat grafting, as pioneered by Khouri et al. in 2015. This approach combines the use of the Brava system, an external tissue expander, with autologous fat grafting to optimize fat graft survival and enhance breast reconstruction outcomes. Brava creates a favorable environment by external expansion, promoting the growth of new blood vessels and improving the survival of transplanted fat cells. Another regenerative alternative that has gained attention is percutaneous aponeurotomy and lipofilling (PAL). Introduced by Khouri et al. in 2013, PAL involves the release of the fibrous tissue using a needle, followed by lipofilling to regenerate a breast tissue. This technique offers a less invasive option for breast reconstruction compared to traditional flap reconstruction techniques, potentially resulting in reduced recovery time and minimized scarring (Khouri et al. 2013).

Table 10.1 shows various techniques which are performed during breast reconstruction. However, a critical consideration in fat grafting procedures is the fate of adipocytes after non-vascularized fat grafting. Eto et al. (2012) conducted a study on this subject, providing evidence of early adipocyte death and subsequent replacement, highlighting the need for further research to optimize fat graft retention and long-term outcomes. Advancements in the combined use of enhanced stromal vascular fraction and platelet-rich plasma (PRP) have shown promising results in improving fat graft maintenance in breast reconstruction (Gentile et al. 2012). Translational studies have indicated that the inclusion of these adjuncts enhances fat graft survival and tissue integration, potentially contributing to better overall reconstruction outcomes. Furthermore, Garza et al. (2021) explored postmastectomy breast reconstruction with autologous fat grafts combined with fibroblast-seeded biological scaffolds. This comparative study demonstrated the potential of scaffolds in enhancing fat graft retention and tissue regeneration, opening up exciting possibilities for further enhancing the success of breast reconstruction procedures. Injectable dermal fillers have also played a significant role in cosmetic and reconstructive procedures, and understanding their characteristics and applications can aid in optimizing treatment plans for individual patient needs (Sánchez-Carpintero et al. 2010). As the field of breast reconstruction continues to evolve, preclinical and clinical studies play a pivotal role in guiding future advancements (Shih et al. 2020).

Table 10.1 Techniques used for breast reconstruction

Technique	Description	Advantages	Disadvantages
Tissue-engineered breast reconstruction with Brava-assisted fat grafting	This approach combines the use of the Brava system, an external tissue expander, with autologous fat grafting to optimize fat graft survival and enhance breast reconstruction outcomes. Brava creates a favorable environment by external expansion, promoting the growth of new blood vessels and improving the survival of transplanted fat cells	More natural and minimally invasive approach than traditional flap reconstruction techniques	Requires multiple procedures
Percutaneous aponeurotomy and lipofilling (PAL)	This technique involves the release of fibrous tissue using a needle, followed by lipofilling to regenerate the breast tissue. This technique offers a less invasive option for breast reconstruction compared to traditional flap reconstruction techniques, potentially resulting in reduced recovery time and minimized scarring	Less invasive than traditional flap reconstruction techniques	May not be suitable for all patients
Enhanced stromal vascular fraction (SVF) and platelet-rich plasma (PRP)	These adjuncts have shown promising results in improving fat graft maintenance in breast reconstruction. They enhance fat graft survival and tissue integration, potentially contributing to better overall reconstruction outcomes	Can improve fat graft retention and long-term outcomes	Requires additional procedures
Scaffold-assisted fat grafting	This technique uses biological scaffolds to enhance fat graft retention and tissue regeneration. Scaffolds provide a supportive environment for transplanted fat cells and promote the growth of new blood vessels	Can improve fat graft retention and long-term outcomes	Requires additional procedures
Injectable dermal fillers	These fillers have played a significant role in cosmetic and reconstructive procedures. They can be used to augment the breast or to correct contour deformities	Can be used to achieve natural-looking results	May require multiple injections

10.2 Background on Breast Reconstruction

Breast reconstruction is a transformative surgical procedure aimed at restoring the form, symmetry, and psychological well-being of women who have undergone mastectomy or experienced breast deformities. Over the years, advancements in surgical techniques, materials, and patient care have significantly improved the options available for breast reconstruction, offering women more choices and personalized approaches to address their unique needs and preferences. National comparative audits and research studies play a crucial role in assessing the current state of breast reconstruction practices and identifying trends and areas for improvement. The study conducted by Ranganath et al. (2023) examined surgery for early breast cancer, including audit data on surgery for screen-detected cancer. Such comprehensive audits help evaluate the quality and outcomes of breast cancer surgery, providing valuable insights into the effectiveness of breast reconstruction procedures (Ranganath et al. 2023).

Table 10.2 shows recent advances and background of breast reconstruction. In the United States, there has been a paradigm shift in breast reconstruction, with an increasing trend toward using breast implants. The study by Albornoz et al. (2013) highlights this shift, which is attributed to advances in implant technology, improved surgical techniques, and increased patient preference for more immediate reconstruction options. One of the challenges in implant-based breast reconstruction is the potential for implant-associated breast deformities, which can impact the aesthetic outcome and patient satisfaction. Acellular dermal matrix (ADM) has emerged as a valuable tool in addressing these challenges (Albornoz et al. 2013). The research by Spear et al. (2011) explores the use of ADM to provide additional support and coverage for implants, reducing the risk of deformities and improving the overall reconstruction results. While implant-based breast reconstruction remains a popular choice, it is essential to recognize the risk factors and potential complications

Table 10.2 Background of breast reconstruction

Topic	Key points
Breast reconstruction	A transformative surgical procedure aimed at restoring the form, symmetry, and psychological well-being of women who have undergone mastectomy or experienced breast deformities
Advancements in breast reconstruction	National comparative audits and research studies have significantly improved the options available for breast reconstruction, offering women more choices and personalized approaches to address their unique needs and preferences
Implant-based breast reconstruction	A popular choice, but it is essential to recognize the risk factors and potential complications associated with this approach
Autologous fat grafting	A natural and minimally invasive approach that has gained popularity in recent years
Fat grafting and breast cancer recurrence	There is no evidence that lipofilling of the breast increases the risk of breast cancer recurrence
Diffusion and perfusion dynamics of fat grafting	Understanding these factors is crucial for optimizing outcomes in breast reconstruction
Impact of breast-conserving therapy and microcalcifications in breast reconstruction	These factors can have implications for patient outcomes and reconstruction possibilities

associated with this approach (Spear et al. 2011). The study by Alhindi et al. (2023) investigates these aspects, helping surgeons identify patients who may be at higher risk and tailor their surgical plans accordingly, leading to better outcomes and reduced complications.

In recent years, autologous fat grafting has gained popularity as a reconstructive option, offering a natural and minimally invasive approach. Studies evaluating the applicability and safety of autologous fat grafting, such as the one conducted by Claro et al. (2012), have shown promising results, providing evidence of its feasibility and effectiveness in breast reconstruction. One concern regarding fat grafting in breast reconstruction has been the fear of increasing the risk of cancer recurrence (Claro et al. 2012). However, reassuring evidence from studies like the one by Kronowitz et al. (2016) indicates that lipofilling of the breast does not increase the risk of breast cancer recurrence, easing concerns for patients and surgeons alike. Understanding the diffusion and perfusion dynamics of fat grafting is crucial for optimizing outcomes in breast reconstruction (Kronowitz et al. 2016). The study sheds light on the key factors influencing fat graft survival, which can guide surgeons in achieving better results and developing improved techniques (Khouri et al. 2014). Apart from surgical aspects, studies exploring the impact of breast-conserving therapy and the role of microcalcifications in breast reconstruction, as discussed by Azam et al. (2021), offer insights into the broader implications of breast surgery on patient outcomes and reconstruction possibilities.

10.3 Understanding Bio-Scaffolds

Tissue engineering is a transformative field that aims to develop functional tissue substitutes to replace or regenerate damaged or lost tissues. This interdisciplinary approach involves the use of biological principles, engineering techniques, and biomaterials to create scaffolds that mimic the extracellular matrix (ECM) and provide a supportive environment for cell attachment, growth, and differentiation. In their research, they introduced the concept of tissue engineering as a novel approach to address the limitations of traditional tissue transplantation and organ transplantation. They proposed the use of biodegradable polymer scaffolds seeded with cells, which would eventually lead to the formation of a functional tissue (Langer and Vacanti 1993). Atala (2007) expanded on this concept by discussing the engineering of tissues, organs, and cells. The review emphasizes the importance of selecting appropriate biomaterials for scaffold fabrication and highlights various strategies to engineer tissues with the desired structure and function (Atala 2007).

Additive manufacturing, commonly known as 3D printing, has revolutionized tissue engineering by enabling precise and patient-specific scaffold fabrication. Melchels et al. (2012) provide an in-depth review of additive manufacturing techniques for tissues and organs. They discuss various biomaterials used in 3D printing and how these techniques can be applied to create complex and customized scaffolds, allowing for more effective tissue regeneration. Porous scaffold design is a critical aspect of tissue engineering, as it influences nutrient diffusion, cell infiltration, and tissue integration (Melchels et al. 2012). Hollister (2005) explores the design principles and fabrication methods for porous scaffolds. The manipulation of pore size, shape, and interconnectivity can influence cell behavior and tissue formation. Creating instructive extracellular microenvironments is another essential approach in tissue engineering (Hollister 2005). Lutolf and Hubbell (2005) delve into the use of synthetic biomaterials that can serve as ECM-mimicking scaffolds. These synthetic biomaterials can be engineered to release bioactive factors, control cell behavior, and direct tissue morphogenesis, facilitating the regeneration process (Lutolf and Hubbell 2005).

Table 10.3 explains recent advancement in bioscaffold. Whole-organ tissue engineering is an exciting avenue that involves decellularizing organs or tissues to remove cellular components, leaving behind the ECM. Badylak et al. (2011) review the decellularization and recellularization process and how it can be utilized to create functional tissues. By repopulating the ECM with new cells, it is possible to regenerate the entire organs or tissues. Natural polymers derived from human tissues are also explored as scaffolds in tissue engineering (Badylak et al. 2011). Ma (2004) discusses the use of biomaterials like collagen, hyaluronic acid, and fibrin, which offer biocompatibility and can promote cell adhesion and tissue regeneration. These biomaterials have shown promise in various tissue engineering applications.

Table 10.3 Recent studies in bio-scaffold

References	Key points
Bio-scaffolds	Three-dimensional matrices that provide a supportive environment for cell attachment, growth, and differentiation
Langer and Vacanti (1993)	Introduced the concept of tissue engineering as a novel approach to address the limitations of traditional tissue transplantation and organ transplantation
Atala (2007)	Discussed the engineering of tissues, organs, and cells, emphasizing the importance of selecting appropriate biomaterials for scaffold fabrication
Melchels et al. (2012)	Provided an in-depth review of additive manufacturing techniques for tissues and organs, discussing various biomaterials used in 3D printing and how these techniques can be applied to create complex and customized scaffolds
Hollister (2005)	Explored the design principles and fabrication methods for porous scaffolds, discussing how the manipulation of pore size, shape, and interconnectivity can influence cell behavior and tissue formation
Lutolf and Hubbell (2005)	Delved into the use of synthetic biomaterials that can serve as ECM-mimicking scaffolds, which can be engineered to release bioactive factors, control cell behavior, and direct tissue morphogenesis
Badylak et al. (2011)	Reviewed the decellularization and recellularization process and how it can be utilized to create functional tissues, by repopulating the ECM with new cells
Ma (2004)	Discussed the use of biomaterials like collagen, hyaluronic acid, and fibrin, which offer biocompatibility and can promote cell adhesion and tissue regeneration
Chen and Liu (2016)	Elaborate on the use of human-derived biomaterials, such as decellularized ECM and cell-based constructs, which can improve the safety and efficacy of tissue-engineered products

Advancing biomaterials of the human origin is a key focus in tissue engineering research (Ma 2004). Chen and Liu (2016) elaborate on the use of human-derived biomaterials, such as decellularized ECM and cell-based constructs, which can improve the safety and efficacy of tissue-engineered products.

10.4 Mechanism of Action

Zhang et al. (2005) investigated the surface biocompatibility of electrospun PCL-collagen nanofibers using fibroblasts. They found that the nanofibers had good biocompatibility and promoted fibroblast adhesion and proliferation. The biocompatibility of the nanofibers was attributed to the presence of collagen, which is a naturally occurring biomaterial that is well-accepted by cells (Zhang et al. 2005). Kaji et al. (2011) engineered systems for the generation of patterned co-cultures for controlling cell-cell interactions. They used PCL-collagen nanofibers to create patterned scaffolds that could be used to culture different types of cells in close proximity. The patterned scaffolds allowed the cells to interact with each other in a controlled manner, which could be used to study cell-cell signaling and other biological processes (Kaji et al. 2011).

Phadke et al. (2013) investigated the effect of biodegradable scaffold architecture on the osteogenic differentiation of human mesenchymal stem cells. They found that the morphology and architecture of the scaffold could influence the differentiation of stem cells into bone-forming cells. The PCL-collagen nanofibers used in this study had a porous structure that was favorable for cell adhesion and proliferation. This porous structure also allowed the cells to interact with each other and with the surrounding environment, which promoted osteogenic differentiation. The mechanism of action of PCL-collagen nanofibers is based on their biocompatibility, ability to promote cell adhesion and proliferation, and ability to control cell-cell interactions. These properties make PCL-collagen nanofibers a promising material for use in tissue engineering and regenerative medicine (Phadke et al. 2013).

Table 10.4 shows various mechanisms used in breast reconstruction. Zhang et al. (2012) found that EPCs were recruited to the site of a craniofacial bone defect and formed new blood vessels. They also found that growth factors such as VEGF and FGF were involved in angiogenesis (Zhang et al. 2012). Burgess et al. (2021) found

Table 10.4 Mechanism of action

Mechanism	Description
Biocompatibility	The ability of the material to be accepted by the body without causing any harm
Cell adhesion	The ability of cells to attach to the surface of the material
Cell proliferation	The ability of cells to divide and multiply
Cell-cell interactions	The ability of cells to interact with each other
Osteogenic differentiation	The process by which cells differentiate into bone-forming cells
Angiogenesis	The process of blood vessel formation

that diabetic wounds had impaired angiogenesis, despite having normal blood flow. This suggests that angiogenesis is a complex process that is influenced by multiple factors (Burgess et al. 2021). Cowan et al. (2005) found that bone morphogenetic protein 2 (BMP-2) and retinoic acid could accelerate angiogenesis and bone formation in vivo. This suggests that these growth factors could be used to promote vascularization in craniofacial bone tissue engineering (Cowan et al. 2005). Tan et al. (2013) found that ASCs in PCAL/CA could improve bone tissue formation in a rat model of bone defect. They also found that the addition of growth factors such as BMP-2 could further improve bone formation (Tan et al. 2013). Researchers found that platelet-rich plasma (PRP) could improve the survival and differentiation of ASCs in PCAL/CA. PRP is a blood product that contains growth factors and other molecules that promote tissue regeneration (Zhang et al. 2020a).

10.5 Types of Bio-Scaffolds Used in Breast Reconstruction

Extracellular matrix scaffolds are derived from natural materials, such as collagen, elastin, and glycosaminoglycans. These scaffolds are designed to mimic the extracellular matrix of native tissue, providing a supportive environment for cells to grow and function. Alginate hydrogels are synthetic scaffolds made from alginate, a naturally occurring polysaccharide. Alginate hydrogels are biocompatible and biodegradable, making them a good option for breast reconstruction. Hodde et al. (2007) studies the effects of sterilization on an extracellular matrix scaffold. Leong et al. (2003) describes the use of solid freeform fabrication to create three-dimensional scaffolds for tissue engineering. Rowley et al. (1999) investigate the use of alginate hydrogels as synthetic extracellular matrix materials.

Extracellular matrix scaffolds are derived from natural materials, such as collagen, elastin, and glycosaminoglycans. These scaffolds are designed to mimic the extracellular matrix of the native tissue, providing a supportive environment for cells to grow and function. Alginate hydrogels are synthetic scaffolds made from alginate, a naturally occurring polysaccharide. Alginate hydrogels are biocompatible and biodegradable, making them a good option for breast reconstruction. Polyglycolic acid (PGA) is a synthetic polymer that is biodegradable and biocompatible. PGA scaffolds are often used to support the growth of a new tissue in breast reconstruction. Polylactic acid (PLA) is another synthetic polymer that is biodegradable and biocompatible. PLA scaffolds are also used in breast reconstruction (Leong et al. 2003; Rowley et al. 1999).

Table 10.5 shows different types of bio-scaffolds. Hyaluronic acid is a naturally occurring polysaccharide that is used to make hydrogels. Hyaluronic acid hydrogels are biocompatible and biodegradable, making them a good option for breast reconstruction. Decellularized scaffolds are made from the native tissue that has been decellularized, meaning that the cells have been removed. Decellularized scaffolds provide a structural support for new tissue growth, but they also contain the extracellular matrix proteins that can help guide cell behavior. Daly et al. (2017) discuss the use of 3D bioprinting to create cartilage and osteochondral constructs. This

Table 10.5 Types of scaffolds

Type	Description
Extracellular matrix scaffolds	Derived from natural materials, such as collagen, elastin, and glycosaminoglycans. Mimic the extracellular matrix of native tissue, providing a supportive environment for cells to grow and function
Alginate hydrogels	Synthetic scaffolds made from alginate, a naturally occurring polysaccharide. Biocompatible and biodegradable
Polyglycolic acid (PGA)	Synthetic polymer that is biodegradable and biocompatible. Often used to support the growth of a new tissue in breast reconstruction
Polylactic acid (PLA)	Another synthetic polymer that is biodegradable and biocompatible. Also used in breast reconstruction
Hyaluronic acid	Naturally occurring polysaccharide that is used to make hydrogels. Biocompatible and biodegradable
Decellularized scaffolds	Made from the native tissue that has been decellularized, meaning that the cells have been removed. Provide a structural support for new tissue growth, but they also contain the extracellular matrix proteins that can help guide cell behavior

technique could potentially be used to create bio-scaffolds for breast reconstruction (Daly et al. 2017). Murphy et al. (2002) describe a technique for improving the pore interconnectivity of tissue engineering scaffolds. This could be important for breast reconstruction, as it would allow for better cell infiltration and nutrient diffusion (Murphy et al. 2002). Li et al. (2006) investigate the use of co-electrospinning to create tissue engineering scaffolds. This technique could be used to create bio-scaffolds with a hierarchical structure that mimics the extracellular matrix of the native tissue. This could be beneficial for breast reconstruction, as it would provide a more supportive environment for new tissue growth (Li et al. 2006).

Sodian et al. (2000) discuss the use of tissue-engineered Tri leaflet heart valves. The authors used a bio-scaffold made from decellularized porcine heart valves to support the growth of a new tissue. This is a promising approach for the development of tissue-engineered heart valves, but it has not yet been used in breast reconstruction (Sodian et al. 2000). Wöltje et al. (2018) investigates the use of human epidermal growth factor (EGF) in wound healing and burn treatments. EGF is a protein that promotes cell growth and differentiation. The authors used a bio-scaffold made from silkworm silk to deliver EGF to the wound site. This is a promising approach for the treatment of wounds and burns, but it has not yet been used in breast reconstruction (Wöltje et al. 2018).

10.6 Clinical Applications

Acellular dermal matrix (ADM) was used for non-extensive rapid breast restoration, according to Salzberg (2006). They discovered that ADM was an excellent alternative for this surgery and produced good aesthetic effects (Salzberg 2006). The literature on fat grafting for breast reconstruction was reviewed by Khouri et al. in 2014. They discovered that fat grafting can be an effective method for enhancing the volume and contour of the breast after reconstruction. They did add that there are dangers associated with fat grafting, such as the possibility of infection and fat necrosis (Khouri et al. 2014).

A review of the literature on autologous fat transfer for breast reconstruction was conducted by Caviggioli et al. in 2017. They discovered that fat grafting is a method that can be used to enhance the cosmetic outcomes of breast reconstruction. Non-expansive instant breast reconstruction is possible with ADM. This kind of breast reconstruction is carried out concurrently with the surgery. The breast mound is built up using ADM, which can then be covered with an implant or autologous tissue by the surgeon. Enhancing the volume and contour of the rebuilt breast using fat grafting is possible. This procedure involves injecting fat into the reconstructed breast from another area of the body, such as the thighs or the belly. An appearance that is more natural-looking can be achieved by fat grafting (Caviggioli et al. 2011).

Table 10.6 shows various clinical applications. Improvement of breast symmetry- Autologous fat transfer can be used to improve the symmetry of the breasts after mastectomy. This is especially important in patients who have undergone radiation therapy, which can cause the breasts to become smaller and less symmetrical. Volume augmentation- Autologous fat transfer can be used to augment the volume of the reconstructed breast. This can be done in patients who have had a mastectomy and do not have enough tissue to create a natural-looking breast shape. Correction of contour defects- Autologous fat transfer can be used to correct contour defects in the reconstructed breast, such as dents or ripples. Improvement of skin quality-Autologous fat transfer can be used to improve the skin quality of the reconstructed breast. This is because fat cells contain growth factors that can help heal the skin and make it look more natural. Reduction of scarring: Autologous fat transfer can be used to reduce the scarring from a mastectomy. This is because fat cells can help fill in the space around the scar, making it less noticeable.

Table 10.6 Clinical applications

Application	ADM	Fat grafting
Non-expansive immediate breast reconstruction	Yes	No
Volume augmentation	No	Yes
Correction of contour defects	Yes	Yes
Improvement of skin quality	No	Yes
Reduction of scarring	No	Yes
Orbital floor reconstruction	Yes	No
Midface augmentation	Yes	No
Cranioplasty	Yes	No
Temporomandibular joint (TMJ) reconstruction	Yes	No
Cleft lip and palate repair	Yes	No

Autologous fat transfer is not without risks. Some of the potential risks include:

1. Fat necrosis- This is the death of fat cells, which can cause lumps or bumps in the breast.
2. Infection- This is a rare but serious complication of fat transfer.
3. Seroma- This is a collection of fluid under the skin, which can occur after fat transfer.
4. Asymmetry- This can occur if the fat is not evenly distributed in the breast.

Overall, autologous fat transfer is a safe and effective procedure for improving the appearance of the reconstructed breast. However, it is important to discuss the risks and benefits of this procedure with your doctor before making a decision (Caviggioli et al. 2011).

The study by Schop et al. (2021) found that autologous fat transfer was associated with significant improvements in the quality of life in patients who had undergone breast reconstruction after radiation therapy. Delay et al. (2009) reported on the results of 880 fat grafting procedures for breast reconstruction. The authors found that fat grafting was a safe and effective technique that produced good aesthetic results (Delay et al. 2009). The study by Claro et al. (2012) evaluated the applicability and safety of autologous fat for breast reconstruction. The authors found that fat grafting was a safe and effective technique that could be used to improve the appearance of the reconstructed breast (Claro et al. 2012).

Orbital floor reconstruction- ADM can be used to reconstruct the orbital floor, which is the bone that supports the eye. This is a common procedure after orbital trauma or surgery. ADM can provide a structural support to the orbital floor and help prevent complications such as vision loss or diplopia (double vision). Midface augmentation: ADM can be used to augment the midface, which is the area of the face between the eyes and the nose. This can be done to improve facial symmetry or to correct age-related changes. ADM can provide volume and structure to the midface and help create a more youthful appearance.

Cranioplasty- ADM can be used to repair defects in the skull. This can be done after trauma, surgery, or tumor removal. ADM can provide a structural support to the skull and help prevent complications such as brain injury or infection. Temporomandibular joint (TMJ) reconstruction: ADM can be used to reconstruct the TMJ, which is the joint that connects the jaw to the skull. This can be done to treat TMJ disorders such as pain, clicking, or locking. ADM can provide a structural support to the TMJ and help improve function. Cleft lip and palate repair: ADM can be used to repair the cleft lip and palate, which are birth defects that occur when the lip or palate does not develop properly. ADM can provide a structural support to the lip and palate and help improve the appearance of the defect.

ADM is not without risks. Some of the potential risks include:

Infection- This is a rare but serious complication of ADM use.

Rejection- The body may reject the ADM, which can lead to inflammation and scarring.

Adhesions- The ADM may adhere to surrounding tissues, which can restrict movement or cause pain (Claro et al. 2012).

Tian et al. (2018) investigated the role of vascularization in craniofacial bone tissue engineering. The authors found that ADM can promote angiogenesis, which is the growth of new blood vessels. This is important for the survival and integration of bone grafts (Tian et al. 2018). The study by Kim et al. (2017) reported on the use of ADM for orbital floor reconstruction. The authors found that ADM was a safe and effective material that provided good functional and aesthetic results (Kim et al. 2017).

10.7 Factors Influencing Bio-Scaffold Selection

Type of tissue to be repaired or regenerated: The type of tissue to be repaired or regenerated is one of the most important factors to consider when selecting a bio-scaffold. Different tissues have different mechanical and biological properties, so the bio-scaffold must be able to provide the necessary support and environment for tissue regeneration. For example, bone tissue requires a bio-scaffold that is strong and durable, while skin tissue requires a bio-scaffold that is flexible and porous (Myeroff and Archdeacon 2011; Parenteau-Bareil et al. 2010; Trottier et al. 2008).

Table 10.7 explains various factors for bio-scaffold selection. The size and shape of the defect is another important factor to consider when selecting a bio-scaffold. The bio-scaffold must be able to fill the defect and provide adequate support for tissue regeneration.

The availability of the biomaterial is also an important factor to consider. Some biomaterials are more readily available than others, and the cost of the biomaterial may also be a factor.

The potential risks and complications of the biomaterial are also important to consider. Some biomaterials may have a higher risk of infection or rejection than others.

Patient-specific factors such as age, health status, and allergies may also influence the choice of bio-scaffold.

In addition to the factors mentioned above, other factors that may influence bio-scaffold selection include the following:

1. Biocompatibility- The bio-scaffold must be biocompatible, meaning that it must not elicit an immune response from the body.
2. Biodegradability- The bio-scaffold must be biodegradable, meaning that it must break down over time and be replaced by new tissue.
3. Mechanical properties- The bio-scaffold must have the appropriate mechanical properties to support the tissue regeneration process.
4. Tissue-guiding properties- The bio-scaffold must have the ability to direct the growth and differentiation of cells into the desired tissue type.
5. Surface properties- The surface properties of the bio-scaffold can influence cell adhesion, proliferation, and differentiation.

Table 10.7 Factors for bio-scaffold selection

Factor	Description
Type of tissue to be repaired or regenerated	The type of tissue to be repaired or regenerated is one of the most important factors to consider when selecting a bio-scaffold. Different tissues have different mechanical and biological properties, so the bio-scaffold must be able to provide the necessary support and environment for tissue regeneration. For example, the bone tissue requires a bio-scaffold that is strong and durable, while the skin tissue requires a bio-scaffold that is flexible and porous
Size and shape of the defect	The size and shape of the defect is another important factor to consider when selecting a bio-scaffold. The bio-scaffold must be able to fill the defect and provide adequate support for tissue regeneration
Availability of the biomaterial	The availability of the biomaterial is also an important factor to consider. Some biomaterials are more readily available than others, and the cost of the biomaterial may also be a factor
Potential risks and complications of the biomaterial	The potential risks and complications of the biomaterial are also important to consider. Some biomaterials may have a higher risk of infection or rejection than others
Patient-specific factors	Patient-specific factors such as age, health status, and allergies may also influence the choice of bio-scaffold
Biocompatibility	The bio-scaffold must be biocompatible, meaning that it must not elicit an immune response from the body
Biodegradability	The bio-scaffold must be biodegradable, meaning that it must break down over time and be replaced by new tissue
Mechanical properties	The bio-scaffold must have the appropriate mechanical properties to support the tissue regeneration process
Tissue-guiding properties	The bio-scaffold must have the ability to direct the growth and differentiation of cells into the desired tissue type
Surface properties	The surface properties of the bio-scaffold can influence cell adhesion, proliferation, and differentiation

The best bio-scaffold for a particular application will depend on a number of factors, including the specific needs of the patient and the goals of the treatment (Saldin et al. 2017; Yamada et al. 2021; Yu et al. 2013; Zhang and Wang 2010; Baptista et al. 2014).

10.8 Preoperative Preparation and Surgical Technique

The study by Spear et al. (2005) on fat injection to correct contour deformities in the reconstructed breast mentions that the patient should be evaluated for any medical conditions that could affect the surgery, such as diabetes or high blood pressure. The patient should also stop smoking and taking aspirin or other blood thinners prior to surgery (Spear et al. 2005).

Fang et al. (2021) on the rational use of adipose-derived stromal cells in breast reconstruction mentions that the patient should have a thorough physical examination and laboratory tests to assess their overall health. The patient should also be counselled about the risks and benefits of the procedure (Fang et al. 2021).

Table 10.8 Preoperative preparation and surgical technique

Step	Description
Preoperative preparation	A thorough medical evaluation to assess the patient's overall health, counselling about the risks and benefits of the procedure, stopping smoking and taking blood thinners prior to surgery, and following any specific instructions from the surgeon
Surgical technique	Harvesting the cells or tissues that will be used for the graft, preparing the recipient site, implanting the graft, and closing the incision

Table 10.8 explains preoperative preparation and surgical technique. Preoperative preparation for any surgical procedure should include:

- A thorough medical evaluation to assess the patient's overall health
- Counselling about the risks and benefits of the procedure
- Stopping smoking and taking blood thinners prior to surgery
- Following any specific instructions from the surgeon

The surgical technique will vary depending on the specific type of tissue engineering being performed. Common steps may include-

- Harvesting the cells or tissues that will be used for the graft
- Preparing the recipient site
- Implanting the graft
- Closing the incision (Wang et al. 2021; Zhang et al. 2019; Boerckel et al. 2014; Abbasi et al. 2020; Luigi et al. 2022; Chen et al. 2022)

10.9 Clinical Outcomes and Complications

The research by Margulies and Salzberg (2019) reports on the early experience with porcine acellular dermal matrix (Strattice) in primary prosthetic breast reconstruction. The authors found that Strattice was an effective and safe material for breast reconstruction, with low rates of complications (Margulies and Salzberg 2019). Salzberg et al. (2016) report on a 13-year experience with acellular dermal matrix-assisted direct-to-implant breast reconstruction. The authors found that this technique was associated with low rates of capsular contracture but higher rates of seroma formation (Salzberg et al. 2016). Luo et al. (2022) presents the US consensus recommendations for the use of acellular dermal matrix for soft tissue reconstruction with breast implants. The authors recommend that Strattice be used in conjunction with breast implants for primary breast reconstruction, but not for secondary breast reconstruction (Luo et al. 2022).

Table 10.9 shows different complications during breast reconstruction. Wells et al. (2022) compares the outcomes of two-stage prosthetic breast reconstruction using a photographic scale to the traditional method of tissue expansion and implant breast reconstruction in a Chinese population. The authors found that the two-stage

Table 10.9 Complications in breast reconstruction

Outcome	Description
Capsular contracture	This is a complication of breast reconstruction that occurs when the scar tissue around the implant becomes tight and constricts the implant. This can cause pain, deformity, and decreased sensation in the breast
Seroma	This is a collection of fluid under the skin that can occur after breast reconstruction. It is usually harmless and resolves on its own, but it can sometimes require drainage
Infection	This is a serious complication of breast reconstruction that can occur if bacteria enter the surgical site. Infection can cause pain, swelling, redness, and warmth in the breast. It can also lead to the need for additional surgery
Bleeding	This is a common complication of breast reconstruction that can occur during or after surgery. It is usually minor and can be controlled with pressure. However, it can sometimes be more serious and require additional surgery
Nerve damage	This is a rare complication of breast reconstruction that can occur if nerves are injured during surgery. Nerve damage can cause numbness, tingling, or pain in the breast
Symmetry	This is the goal of breast reconstruction to achieve symmetry between the two breasts. However, it is not always possible to achieve perfect symmetry
Patient satisfaction	This is the degree to which a patient is satisfied with their breast reconstruction. Patient satisfaction is influenced by a number of factors, including the appearance of the breast, the level of pain, and the overall experience of the surgery

technique was associated with better aesthetic outcomes than the traditional technique (Wells et al. 2022). Olivas-Menayo and Berniz (2021) describes a suture technique for correcting inverted nipples. The authors found that this technique was effective in correcting inverted nipples in the majority of patients (Olivas-Menayo and Berniz 2021). Vardanian et al. (2011) compares the outcomes of implant-based immediate breast reconstruction with and without acellular dermal matrix. The authors found that the use of acellular dermal matrix was associated with lower rates of capsular contracture but higher rates of seroma formation (Vardanian et al. 2011).

The research by Krimmel et al. (2003) reports on a 20-year follow-up of microvascular transfer of the composite iliac crest and rib to the chest wall. The authors found that this technique was effective in reconstructing the breast in patients with complex defects (Krimmel et al. 2003). Colakoglu et al. (2011) examine the impact of complications on patient satisfaction in breast reconstruction. The authors found that patients who experienced complications were less satisfied with their breast reconstruction than patients who did not experience complications (Colakoglu et al. 2011).

10.10 Future Directions and Research

The advancement of novel biomaterials that possess enhanced biocompatibility and biodegradability, while closely emulating the natural properties of tissues, stands as a significant area of focus. Additionally, the exploration of innovative techniques for precisely delivering cells and growth factors to the afflicted site in cases of injury or disease is paramount. The utilization of stem cells for the purpose of tissue and organ regeneration represents another promising avenue. Furthermore, the progress in 3D printing technologies for fabricating tissue analogs is a notable area of development. The incorporation of computational modeling holds immense potential in refining the design and optimization of scaffolds and constructs used in tissue engineering. Likewise, the establishment of new animal models for evaluating tissue engineering technologies is crucial (Camponogara et al. 2022; Patrachari et al. 2012; Kolambkar et al. 2011). The transition of tissue engineering innovations from laboratory settings to clinical applications is a critical phase. These outlined challenges and opportunities in tissue engineering exemplify the dynamic nature of the field, with substantial potential for groundbreaking treatments across a spectrum of diseases and injuries (Dearman et al. 2021).

Table 10.10 explains future research in bio-scaffold and tissue engineering. Examples of specific studies being conducted in these fields include:

New biomaterials are being created by researchers that are based on natural substances like collagen and hyaluronic acid. These materials can better imitate the natural features of tissues since they are more biocompatible and biodegradable than manufactured materials. New techniques are being developed by researchers to transfer cells and growth factors to the location of sickness or injury. These techniques make use of cell-sheet technology, nanoparticles, and microcarriers.

In order to regenerate tissues and organs, scientists are employing stem cells. Because they can develop into numerous cell types, stem cells are perfect for tissue engineering applications. For the purpose of producing tissue analogs, researchers are exploring new 3D printing technologies. These technologies make it possible to build scaffolds with intricate designs that resemble the architecture of tissues in their native state. To enhance the design and optimization of tissue engineering scaffolds and constructs, researchers are embracing computational modeling. The behavior of cells and tissues in response to various environmental variables can be predicted using computational models. To test tissue engineering methods, researchers are creating new animal models. For evaluating the security and effectiveness of novel tissue engineering therapies, animal models are crucial. To move tissue engineering technology from the lab to the clinic, researchers are working. Although there has been substantial improvement recently, this is still a huge challenge (Gentile et al. 2020; Pan and Ding 2012; Zhang et al. 2020b; Pati et al. 2014).

Table 10.10 Future research

Area	Future directions	Research
New biomaterials	Development of new biomaterials that are more biocompatible and biodegradable and that can better mimic the natural properties of tissues	Researchers are developing new biomaterials based on natural materials, such as collagen and hyaluronic acid. They are also developing new methods for modifying the surface properties of biomaterials to improve cell adhesion and growth
New methods for delivering cells and growth factors	Development of new methods for delivering cells and growth factors to the site of injury or disease	Researchers are developing new methods for delivering cells and growth factors, such as using nanoparticles, microcarriers, and cell-sheet technology. They are also developing new ways to control the release of cells and growth factors over time
Use of stem cells to regenerate tissues and organs	Study of how to use stem cells to repair damaged tissues and organs and to create new tissues and organs	Researchers are using stem cells to regenerate tissues and organs. They are studying how to use stem cells to differentiate into a variety of cell types and how to control the growth and differentiation of stem cells
Development of new 3D printing technologies for creating tissue analogues	Development of new 3D printing technologies for creating tissue analogues	Researchers are developing new 3D printing technologies that allow for the creation of scaffolds with complex structures that mimic the natural architecture of tissues. They are also developing new ways to print cells and growth factors into 3D scaffolds
Use of computational modeling to improve the design and optimization of tissue engineering scaffolds and constructs	Use of computational modeling to predict the behavior of cells and tissues in response to different environmental conditions	Researchers are using computational modeling to design new scaffolds that are more effective at promoting cell growth and differentiation. They are also using computational modeling to optimize the design of tissue engineering scaffolds and constructs
Development of new animal models for testing tissue engineering technologies	Development of animal models that are more closely matched to human tissues and diseases	Researchers are developing new animal models for testing tissue engineering technologies. They are working to develop animal models that are more closely matched to human tissues and diseases
Translation of tissue engineering technologies from the laboratory to the clinic	Translation of tissue engineering technologies from the laboratory to the clinic	Researchers are working to translate tissue engineering technologies from the laboratory to the clinic. This is a major challenge, but there has been significant progress in recent years

10.11 Conclusion

Breast reconstruction is a vital procedure for women who have undergone mastectomy or have breast deformities. Autologous fat grafting is a natural and minimally invasive approach, but fat survival can be a challenge. Bio-scaffolds can be used to optimize the outcomes of breast reconstruction by providing a supportive environment for transplanted fat. Bio-scaffolds are three-dimensional matrices that can be made from natural or synthetic materials. They promote tissue regeneration by stimulating angiogenesis, enhancing tissue growth, and providing mechanical support. This can improve fat graft survival and aesthetic results. There are many different types of bio-scaffolds available, and the choice of scaffold will depend on the specific patient and the goals of reconstruction. Some factors to consider include the patient's age, health status, and body habitus, as well as the size and shape of the breast defect. The use of bio-scaffolds in breast reconstruction is still an evolving field, but there is growing evidence to support their efficacy. Research is needed to further optimize the use of bio-scaffolds in breast reconstruction and to develop new and improved materials. This research has provided plastic surgeons with a comprehensive overview of the role of bio-scaffolds in breast reconstruction. By understanding the mechanisms of action, clinical applications, and factors influencing selection, surgeons can make informed decisions about the use of bio-scaffolds to improve outcomes for their patients.

Acknowledgments Authors would like to thank Dr. Manik Deosarkar, HOD Department of Chemical Engineering, Vishwakarma Institute of Technology, Pune, for their constant support and help.

Author Contributions Acquisition of data: Mitesh Ikar. Writing of the manuscript: Mitesh Ikar. Supervision: Sunil Sable.

References

Abbasi N, Hamlet S, Love RM, Nguyen N-T (2020) Porous scaffolds for bone regeneration. J Sci Adv Mater Devices 5(1):1–9. https://doi.org/10.1016/j.jsamd.2020.01.007. ISSN 2468-2179

Albornoz CR, Bach PB, Mehrara BJ, Disa JJ, Pusic AL, McCarthy CM, Cordeiro PG, Matros E (2013) A paradigm shift in U.S. breast reconstruction: increasing implant rates. Plast Reconstr Surg 131(1):15–23. https://doi.org/10.1097/PRS.0b013e3182729cde. PMID: 23271515

Alhindi N, Bamakhrama B, Alzahrani A et al (2023) Risk factors of implant loss and complications post-implant based breast reconstruction: a meta-analysis. Eur J Plast Surg 46:865. https://doi.org/10.1007/s00238-023-02077-x

Atala A (2007) Engineering tissues, organs and cells. J Tissue Eng Regen Med 1(2):83–96. https://doi.org/10.1002/term.18. PMID: 18038397

Azam S, Eriksson M, Sjölander A et al (2021) Mammographic microcalcifications and risk of breast cancer. Br J Cancer 125:759–765. https://doi.org/10.1038/s41416-021-01459-x

Badylak SF, Taylor D, Uygun K (2011) Whole-organ tissue engineering: decellularization and recellularization of three-dimensional matrix scaffolds. Annu Rev Biomed Eng 13:27–53. https://doi.org/10.1146/annurev-bioeng-071910-124743. PMID: 21417722

Baptista LS, do Amaral RJ, Carias RB et al (2014) Decellularized bovine pericardium: natural biomaterial for cardiac surgery preclinical investigation. Artif Organs 38(2):E45–E54. https://doi.org/10.1111/aor.12251

Boerckel JD, Mason DE, McDermott AM, Alsberg E (2014) Microcomputed tomography: approaches and applications in bioengineering. Stem Cell Res Ther 5(6):144. https://doi.org/10.1186/scrt534. PMID: 25689288; PMCID: PMC4290379

Burgess JL, Wyant WA, Abdo Abujamra B, Kirsner RS, Jozic I (2021) Diabetic wound-healing science. Medicina (Kaunas) 57(10):1072. https://doi.org/10.3390/medicina57101072. PMID: 34684109; PMCID: PMC8539411

Camponogara F, Zanotti F, Trentini M, Tiengo E, Zanolla I, Pishavar E, Soliani E, Scatto M, Gargiulo P, Zambito Y, De Luca S, Ferroni L, Zavan B (2022) Biomaterials for regenerative medicine in Italy: brief state of the art of the principal research centers. Int J Mol Sci 23(15):8245. https://doi.org/10.3390/ijms23158245. PMID: 35897825; PMCID: PMC9368060

Caviggioli F, Maione L, Forcellini D, Klinger F, Klinger M (2011) Autologous fat graft in post-mastectomy pain syndrome. Plast Reconstr Surg 128(2):349–352. https://doi.org/10.1097/PRS.0b013e31821e70e7. PMID: 21788826

Chen FM, Liu X (2016) Advancing biomaterials of human origin for tissue engineering. Prog Polym Sci 53:86–168. https://doi.org/10.1016/j.progpolymsci.2015.02.004. Epub 2015 Mar 28. PMID: 27022202; PMCID: PMC4808059

Chen Y, Dong X, Shafiq M et al (2022) Recent advancements on three-dimensional electrospun nanofiber scaffolds for tissue engineering. Adv Fiber Mater 4:959–986. https://doi.org/10.1007/s42765-022-00170-7

Claro F Jr, Figueiredo JC, Zampar AG, Pinto-Neto AM (2012) Applicability and safety of autologous fat for reconstruction of the breast. Br J Surg 99(6):768–780. https://doi.org/10.1002/bjs.8722. Epub 2012 Apr 4. PMID: 22488516

Colakoglu S, Khansa I, Curtis MS, Yueh JH, Ogunleye A, Haewyon C, Tobias AM, Lee BT (2011) Impact of complications on patient satisfaction in breast reconstruction. Plast Reconstr Surg 127(4):1428–1436. https://doi.org/10.1097/PRS.0b013e318208d0d4. PMID: 21460651

Cowan CM, Aalami OO, Shi YY, Chou YF, Mari C, Thomas R, Quarto N, Nacamuli RP, Contag CH, Wu B, Longaker MT (2005) Bone morphogenetic protein 2 and retinoic acid accelerate in vivo bone formation, osteoclast recruitment, and bone turnover. Tissue Eng 11(3–4):645–658. https://doi.org/10.1089/ten.2005.11.645. PMID: 15869441

Daly AC, Freeman FE, Gonzalez-Fernandez T, Critchley SE, Nulty J, Kelly DJ (2017) 3D bioprinting for cartilage and osteochondral tissue engineering. Adv Healthc Mater 6(22). https://doi.org/10.1002/adhm.201700298. Epub 2017 Aug 14. PMID: 28804984

Dearman BL, Boyce ST, Greenwood JE (2021) Advances in skin tissue bioengineering and the challenges of clinical translation. Front Surg 8:640879. https://doi.org/10.3389/fsurg.2021.640879. PMID: 34504864; PMCID: PMC8421760

Delay E, Garson S, Tousson G, Sinna R (2009) Fat injection to the breast: technique, results, and indications based on 880 procedures over 10 years. Aesthet Surg J 29(5):360–376. https://doi.org/10.1016/j.asj.2009.08.010. PMID: 19825464

Eto H, Kato H, Suga H, Aoi N, Doi K, Kuno S, Yoshimura K (2012) The fate of adipocytes after nonvascularized fat grafting: evidence of early death and replacement of adipocytes. Plast Reconstr Surg 129(5):1081–1092. https://doi.org/10.1097/PRS.0b013e31824a2b19. PMID: 22261562

Fang J, Chen F, Liu D et al (2021) Adipose tissue-derived stem cells in breast reconstruction: a brief review on biology and translation. Stem Cell Res Ther 12:8. https://doi.org/10.1186/s13287-020-01955-6

Garza R III, Ochoa O, Chrysopoulo M (2021) Post-mastectomy breast reconstruction with autologous tissue: current methods and techniques. Plast Reconstr Surg Glob Open 9(2):e3433. https://doi.org/10.1097/GOX.0000000000003433. PMID: 33680677; PMCID: PMC7929567

Gentile P, Orlandi A, Scioli MG, Di Pasquali C, Bocchini I, Curcio CB, Floris M, Fiaschetti V, Floris R, Cervell V (2012) A comparative translational study: the combined use of enhanced stromal vascular fraction and platelet-rich plasma improves fat grafting maintenance in breast reconstruction. Stem Cells Transl Med 1(4):341–351. https://doi.org/10.5966/sctm.2011-0065. Epub 2012 Apr 13. PMID: 23197813; PMCID: PMC3659694

Gentile P, Kothari A, Casella D, Calabrese C (2020) Fat graft enhanced with adipose-derived stem cells in aesthetic breast augmentation: clinical, histological, and instrumental evaluation. Aesthet Surg J 40(9):962–977. https://doi.org/10.1093/asj/sjz292

Hodde J, Janis A, Ernst D, Zopf D, Sherman D, Johnson C (2007) Effects of sterilization on an extracellular matrix scaffold: part I. Composition and matrix architecture. J Mater Sci Mater Med 18(4):537–543. https://doi.org/10.1007/s10856-007-2300-x. PMID: 17546412

Hollister S (2005) Porous scaffold design for tissue engineering. Nat Mater 4:518–524. https://doi.org/10.1038/nmat1421

Kaji H, Camci-Unal G, Langer R, Khademhosseini A (2011) Engineering systems for the generation of patterned co-cultures for controlling cell-cell interactions. Biochim Biophys Acta 1810(3):239–250. https://doi.org/10.1016/j.bbagen.2010.07.002. Epub 2010 Jul 23. PMID: 20655984; PMCID: PMC3026923

Khouri RK, Smit JM, Cardoso E, Pallua N, Lantieri L, Mathijssen IMJ, Khouri RK Jr, Rigotti G (2013) Percutaneous aponeurotomy and lipofilling: a regenerative alternative to flap reconstruction? Plast Reconstr Surg 132(5):1280–1290. https://doi.org/10.1097/PRS.0b013e3182a4c3a9. PMID: 23924652

Khouri RK Jr, Khouri RE, Lujan-Hernandez JR, Khouri KR, Lancerotto L, Orgill DP (2014) Diffusion and perfusion: the keys to fat grafting. Plast Reconstr Surg Glob Open 2(9):e220. https://doi.org/10.1097/GOX.0000000000000183. PMID: 25426403; PMCID: PMC4229279

Khouri RK et al (2015) Tissue-engineered breast reconstruction with Brava-assisted fat grafting: a 7-year, 488-patient, multicenter experience. Plast Reconstr Surg 135:643–658

Kim J, Lew DH, Roh TS, Lee WJ (2017) Use of acellular allogenic dermal matrix (MegaDerm) in orbital wall reconstruction: a comparison with absorbable mesh plate and porous polyethylene. J Craniofac Surg 28(7):e644–e649. https://doi.org/10.1097/SCS.0000000000003782. PMID: 28834839

Kolambkar YM, Boerckel JD, Dupont KM, Bajin M, Huebsch N, Mooney DJ, Hutmacher DW, Guldberg RE (2011) Spatiotemporal delivery of bone morphogenetic protein enhances functional repair of segmental bone defects. Bone 49(3):485–492. https://doi.org/10.1016/j.bone.2011.05.010. Epub 2011 May 18. PMID: 21621027; PMCID: PMC3143222

Krimmel M, Hoffmann J, Zerfowski M, Reinert S (2003) Central resorption in an iliac crest transplant with microvascular anastomosis–report of 2 cases. J Craniofac Surg 31(3):176–178. https://doi.org/10.1016/S1010-5182(03)00021-0. ISSN 1010-5182

Kronowitz SJ, Mandujano CC, Liu J, Kuerer HM, Smith B, Garvey P, Jagsi R, Hsu L, Hanson S, Valero V (2016) Lipofilling of the breast does not increase the risk of recurrence of breast cancer: a matched controlled study. Plast Reconstr Surg 137(2):385–393. https://doi.org/10.1097/01.prs.0000475741.32563.50. PMID: 26818270

Langer R, Vacanti JP (1993) Tissue engineering. Science 260(5110):920–926. https://doi.org/10.1126/science.8493529. PMID: 8493529

Leong KF, Cheah CM, Chua CK (2003) Solid freeform fabrication of three-dimensional scaffolds for engineering replacement tissues and organs. Biomaterials 24(13):2363–2378. https://doi.org/10.1016/s0142-9612(03)00030-9. PMID: 12699674

Li M, Mondrinos MJ, Chen X, Gandhi MR, Ko FK, Lelkes PI (2006) Co-electrospun poly(lactide-co-glycolide), gelatin, and elastin blends for tissue engineering scaffolds. J Biomed Mater Res A 79(4):963–973. https://doi.org/10.1002/jbm.a.30833. PMID: 16948146

Luigi MD, Guo Z, An L, Armstrong JN, Zhou C, Ren S (2022) RSC Adv 12:21213–21222

Luo J, Moss WD, Pires GR, Rhemtulla IA, Rosales M, Stoddard GJ, Agarwal JP, Kwok AC (2022) A nationwide analysis evaluating the safety of using acellular dermal matrix with tissue expander-based breast reconstruction. Arch Plast Surg 49(6):716–723. https://doi.org/10.1055/s-0042-1758638. PMID: 36523919; PMCID: PMC9747277

Lutolf M, Hubbell J (2005) Synthetic biomaterials as instructive extracellular microenvironments for morphogenesis in tissue engineering. Nat Biotechnol 23:47–55. https://doi.org/10.1038/nbt1055

Ma P (2004) Scaffolds for tissue fabrication. Mater Today 7:30–40. https://doi.org/10.1016/S1369-7021(04)00233-0

Margulies IG, Salzberg CA (2019) The use of acellular dermal matrix in breast reconstruction: evolution of techniques over 2 decades. Gland Surg 8(1):3–10. https://doi.org/10.21037/gs.2018.10.05. PMID: 30842922; PMCID: PMC6378261

Melchels FPW, Domingos MAN, Klein TJ, Malda J, Da Silva Bartolo PJ, Hutmacher DW (2012) Additive manufacturing of tissues and organs. Prog Polym Sci 37(8):1079–1104. https://doi.org/10.1016/j.progpolymsci.2011.11.007

Murphy WL, Dennis RG, Kileny JL, Mooney DJ (2002) Salt fusion: an approach to improve pore interconnectivity within tissue engineering scaffolds. Tissue Eng 8(1):43–52. https://doi.org/10.1089/107632702753503045. PMID: 11886653

Myeroff C, Archdeacon M (2011) Autogenous bone graft: donor sites and techniques. J Bone Joint Surg Am 93(23):2227–2236. https://doi.org/10.2106/JBJS.J.01513. PMID: 22159859

Olivas-Menayo J, Berniz C (2021) Inverted nipple correction techniques: an algorithm based on scientific evidence, patients' expectations and potential complications. Aesth Plast Surg 45:472–480. https://doi.org/10.1007/s00266-020-01909-6

Pan Z, Ding J (2012) Poly(lactide-co-glycolide) porous scaffolds for tissue engineering and regenerative medicine. Interface Focus 2(3):366–377. https://doi.org/10.1098/rsfs.2011.0123

Parenteau-Bareil R, Gauvin R, Berthod F (2010) Collagen-based biomaterials for tissue engineering applications. Materials (Basel) 3(3):1863–1887. https://doi.org/10.3390/ma3031863. PMCID: PMC5445871

Pati F, Jang J, Ha DH, Won Kim S, Rhie JW, Shim JH, Kim DH, Cho DW (2014) Printing three-dimensional tissue analogues with decellularized extracellular matrix bioink. Nat Commun 5:3935. https://doi.org/10.1038/ncomms4935. PMID: 24887553; PMCID: PMC4059935

Patrachari AR, Podichetty JT, Madihally SV (2012) Application of computational fluid dynamics in tissue engineering. J Biosci Bioeng 114(2):123–132. https://doi.org/10.1016/j.jbiosc.2012.03.010. ISSN 1389-1723

Phadke A, Hwang Y, Kim SH, Kim SH, Yamaguchi T, Masuda K, Varghese S (2013) Effect of scaffold microarchitecture on osteogenic differentiation of human mesenchymal stem cells. Eur Cell Mater 25:114–129. https://doi.org/10.22203/ecm.v025a08. PMID: 23329467; PMCID: PMC4387005

Ranganath R, Hui M, Uppin S et al (2023) An audit of breast conservation surgery for breast cancer—a 7-year experience from southern India. Indian J Surg Oncol 14:492–496. https://doi.org/10.1007/s13193-022-01532-8

Rowley JA, Madlambayan G, Mooney DJ (1999) Alginate hydrogels as synthetic extracellular matrix materials. Biomaterials 20(1):45–53. https://doi.org/10.1016/s0142-9612(98)00107-0. PMID: 9916770

Saldin LT, Cramer MC, Velankar SS, White LJ, Badylak SF (2017) Extracellular matrix hydrogels from decellularized tissues: structure and function. Acta Biomater 49:1–15. https://doi.org/10.1016/j.actbio.2016.11.068. Epub 2016 Dec 1. PMID: 27915024; PMCID: PMC5253110

Salzberg CA (2006) Nonexpansive immediate breast reconstruction using human acellular tissue matrix graft (AlloDerm). Ann Plast Surg 57(1):1–5. https://doi.org/10.1097/01.sap.0000214873.13102.9f. PMID: 16799299

Salzberg CA, Ashikari AY, Berry C, Hunsicker LM (2016) Acellular dermal matrix-assisted direct-to-implant breast reconstruction and capsular contracture: a 13-year experience. Plast Reconstr Surg 138(2):329–337. https://doi.org/10.1097/PRS.0000000000002331. PMID: 27064232

Sánchez-Carpintero I, Candelas D, Ruiz-Rodríguez R (2010) Dermal fillers: types, indications, and complications. Actas Dermosifiliogr (Engl Ed) 101(5):381–393. https://doi.org/10.1016/S1578-2190(10)70660-0. ISSN 1578-2190

Schop SJ, Joosen MEM, Wolswijk T, Heuts EM, van der Hulst RRWJ, Piatkowski de Grzymala AA (2021) Quality of life after autologous fat transfer additional to prosthetic breast reconstruction in women after breast surgery: a systematic review. Eur J Surg Oncol 47(4):772–777. https://doi.org/10.1016/j.ejso.2020.10.021. ISSN 0748-7983

Shih L, Davis MJ, Winocour SJ (2020) The science of fat grafting. Semin Plast Surg 34(1):5–10. https://doi.org/10.1055/s-0039-3402073. Epub 2020 Feb 15. PMID: 32071573; PMCID: PMC7023968

Sodian R, Hoerstrup SP, Sperling JS, Daebritz S, Martin DP, Moran AM, Kim BS, Schoen FJ, Vacanti JP, Mayer JE Jr (2000) Early in vivo experience with tissue-engineered trileaflet heart valves. Circulation 102(19 Suppl 3):III22–III29. https://doi.org/10.1161/01.cir.102.suppl_3.iii-22. PMID: 11082357

Spear SL, Wilson HB, Lockwood MD (2005) Fat injection to correct contour deformities in the reconstructed breast. Plast Reconstr Surg 116(5):1300–1305. https://doi.org/10.1097/01.prs.0000181509.67319.cf. PMID: 16217471

Spear SL, Seruya M, Clemens MW, Teitelbaum S, Nahabedian MY (2011) Acellular dermal matrix for the treatment and prevention of implant-associated breast deformities. Plast Reconstr Surg 127(3):1047–1058. https://doi.org/10.1097/PRS.0b013e31820436af. PMID: 21088648

Tan Q, Lui PP, Lee YW (2013) In vivo identity of tendon stem cells and the roles of stem cells in tendon healing. Stem Cells Dev 22(23):3128–3140. https://doi.org/10.1089/scd.2013.0073. Epub 2013 Aug 3. PMID: 23815595; PMCID: PMC3857013

Tian T, Zhang T, Lin Y, Cai X (2018) Vascularization in craniofacial bone tissue engineering. J Dent Res 97(9):969–976. https://doi.org/10.1177/0022034518767120

Trottier V, Marceau-Fortier G, Germain L, Vincent C, Fradette J (2008) IFATS collection: using human adipose-derived stem/stromal cells for the production of new skin substitutes. Stem Cells 26(10):2713–2723. https://doi.org/10.1634/stemcells.2008-0031. Epub 2008 Jul 10. PMID: 18617689

Vardanian AJ, Clayton JL, Roostaeian J, Shirvanian V, Da Lio A, Lipa JE, Crisera C, Festekjian JH (2011) Comparison of implant-based immediate breast reconstruction with and without acellular dermal matrix. Plast Reconstr Surg 128(5):403e–410e. https://doi.org/10.1097/PRS.0b013e31822b6637. PMID: 22030500

Wang Q, Ma Z, Wang Y et al (2021) Fabrication and characterization of 3D printed biocomposite scaffolds based on PCL and zirconia nanoparticles. Bio-Des Manuf 4:60–71. https://doi.org/10.1007/s42242-020-00095-3

Wells MW, Chang IA, Gatherwright JR, Festekjian JH, Delong MR (2022) Postsurgical outcomes with meshes for two-stage prosthetic breast reconstruction in 20,817 patients. Plast Reconstr Surg Glob Open 10(12):e4699. https://doi.org/10.1097/GOX.0000000000004699. PMID: 36518688; PMCID: PMC9742085

Wöltje M, Böbel M, Bienert M, Neuss S, Aibibu D, Cherif C (2018) Functionalized silk fibers from transgenic silkworms for wound healing applications: surface presentation of bioactive epidermal growth factor. J Biomed Mater Res A 106(10):2643–2652. https://doi.org/10.1002/jbm.a.36458. Epub 2018 Sep 8. PMID: 29790257

Yamada S, Behfar A, Terzic A (2021) Regenerative medicine clinical readiness. Regen Med 16(3):309–322. https://doi.org/10.2217/rme-2020-0178. Epub 2021 Feb 24. PMID: 33622049; PMCID: PMC8050983

Yu Y, Zhang Y, Martin JA, Ozbolat IT (2013) Evaluation of cell viability and functionality in vessel-like bioprintable cell-laden tubular channels. J Biomech Eng 135(9):91011. https://doi.org/10.1115/1.4024575. PMID: 23719889; PMCID: PMC3708706

Zhang J, Wang JH (2010) Platelet-rich plasma releasate promotes differentiation of tendon stem cells into active tenocytes. Am J Sports Med 38(12):2477–2486. https://doi.org/10.1177/0363546510376750. Epub 2010 Aug 27. PMID: 20802092

Zhang YZ, Venugopal J, Huang ZM, Lim CT, Ramakrishna S (2005) Characterization of the surface biocompatibility of the electrospun PCL-collagen nanofibers using fibroblasts. Biomacromolecules 6(5):2583–2589. https://doi.org/10.1021/bm050314k. PMID: 16153095

Zhang W, Zhu C, Wu Y, Ye D, Wang S, Zuo G (2012) Vascularization in craniofacial bone tissue engineering. Oral Surg Oral Med Oral Pathol Oral Radiol 113(1):e1–e8. https://doi.org/10.1016/j.tripleo.2011.08.007

Zhang W, Ullah I, Lei Shi Y, Zhang HO, Zhou J, Ullah MW, Zhang X, Li W (2019) Fabrication and characterization of porous polycaprolactone scaffold via extrusion-based cryogenic 3D printing for tissue engineering. Mater Des 180:107946. https://doi.org/10.1016/j.matdes.2019.107946. ISSN 0264-1275

Zhang J, Zhang J, Zhang N, Li T, Zhou X, Jia J, Liang Y, Sun X, Chen H (2020a) The effects of platelet-rich and platelet-poor plasma on biological characteristics of BM-MSCs in vitro. Anal Cell Pathol (Amst) 2020:8546231. https://doi.org/10.1155/2020/8546231. PMID: 32908815; PMCID: PMC7471809

Zhang J, Liu Y, Chen Y, Yuan L, Liu H, Wang J, Liu Q, Zhang Y (2020b) Adipose-derived stem cells: current applications and future directions in the regeneration of multiple tissues. Stem Cells Int 2020:8810813. https://doi.org/10.1155/2020/8810813. PMID: 33488736; PMCID: PMC7787857

Toxicological and Regulatory Aspects of Natural Product Based Bio-Scaffold

11

Tridip Boruah, Twinkle Chetia, Romesh Borgohain, Himu Roy, Morjina Parveen, Rinki Bharali, and Susuma Paul

Abstract

Natural product-based bio-scaffold is a structural framework made from natural sources such as plants, fungi, animals or microbes that can be used as a foundation or template to create bioactive substances that have the potential to be used in agriculture, medicine or other industries. Natural products like these provide useful bioactive components, but it is important to know their safety profile to avoid health and environmental hazards. The toxicology of natural product-based bio-scaffolds encompasses the examination of the possible negative consequences and toxicity related to the utilisation of natural component derivatives in many domains, including medicine, agriculture and cosmetics. In order to optimise the advantages of these substances and mitigate possible negative consequences on human health and the environment, it is imperative to comprehend the potential hazards associated with them and effectively apply suitable safety protocols. There is a convoluted web of relationships between national agencies, industry stakeholders and international norms that governs bio-scaffolds based on natural products. The regulatory framework was created to strike a balance between the competing needs of fostering innovation and expanding people's access to helpful products and safeguarding the public's health and the environment. This chapter aims to consolidate the pertinent literature within a unified framework, with the intention of facilitating future study in this field.

T. Boruah (✉) · M. Parveen · R. Bharali · S. Paul
Department of Botany, Madhab Choudhury College, Barpeta, Assam, India
e-mail: tridip_boruah@gauhati.ac.in

T. Chetia · H. Roy
Department of Botany, Cotton University, Guwahati, Assam, India

R. Borgohain
Department of Chemistry, Birjhora Mahavidyalaya, Bongaigaon, Assam, India

M. Kumar et al. (eds.), *Natural Product Inspired Scaffolds*,
https://doi.org/10.1007/978-981-97-3111-4_11

Keywords
Toxicity · Natural products · Tissue engineering · Chitosan · Inflammation

11.1 Introduction

Natural product-based bio-scaffolds are an interesting combination of biology and materials science, taking advantage of the natural characteristics of compounds derived from living organisms to help with tissue engineering and regenerative medicine (Dromel and Singh 2021; Mohanty et al. 2022). Collagen holds a key position within the field of biomaterials. Collagen, derived from connective tissues in animals, exhibits a high degree of versatility as a scaffold material owing to its structural resemblance to endogenous tissues. The material is widely utilised in various applications, spanning from the transplantation of the skin to the regeneration of bones. It provides a biocompatible structure that facilitates the adhesion and proliferation of cells (Bhadra et al. 2021). Fibrin, a protein that plays a crucial role in the physiological process of haemostasis, offers an intriguing natural bio-scaffold. Fibrin scaffolds, which are derived from blood plasma, play a significant role in the process of wound healing and tissue regeneration. Fibrin plays a pivotal role in promoting tissue repair by creating a temporary and supportive matrix that facilitates essential cellular processes (Anitua et al. 2019). Chitosan, a biopolymer renowned for its remarkable biocompatibility, is obtained from the exoskeleton of crustaceans. Chitosan-based scaffolds have been utilised in the fields of wound healing and tissue engineering due to its capacity to enhance cell adhesion and regulate the inflammatory response (Gholap et al. 2023). Hyaluronic acid, a glycosaminoglycan that is present in high quantities within the extracellular matrix, presents a compelling natural substance for the development of bio-scaffolds. The applications of hyaluronic acid range from facilitating joint lubrication to supporting tissue engineering processes. In the field of tissue engineering, hyaluronic acid scaffolds play a crucial role in establishing an environment of favourable conditions that promotes cell proliferation and differentiation (Ruiz et al. 2019). Alginate, derived from brown seaweed, is utilised because of its capacity to generate hydrogels. Alginate-based scaffolds have been extensively utilised in numerous fields such as medication administration, wound dressings and tissue engineering. These scaffolds offer a three-dimensional structure that effectively supports cellular processes (Abasalizadeh et al. 2020). Silk fibroin, derived from the natural protein fibres of silk, has recently gained prominence as a bio-scaffold material due to its distinctive characteristics. The utilisation of silk fibroin in tissue engineering encompasses various domains, such as the restoration of the skin and the mending of the nerve tissue. These applications leverage the advantageous properties of silk fibroin, namely, its biocompatibility and mechanical robustness (Chen et al. 2023a, b). The aforementioned examples highlight the wide range of natural ingredients that function as the foundation for bio-scaffolds. Researchers strive to replicate and augment the complex interaction between cells and their surroundings by incorporating them into biomaterials. This endeavour holds the potential to drive progress in the fields of tissue engineering and regenerative medicine.

Bio-scaffolds, despite its significant potential in the fields of tissue engineering and regenerative medicine, have a number of problems that necessitate resolution in order to achieve successful implementation. One notable obstacle is to the concept of immunogenicity. Certain biomaterials used as scaffolds have the ability to elicit immunological responses, resulting in inflammation and the possibility of rejection. There are ongoing endeavours to change materials with the aim of decreasing immunogenicity and improving biocompatibility (Kasravi et al. 2023). The rate of biodegradation is a further vital concern. It is crucial to achieve a balance between scaffold degradation and tissue regeneration. Excessive degradation at a rapid rate might undermine the overall structural integrity, whereas gradual degradation can hinder the process of tissue integration. To address this issue, researchers concentrate on optimising the composition and structure of bio-scaffolds (Collins et al. 2021). Another obstacle that must be overcome is posed by the mechanical qualities of bio-scaffolds, particularly in load-bearing tissues. Improving the mechanical strength of a material by increasing the amount of reinforcing material it contains or making modifications to the fabrication process is an important field of research (Saiz et al. 2013). The creation of blood vessels inside the bio-scaffold, known as vascularisation, is a difficult problem to solve. Reduced nutrition and oxygen delivery to scaffold-bound cells result from inadequate vascularisation. New approaches, such as the addition of angiogenic agents, are being developed to enhance this quality (Lopes et al. 2021). It is difficult to achieve a uniform dispersion of cells within the bio-scaffold and to direct their differentiation. Attempts are made to improve cell adhesion, migration and differentiation by modifying surfaces, applying bioactive coatings and including signalling molecules (Boyetey et al. 2023). Maintaining the bio- and structural integrity of bio-scaffolds during sterilisation is a technological difficulty. Sterilisation methods that are both effective and compatible with various biomaterials are essential for therapeutic use (Tupone et al. 2021). Current research endeavours are focused on the identification and mitigation of factors that contribute to the degradation or instability of bio-scaffolds over prolonged durations (Okolie et al. 2023). Clinical use requires regulatory permission, which is a lengthy and stringent process. In order to streamline and speed up the approval process for bio-scaffold technologies, researchers, doctors and regulatory agencies must work together. Unlocking the full potential of bio-scaffolds in regenerative medicine will require a multidisciplinary approach to overcoming these obstacles through continuous research and collaboration between scientists, engineers, clinicians and regulatory authorities. This chapter is a comprehensive endeavour to gather all reliable published materials and organise them under a unified title, with the aim of facilitating future advanced research.

11.2 Chemical Composition of Natural Product-Based Bio-Scaffolds

Examples of various chemical components used for the preparation of natural product-based bio-scaffold and its functions are given in Table 11.1.

11.2.1 Collagen-Based Bio-Scaffolds

These are tridimensional formations predominantly consisting of collagen, a fibrous protein that serves as a prominent constituent of the extracellular matrix in numerous tissues. The bio-scaffolds have been engineered with the intention of emulating the inherent microenvironment of cells and tissues, so furnishing a conducive framework for cellular proliferation, movement and the process of tissue restoration. Collagen serves as the predominant structural protein inside the extracellular matrix of connective tissues. Bio-scaffolds made from collagen can be obtained from several sources, such as porcine, bovine or human tissues. The selection of the source material has the potential to impact the biocompatibility and immunogenicity of the scaffold. The collagen fibres exhibit a structural arrangement that mimics the original extracellular matrix, so serving as a natural scaffold characterised by a fibrous network. This architectural framework offers mechanical reinforcement and promotes cellular adhesion and movement (Khanna et al. 2021). Bio-scaffolds made of collagen scaffolds are extensively employed in the field of tissue engineering for the purpose of regenerating tissues that have been injured or destroyed. These techniques have been utilised to fix the bone, skin, cartilage, blood vessels and many other types of tissue. Cells have the ability to identify collagen as a native substrate, hence facilitating the processes of cell attachment and proliferation. The scaffold creates an ideal environment for cellular organisation and differentiation (Rai et al. 2017). Collagen-based scaffolds are commonly engineered with the intention of being susceptible to biodegradation. Over a period of time, the scaffolds undergo degradation within the body, facilitating the gradual replacement of the scaffold by the freshly generated tissue. The pace of deterioration can be adjusted to align with the natural tissue regeneration process (Goonoo et al. 2013). The stability and mechanical qualities of collagen scaffolds can be enhanced through the process of chemical or physical cross-linking. Cross-linking has a crucial role in regulating the rate of degradation and enhancing the overall functionality of the scaffold (Cornwell et al. 2007). Current research endeavours are centred around enhancing the mechanical characteristics, bioactivity and integration capabilities of scaffolds composed of collagen. The integration of sophisticated technologies, such as 3D printing, contributes to the improvement of precision and customisation capabilities. The utilisation of collagen-based bio-scaffolds has a significant potential in the realm of regenerative medicine and tissue engineering owing to their inherent biocompatibility, bioactivity and capacity to facilitate cellular proliferation and tissue regeneration (Bisht et al. 2021). Ongoing research is essential, however, in order to resolve obstacles and optimise their use in a variety of therapeutic applications.

Table 11.1 Chemical make-up of natural product-based bio-scaffold

S. no.	Name of bio-scaffold	Chemical composition	Functions	References
1.	Collagen	Glycine, prolin and hydroxyproline	Ligament repair and regeneration	Glowacki and Mizuno (2008)
2.	Laminin	One α and one β chain	Cell adhesion, bone tissue engineering	Iorio et al. (2015)
3.	Chitosan	(β1–4)-linked residues of *N*-acetyl-2-amino-2--deoxy-D-glucose (glucosamine, GlcN) and 2-amino-2-deoxy-D--glucose (*N*-acetyl glucosamine, GlcNAc) residues	Supports the attachment and proliferation of bone-forming osteoblast cells	Rodriguez-Vazquez et al. (2015)
4.	Alginate	1,4-linked β-D-mannuronic acid M and 1,4-α-I-guluronic acid G residues	Cell encapsulation, cell adhesion, immunomodulation	Zhang et al. (2023)
5.	Hydrogel	Hyaluronic acid, laminin, fibronectin and collagen	Cell adhesion, proliferation; transport gases, nutrients and growth factors	Zhou et al. (2022)
6.	Cellulose	Polysaccharides	Fixes muscle tissue	Varma and Iravani (2022)
7.	Porous scaffolds	Polysaccharides, polynucleotides and lipids	Control cell functions and guide regeneration of the functional tissue and organs	Alizadeh et al
8.	Ceramic scaffolds	Hydroxyapatite (HA) and β-tri-calcium phosphate (TCP)	Regenerate damaged tissues	Huang and Best (2022)
9.	Fibrous scaffold	Polysaccharides, alginate and proteins	Facilitates the formation of artificial functional tissues	Tang et al
10.	Microsphere scaffolds	Polylactic and poly(3-hydroxybutyrate-co-3-hydroxyvalerate)	Maintain the high level of interconnectivity among pores	Singh et al. (2008)
11.	Hyaluronic acid	Repeating disaccharides [β-1,4-D-glucuronic acid (known as uronic acid) and β-1,3-*N*-acetyl-D-glucosamide]	Provides an open, hydrated structure for the passage of nutrients	Ding et al. (2023)

(continued)

Table 11.1 (continued)

S. no.	Name of bio-scaffold	Chemical composition	Functions	References
12.	Polymer scaffolds	Poly(lactic acid) (PLA), poly(glycolic acid) (PGA) and their copolymer poly(lactic-co-glycolic acid) (PLGA)	Cell attachment, improve untoward issues associated with hydrophilicity and biodegradability	Kim et al. (2019)
13.	Magnesium scaffolds	WZM211(MgY 2 wt%, Zn 1 wt%, Mn 1 wt%) fibres	Bone fracture healing capacity	Yazdimamaghani et al. (2017)
14.	Fibroin	Recurrent amino acid sequence (Gly-Ser-Gly-Ala-Gly--Ala)	Cartilage, corneal, neural, skin tissue engineering	Krishani et al. (2023)
15.	Keratin	Various types of keratins, keratin-associated proteins (KFAPs) and enzymes	Vascular, skin, urethral, nerve, bone tissue engineering	Krishani et al. (2023)
16.	Elastin	Prolin, glycine, desmosine and isodesmosine	Cartilage, cardiac, vascular tissue engineering	Krishani et al. (2023)
17.	Gelatin	Arg-Gly-Asp (RGD) peptide sequence	Cell adhesion, proliferation and differentiation	Krishani et al. (2023)
18.	Fibrin	Amylose (a linear polymer linked by $\alpha(1–4)$ linkages) is connected to amylopectin (highly branched polymer) by $\alpha(1–6)$ linkages	Cell adhesion, growth, proliferation, differentiation and wound healing capacity	Krishani et al. (2023)
19.	Chitin	Repeated units of 2-(acetylamino)-2-deoxy-D-glucose	Helps in cell adhesion, differentiation and migration in tissue engineering	Krishani et al. (2023)
20.	Pullulan	Repeated maltotriose units connected by $\alpha(1–6)$ linkages	Plays a vital role in tissue engineering due to its adjustable property, biocompatibility, biodegradability and adhesive nature	Krishani et al. (2023)

11.2.2 Fibrin-Based Bio-Scaffolds

These are three-dimensional formations that are mainly made up of fibrin, which is a protein with a fibrous structure that is an essential component in the process by which blood clots. The process of blood clotting results in the formation of a protein called fibrin. Thrombin is responsible for its formation by transforming fibrinogen into thrombin. A mesh-like structure is created by fibrin, which serves as a scaffold for platelets and other cells that are engaged in the process of wound healing. Allowing fibrinogen to be transformed into fibrin by the enzymatic activity of thrombin is the first step in the formation of scaffolds based on fibrin. This process is an analogue of the natural mechanism that causes blood to clot (Ciolacu et al. 2022). Fibrin-based scaffolds have been extensively utilised in tissue engineering endeavours to facilitate the regeneration of diverse tissue types, such as the skin, bone, cartilage and nerve tissue. Additionally, these scaffolds have found utility in the field of regenerative medicine owing to their favourable biocompatibility, bioresorbability and capacity to facilitate cell adhesion and proliferation. Fibrin-based products are additionally employed in the context of wound healing to facilitate the process of tissue repair and regeneration (Ozdil and Aydin 2014). Fibrin serves as a highly favourable substrate for the processes of cell adhesion and migration. This substance facilitates the adhesion and proliferation of diverse cell types that play an integral part in the process of tissue regeneration (Chung and Park 2007). Enzymatic activities in the body, specifically plasmin, spontaneously destroy fibrin scaffolds. The manipulation of fibrinogen and thrombin concentrations during scaffold construction can exert control over the pace of disintegration. Fibrin-based scaffolds can have their mechanical qualities changed by changing the amount of fibrinogen and thrombin or by adding other materials to make the structure stronger. Cross-linking fibrin scaffolds improves their stability and mechanical qualities. Cross-linking agents or procedures can be used to modify the degradation rate and mechanical strength to meet the needs of tissue engineering (Noori et al. 2017). Fibrin has the potential to be integrated with various biomaterials, including polymers and ceramics, resulting in the development of hybrid scaffolds that exhibit enhanced mechanical characteristics and functions. The integration of bioactive molecules: Fibrin-based scaffolds have the capacity to incorporate bioactive compounds, growth factors or medicines, thereby augmenting tissue regeneration or facilitating regulated release for therapeutic applications. FDA approval: The Food and Drug Administration (FDA) of the United States has given its stamp of approval to items containing fibrin for use in a variety of medical applications (Madduri and Gander 2012; Roberts et al. 2020). Fibrin-based bio-scaffolds possess several notable benefits, including exceptional biocompatibility, facilitation of cellular proliferation and a regulated degradation profile. They are especially helpful in situations in which a temporary scaffold is required to direct the regeneration of the tissue and which must afterwards be gradually replaced by the body's native tissue (Fatimi et al. 2022). Current research efforts are persistently investigating strategies to enhance the efficiency of fibrin-based scaffolds in various tissue engineering contexts.

11.2.3 Poly(Lactic-co-Glycolic Acid) (PLGA) Bio-Scaffolds

These materials belong to the category of synthetic polymer-based substances utilised in the fields of tissue engineering and regenerative medicine. PLGA is a copolymer of glycolic acid and lactic acid that has been extensively utilised due to its biocompatibility, adaptability and controllable degradation kinetics. This copolymer consists of monomers of lactic acid and glycolic acid. This particular copolymer is made up of monomers derived from lactic acid and glycolic acid. It is possible to change the proportion of glycolic acid to lactic acid in the polymer chain in order to affect the material's qualities, such as the rate at which it degrades and its mechanical strength. One of the notable benefits associated with poly(lactic-co-glycolic acid) (PLGA) is its capacity to be adjusted in terms of its degradation profile. The manipulation of factors such as the ratio of lactic acid to glycolic acid, molecular weight and crystallinity can exert control on the pace of degradation. The inclusion of this feature is of greatest significance in order to align the degradation of the scaffold with the pace at which tissue regeneration occurs. This characteristic is very important for making sure that the rate of scaffold degradation matches the rate of tissue regeneration (Makadia and Siegel 2011; Gentile et al. 2014). The molecular weight and composition of PLGA can be altered in order to produce the desired changes in the material's mechanical properties. This makes it possible to customise scaffolds so that they can meet the specific mechanical requirements of various tissues (Alamán-Díez et al. 2022). Poly(lactic-co-glycolic acid) (PLGA) has the potential to be employed in the regulation of the release of bioactive molecules, growth factors or pharmaceutical compounds. The inclusion of these compounds within the polymer matrix facilitates the achievement of sustained release over an extended period. Complex and customised scaffolds with unique geometries for tissue engineering applications can be produced through the utilisation of diverse techniques such as 3D printing and electrospinning (Nikolova and Chavali 2019). The utilisation of PLGA scaffolds has been seen in the context of tissue regeneration, encompassing diverse tissues such as the bone, cartilage and nerve. The adaptability of PLGA enables its utilisation in various kinds of tissue and engineering specifications. PLGA bio-scaffolds present a synthetic alternative for tissue engineering applications, affording a substantial level of manipulation over material characteristics. The problems of using PLGA scaffolds and the ways in which they can be optimised for certain tissues are still being investigated (Litowczenko et al. 2022; Liu et al. 2023).

11.2.4 Chitosan-Based Bio-Scaffolds

Chitosan-based bio-scaffolds refer to three-dimensional structures predominantly consisting of chitosan, a naturally occurring polymer sourced from chitin, a substance present in the exoskeletons of crustaceans and insects. Chitosan has garnered a considerable interest within the realm of tissue engineering and regenerative medicine owing to its favourable attributes such as biocompatibility, biodegradability, antibacterial characteristics and capacity to facilitate cell adhesion and proliferation. Chitosan is obtained through the process of deacetylation of chitin, which is a polysaccharide present in the exoskeletons of crustaceans, insects and certain fungi (Federer et al. 2021). Chitosan has notable attributes such as biocompatibility, cationic properties and structural resemblance to glycosaminoglycans present in endogenous tissues. Chitosan facilitates the processes of cell adhesion, migration and proliferation. The positive charge of the entity enables it to engage in interactions with cell membranes and extracellular matrix components that possess a negative charge (Sahranavard et al. 2020). Chitosan exhibits biodegradability, with the rate of degradation being subject to regulation through the manipulation of parameters such as the degree of deacetylation. Enzymes, such as lysozyme, are involved in the degradation of chitosan, leading to the formation of biocompatible by-products. Because of its intrinsic antibacterial qualities, chitosan can be used in situations where preventing the spread of infection is of paramount importance. The cationic character of chitosan is responsible for this quality, as it can cause disruption to the cell membranes of microorganisms. By manipulating parameters including the degree of deacetylation, molecular weight and insertion of reinforcing chemicals, the mechanical characteristics of chitosan-based scaffolds can be fine-tuned (Wang et al. 2020). Chitosan scaffolds possess the capability to be tailored with precise porosity and interconnectivity, hence facilitating improved cell penetration, nutrient diffusion and waste elimination inside the scaffold structure. Chitosan-derived bio-scaffolds have a wide range of potential applications in the field of tissue engineering owing to its inherent natural composition, compatibility with biological systems and adjustable characteristics (Ciolacu et al. 2022). Ongoing research intends to further increase the clinical relevance of chitosan-based scaffolds by further optimising these scaffolds for use in a variety of tissue engineering applications.

11.2.5 Cross-Linking Agents

Cross-linking refers to the chemical or physical connection of polymer chains within a bio-scaffold, resulting in the formation of a three-dimensional network. The process of cross-linking contributes to the enhancement of stability and the establishment of structural integrity inside the scaffold (Chen et al. 2023a, b). The construction of bio-scaffolds can include the use of several cross-linking agents, with the selection of the appropriate cross-linking approach being contingent upon the specific polymer employed and the desired purpose of the scaffolds. This chemical cross-linking agent is frequently employed for the purpose of cross-linking proteins, including collagen (Zeugolis et al. 2009). The process of cross-linking polymer chains is achieved by glutaraldehyde through the formation of covalent bonds with amino groups. Nevertheless, the application of this substance is occasionally constrained as a result of its cytotoxic properties. Genipin, a natural cross-linking agent, is obtained from the fruit of the gardenia plant. It has been found to possess lower cytotoxicity compared to glutaraldehyde. Collagen-based scaffolds are frequently employed for this purpose (Oryan et al. 2018). Carbodiimides, such as EDC/NHS, are commonly used in various scientific and biomedical applications. Carbodiimides, such as 1-ethyl-3-(3-dimethylaminopropyl) carbodiimide (EDC) and *N*-hydroxysuccinimide (NHS), are frequently employed in the process of cross-linking carboxyl and amino groups inside polymer structures, such as collagen. The utilisation of EDC/NHS chemistry is frequently employed in the fabrication of hydrogel scaffolds. Dibenzaldehyde is employed in the process of cross-linking chitosan-based scaffolds, where it establishes Schiff base connections with amino groups (Campiglio et al. 2020). Isocyanates, such as hexamethylene diisocyanate (HMDI), possess the ability to facilitate the cross-linking of polymers that possess hydroxyl groups. Polyurethane-based scaffolds often employ them in a widespread manner (Swapan 2018). Thermal cross-linking refers to the ability of certain polymers to undergo cross-linking with the application of heat. One illustrative instance involves the application of heat to thermoplastics, so enabling intermolecular interactions to facilitate the formation of a network characterised by cross-linkages. Ultraviolet (UV) radiation has the capability to induce the cross-linking process of specific polymers, with a special emphasis on hydrogel systems. The addition of photoinitiators can be employed to enhance the efficiency of the process. The enzyme in question is employed for the enzymatic cross-linking of proteins, such as gelatin and collagen. The process involves the formation of covalent connections between lysine and glutamine residues. These molecules possess both hydrophobic and hydrophilic regions, enabling them to spontaneously form nanofibres through self-assembly. The process of self-assembly efficiently establishes cross-links among the peptide amphiphiles, resulting in the formation of a structurally stable scaffold (Song and Wang 2020; Unal and West 2020). The selection of the cross-linking technique is contingent upon various elements, such as the polymer's nature, the desired characteristics of the scaffold and the intended purpose. It is imperative to acknowledge that an excessive amount of cross-linking has the potential to impact the biocompatibility of the scaffold. Therefore, the optimisation process becomes pivotal in order to get the ideal equilibrium between stability and biofunctionality (Aljohani et al. 2018). Furthermore, in applications where the scaffold will come into contact with cells or tissues, the potential cytotoxicity of particular chemical cross-linkers should be taken into account.

11.3 Application of Natural Product-Based Bio-Scaffold

Bio-scaffolds derived from natural products have diverse applications in numerous disciplines (Fig. 11.1), demonstrating their adaptability in addressing critical issues in medicine and biotechnology. These bio-scaffolds are extremely important in tissue engineering because they provide a three-dimensional scaffolding that is quite similar to the extracellular matrix of natural tissues. Collagen scaffolds for skin regeneration and chitosan scaffolds for cartilage healing are two prominent examples of this phenomenon. These natural materials are ideal for cell adhesion, proliferation and differentiation because of their intrinsic biocompatibility (Rasouli et al. 2021; Bitting et al. 2022). The utilisation of bio-scaffolds derived from natural products greatly enhances the advantages of regenerative therapy. Scaffolds play a significant role in the process of tissue healing and organ replacement, facilitating the regeneration of injured tissues or the complete restoration of organs (Dromel and

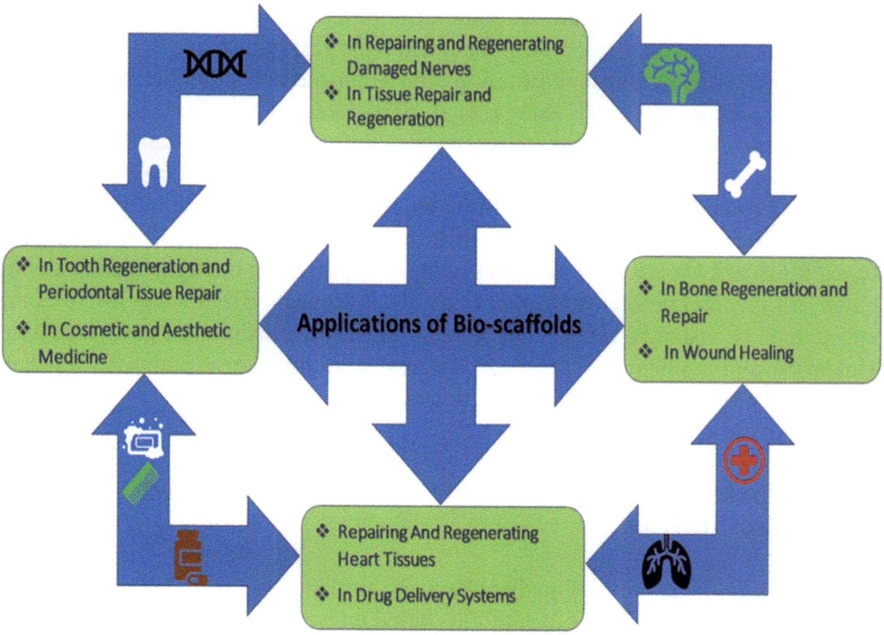

Fig. 11.1 Generalised perspective of bio-scaffold-based applications in the modern day science

Singh 2021). Materials having regenerative qualities, such as hyaluronic acid and silk fibroin, play an important role in tissue repair and regeneration. Bio-scaffolds made from natural materials, such as fibrin and chitosan, are used as wound dressings in wound healing. Their application in this context fosters a favourable environment for tissue repair, which aids in the healing process (Prasathkumar and Sadhasivam 2021). Drug delivery systems also utilise the distinctive characteristics of bio-scaffolds derived from natural products. Alginate and chitosan are commonly employed in the field of drug encapsulation to enable precise regulation and prolonged release of pharmaceutical substances. The porosity nature of these scaffolds enables the controlled and sustained release of medicinal substances over a certain duration (Cong et al. 2018). Bio-scaffolds have found usage in orthopaedics for purposes including bone regeneration and repair. Scaffolds are used to aid in the development and consolidation of bone, and they are often made from natural materials like collagen and hydroxyapatite (Paladini and Pollini 2022). The field of dental tissue engineering investigates the application of bio-scaffolds for the purpose of tooth regeneration and the repair of the periodontal tissue. Materials such as chitosan and collagen seem to be useful in providing a supporting structure that encourages the proliferation of dental cells and tissue regeneration (Moghadam et al. 2021). Natural product-based bio-scaffolds are commonly employed in cosmetic and aesthetic medicine, particularly in procedures such as dermal fillers. Hyaluronic acid-based scaffolds have been found to play a significant role in augmenting volume and diminishing the appearance of wrinkles. The field of neural tissue engineering derives advantages from the exploration of bio-scaffolds derived from natural products, which have the potential for the restoration and rejuvenation of impaired nerves. Silk fibroin scaffolds exhibit a considerable potential in facilitating the proliferation of brain cells (Magaz et al. 2020; Graciela et al. 2023). Applications in cardiovascular medicine investigate the possibility of using bio-scaffolds to mend and regenerate a damaged heart tissue. Scaffolds for the purpose of cardiac tissue engineering can be created using fibrin in addition to other naturally occurring materials. Because of their wide range of potential applications, natural product-based bio-scaffolds are becoming increasingly important in the fight against a wide variety of medical problems (Kumar et al. 2019; Chandy 2021). Ongoing research endeavours persistently aim to enhance and refine the utilisation of these technologies in current applications while also investigating innovative pathways for their implementation in various biomedical scenarios.

11.4 Toxicological Aspects of Natural Product-Based Bio-Scaffold

Particularly in the context of medical devices, implants and drug delivery systems, the toxicological features of bio-scaffolds based on natural products are critical concerns (few examples are given in Table 11.2) in their development and implementation (Abdellatif et al. 2021). Although natural products are commonly acknowledged for their biocompatibility, there are various aspects that can impact their safety profile.

11.4.1 Material Source and Purity

Toxicology considerations in bio-scaffolds derived from natural products entail evaluating the safety and potential risks of the materials used in biomedical applications. This procedure is essential for ensuring that the materials do not injure living organisms and are fit for their intended purpose (Reina et al. 2017). In the case where the bio-scaffold is obtained from botanical sources, it is crucial to ascertain the precise botanical origin. Plants exhibit heterogeneity in the distribution of bioactive molecules among their many anatomical structures, and certain plant species may harbour constituents that possess hazardous properties. The composition of the bio-scaffold can be influenced by the method employed for its extraction from the natural source (Gurib-Fakim 2006). For example, certain solvents may leave residues that could be harmful; therefore it is needed to be ensured that the bio-scaffold is free from contaminants such as heavy metals, pesticides and other chemical impurities after successful extraction was obtained. Analytical techniques like chromatography and spectroscopy can be employed to assess purity of the material followed by the sterilisation process to eliminate the presence of microorganisms, such as bacteria, fungi or endotoxins. Moreover, identification and characterisation of the bioactive compounds present in the natural products are an important part since

Table 11.2 Table displaying the toxic effects of the bio-scaffold to human health

S. no.	Type of bio-scaffold	Human toxicity	References
1.	Collagen based	Eye irritation, allergic reaction, negative effect of wound healing	Dong and Lv (2016)
2.	Gelatin based	Inhibition of blood protein production, much weight loss	Robscheit-Robbins et al. (1944)
3.	Silk fibroin	Peripheral arterial disease, mitochondrial Dysfunction lipid peroxidation, mitochondrial membrane collapse	Naserzadeh et al. (2018)
4.	Cellulose nanofibres	Skin and eye toxicity	Kim et al. (2019)
5.	Chitosan based	Reduced cell viability, proliferation	Matica et al. (2017)
6.	Alginate	Renal toxicity, hypochromic and microcytic anaemia	Braga et al. (2005)
7.	Xanthan gum	Digestive issues, allergies	Dzionek et al. (2022)
8.	Hyaluronic acid based	Allergic reaction, cytotoxic	Jansen et al. (2004)
9.	Chondroitin sulphate	Hepatotoxicity, liver disease	Cerda et al. (2013)
10.	Starch based	Anaemia	Boyd et al
11.	Glucosamine sulphate based	Liver disease	Cerda et al. (2013)
12.	Guar gum	Abdominal discomfort in children	Mortensen et al. (2017)
13.	Pullulan	Hypersensitivity pneumonitis, respiratory allergies	Kumari and Singh (2022)
14.	Actin	Cell necrosis, hepatic necrosis, adult respiratory distress syndrome	William et al
15.	Keratin	Toxic effects on cell growth	Yamauchi et al. (1998)
16.	Laminin	Kidney disease	Abrass et al. (2010)
17.	Zein	Allergic reactions	Pastorello et al. (2009)
18.	Glutenin	Coeliac disease, inflammation	Dewar et al. (2006)
19.	Myosin	Hypertrophic or dilated cardiomyopathy	Alamo et al. (2017)
20.	Gellan gum	Abnormalities in the gut lining	Petre and Bjarnadottir (2019)

some compounds may have therapeutic effects, while others could be toxic; therefore understanding the composition is critical for safety assessment. Apart from this the dose-response relationship of the bio-scaffolds is critical as few substances that are generally safe can be harmful at high concentrations when used in bio-scaffolds. Toxicological studies should explore a range of doses to establish safety levels. Both in vivo (animal studies) and in vitro (cell culture) toxicity testing are essential to evaluate the safety profile of bio-scaffolds. Assessments may include acute toxicity, sub-chronic toxicity and chronic toxicity studies. Again, evaluation of the biocompatibility of the bio-scaffold with the target tissues is essential. This involves assessing the immune response, inflammatory reactions and other biological interactions. Chronic toxicity studies can help identify potential health risks over extended periods. It is important to ensure that the bio-scaffold complies with regulatory standards for biomedical applications. Different countries may have specific regulations regarding the use of natural products in medical devices or therapies; therefore it is utmost necessary to clearly label the product with information on its composition, recommended usage and any potential risks.

11.4.2 Immunogenicity and Allergic Reactions

Immunogenicity and allergic reactions are crucial factors in the development of bio-scaffolds derived from natural products for biomedical applications. These aspects include comprehending how the immune system of the body reacts to the substance and assessing the likelihood of adverse reactions (Yang et al. 2019). If the bio-scaffold contains proteins or peptides, there is a risk of immunogenicity among sensitive patients. Identification and characterisation of these components to assess their potential to elicit an immune response is an essential part of bio-scaffold research. Certain natural products may also contain polysaccharides or glycoproteins that can trigger immune reactions. Conduction of animal studies to assess the immunogenicity of the bio-scaffold is of utmost importance for monitoring the signs of immune response, including the production of antibodies and inflammatory reactions. Performing in vitro assays to assess the potential of the bio-scaffold to stimulate immune cells is essential since these assays can include lymphocyte proliferation assays and cytokine release assays. Some individuals may have pre-existing allergies to certain substances, and exposure to these allergens can lead to allergic reactions. Conduction of patch testing to assess the potential for skin sensitisation or contact dermatitis is an essential testing measure for successful application of bio-scaffold in the human body. This is especially important for materials that may come into direct contact with the skin. Moreover, implementation of proper monitoring protocols during preclinical and clinical trials to detect any signs of immunogenicity or allergic reactions is important to ensure bio-scaffold compatibility. This includes regular assessments of patient health and immune responses.

11.4.3 Biocompatibility and Cytotoxicity

The consideration of biocompatibility and cytotoxicity is of utmost importance during the creation of bio-scaffolds derived from natural products, particularly for their potential use in biomedical applications. These criteria evaluate the extent to which the material interacts with biological systems and its potential to cause adverse effects (Mansur and Costa 2008). The metabolic activity of cells exposed to the substance can be measured using cell viability tests like MTT or Alamar Blue (Rampersad 2012). Elevated concentrations of cytokines may serve as an indicator of an unfavourable inflammatory response. In the event of the bio-scaffold being exposed to blood, it is necessary to evaluate its compatibility with blood components (Lu et al. 2022). Haemolysis and coagulation assays can shed light on potential blood-harming effects. The examination of the bio-scaffold's compatibility with specific tissues is an essential consideration. This is especially crucial for substances intended for implantation or contact with specific organs (Singh et al. 2020). Perform in vivo investigations, such as implantation tests conducted on animal models, to assess the comprehensive biocompatibility of the material. It is recommended to observe and assess indicators of inflammation, tissue integration and foreign body response (Helenius et al. 2006).The use of relevant cell culture models to assess cytotoxicity is essential. Different cell types may exhibit varying sensitivities to the material, so choosing cell lines that are appropriate for the intended application is a major concern. After exploring a range of concentrations, only the desired one is to be selected to establish a dose-response relationship. Understanding of whether cytotoxic effects are concentration-dependent or not, knowledge about differentiation between apoptosis (programmed cell death) and necrosis (uncontrolled cell death), and theoretical understanding of mode of cell death can provide insights into the mechanisms of cytotoxicity associated with bio-scaffolds. Bio-scaffold degradation products should be tested for cytotoxicity. Certain degradation products may have distinct biological effects than the parent substance (Barnes et al. 2016). The exploration of surface changes as a means to improve biocompatibility is worth considering. The application of surface treatments or coatings has the potential to mitigate cytotoxicity and enhance the interaction with biological tissues (Uddin et al. 2015). Researchers can guarantee the safety of bio-scaffolds derived from natural products by thoroughly testing them for biocompatibility and cytotoxicity. It is recommended to engage in collaborative efforts with professionals specialising in the fields of cell biology, toxicology and regulatory affairs in order to ensure the efficient design and accurate interpretation of these investigations.

11.4.4 Inflammatory Responses

Inflammation is essential to the body's defence against injury and illness. Bio-scaffolds made from natural products are artificial scaffolds engineered to promote wound healing and tissue regeneration. A bio-scaffold's ability to interact with the immune system is essential to its performance. There are several natural products that have qualities that can reduce inflammation. Bio-scaffolds that are formed from these natural products have the ability to modify the inflammatory response, which can assist in the upkeep of a healthy equilibrium between pro-inflammatory and anti-inflammatory signals (You et al. 2022). For example, Curcumin, which comes from turmeric, and resveratrol, which can be found in grapes and berries, are two examples of chemicals that are known for their ability to reduce inflammation. The use of such natural products within bio-scaffolds has the potential to assist in the reduction of excessive inflammation (Inchingolo et al. 2022). Natural products frequently exhibit immunomodulatory qualities, hence exerting an influence on the functioning of immune cells. The utilisation of this approach holds potential advantages within the realm of bio-scaffolds for the purpose of immune response regulation. One example of a substance that has been extensively researched for its immunomodulatory properties is ginseng. When included into a biological scaffold, it has the potential to enhance the regulation and equilibrium of the immune response in the context of tissue regeneration (Sarasati et al. 2023). Natural products may be perceived as less foreign by the body, thereby decreasing the likelihood of a foreign body reaction to the implanted bio-scaffold. This can contribute to improved tissue integration and assimilation. Materials derived from chitin in crustacean shells, such as chitosan, have been investigated for their biocompatibility and anti-inflammatory properties (Chung et al. 2017; Gholap et al. 2023). Bioactive compounds such as polyphenols, flavonoids and alkaloids are frequently found in natural products. These chemicals have the potential to modulate signalling pathways that are implicated in the process of inflammation. The integration of these bioactive compounds into bio-scaffolds has the potential to provide a regulated release, thereby exerting an influence on the nearby cellular environment (Schilrreff and Alexiev 2022). The utilisation of bio-scaffolds derived from natural products is now being investigated for a range of clinical purposes, such as wound healing, orthopaedic implants and tissue engineering. The comprehension and enhancement of their interactions with the immune system are crucial for achieving favourable results (Krishani et al. 2023). In conclusion, bio-scaffolds based on natural products have complex inflammatory responses that include immune system regulation, inflammation control and stimulation of repair mechanisms. The integration of natural substances into biomaterial scaffolds shows potential in improving biocompatibility and facilitating effective tissue regeneration. Nevertheless, it is imperative to conduct thorough investigations and rigorous clinical trials in order to gain a comprehensive understanding and effectively exploit the capabilities of these substances in many medicinal contexts.

11.4.5 Haemocompatibility

Haemocompatibility is a crucial factor in the development of bio-scaffolds derived from natural products, especially when these materials come into contact with blood. The establishment of compatibility between blood components is crucial for various medical applications, including implants and equipment intended for medicinal purposes. Bio-scaffolds can acquire favourable haemocompatible characteristics by utilising natural ingredients, which are renowned for their wide range of bioactive components. The fact that certain natural chemicals have an anticoagulant effect is an important factor. Heparin, which comes from both plants and animals, is one example of a substance that has anticoagulant properties. The addition of such components into bio-scaffolds has the potential to reduce the risk of unintended blood clot formation, hence increasing the haemocompatibility of the material as a whole (Weber et al. 2018; Ji et al. 2023). In addition, bio-scaffolds' interactions with blood are heavily dependent on their surface properties. Inherent biocompatibility of natural goods aids in the development of surfaces that are less prone to cause unwanted reactions like clot formation. Chitosan, derived from chitin, is a natural substance that has been extensively studied for its biocompatibility and ability to minimise thrombogenicity. In addition to its role in avoiding clot formation, the minimisation of platelet activation is another vital component of haemocompatibility. Compounds like polyphenols that are found in nature and have antiplatelet qualities can be added to bio-scaffolds to stop platelets from activating and lower the risk of thrombosis (Brovold et al. 2018). The maintenance of haemocompatibility also requires the encouragement of the proliferation of endothelial cells. The endothelial cells that line blood arteries are critically important in preventing the formation of blood clots. A more haemocompatible environment can be achieved with the use of bio-scaffolds that are designed to encourage the growth and activity of endothelial cells and that commonly imitate the extracellular matrix (Yao et al. 2022). Biodegradability is a quality that makes long-term haemocompatibility better. Bio-scaffolds derived from natural sources that undergo gradual degradation without generating detrimental by-products have the advantage of reducing the potential for persistent inflammatory reactions or blood clot formation (Krishani et al. 2023).

The haemocompatibility of bio-scaffolds can be modified using surface modification approaches. To increase these scaffolds' compatibility with human blood, it is possible to graft or coat them with natural materials. However, thorough biological testing is still required (Pankajakshan and Agrawal 2010; Arora et al. 2021). In vitro and in vivo studies should evaluate coagulation time, platelet adhesion and haemolysis, among other parameters. Only through such exhaustive evaluations can the haemocompatibility of bio-scaffolds derived from natural products be comprehensively understood and guaranteed (Weber et al. 2018). In summary, the attainment of haemocompatibility in bio-scaffolds derived from natural products necessitates a sophisticated strategy that integrates the intrinsic characteristics of these materials with meticulous design considerations. The objective is to develop bio-scaffolds that possess the dual capability of facilitating tissue regeneration and establishing a harmonic interaction with the circulatory system, so ensuring the safety and effectiveness of their utilisation in medical contexts.

11.4.6 Organ Toxicity

Organ toxicity is of utmost importance in the development and application of bio-scaffolds derived from natural sources, especially when contemplating their use in medical devices and implants (Bettini et al. 2020). Natural goods are commonly acknowledged for their biocompatibility; yet, it is crucial to address certain criteria to assure their safety, particularly in relation to specific organs. Toxicological concerns into bio-scaffolds extracted from natural product are given in Fig. 11.2. The liver, which serves as a pivotal metabolic centre, can potentially experience alterations due to the metabolic processes involving natural compounds. The investigation of potential harmful metabolites and the assessment of their effects on hepatic cells are of utmost importance, necessitating the use of in vitro and in vivo investigations (Piccinin et al. 2019). Degradation products or unmetabolised components

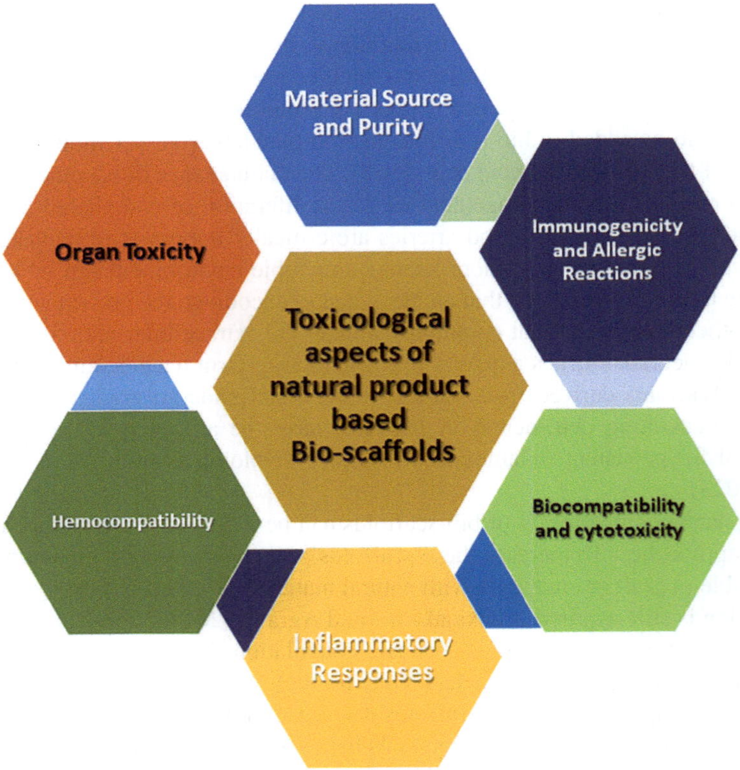

Fig. 11.2 Toxicological insights into bio-scaffolds extracted from natural products

may have an effect on kidney function; thus it's important to examine renal clearance pathways. The safety of bio-scaffolds in connection to kidney health is best evaluated by renal toxicity studies (Feng et al. 2010). The study of cardiotoxicity is crucial due to the potential interaction between bio-scaffold materials and cardiac tissues. The monitoring of electrocardiographic data and the performance of investigations on cardiac cell lines and animal models might yield valuable insights into the potential cardiotoxic consequences (Chintanaphol et al. 2022). Pulmonary consequences are something that need to be taken into consideration, particularly in cases where the bio-scaffold components can be breathed in. The evaluation of the influence on lung tissues can be aided by the study of respiratory exposure as well as the performance of pulmonary toxicity evaluations (Pacurari et al. 2016). Concerns of absorption and discomfort are raised whenever there are interactions with the gastrointestinal tract. It is essential, for the purpose of gaining an understanding of the effects on the digestive system, to conduct research both in vitro and in vivo to evaluate gastrointestinal absorption and local irritation (Winkler et al. 2018). The possibility of neurotoxicity, or detrimental effects on neural tissues, requires investigation. Assessing effects on the nervous system requires both in vitro investigations of neuronal cells and in vivo evaluations. It is crucial to conduct immunotoxicity tests in order to evaluate the effect on immune systems. Assessing the risk of immune system activation and the possibility for detrimental effects on lymph nodes and the spleen is part of this process (Liang et al. 2022). The effects on reproduction and growth need to be carefully looked at, especially when pregnant women might be exposed. Animal models used for reproductive and developmental toxicity studies give us significant details about how chemicals might affect fertility, pregnancy and foetal development (Pentšuk and van der Laan 2009). The issue of endocrine disruption, which results in the occurrence of hormonal imbalances, is a matter of concern that necessitates investigation through both in vitro and in vivo studies, with a particular emphasis on the examination of hormone-producing organs (Vieira et al. 2021). Targeted toxicity studies are necessary to evaluate the localised tissue effects of a bio-scaffold when it is implanted in a specific anatomical setting. These assessments facilitate the comprehension of the localised effects on particular tissues or organs (Nikolova and Chavali 2019). In a nutshell a thorough assessment of the possible organ toxicity of nature-based bio-scaffolds is necessary to guarantee their safety. By integrating in vitro and in vivo investigations, we can learn more about how these materials will behave in the body as a whole, and as a result, we can better tailor them to ensure they are safe for use in healthcare facilities. The involvement of toxicologists and adherence to regulatory criteria are essential components of this process.

11.5 Regulatory Aspects of Natural Product-Based Bio-Scaffold

The regulations pertaining to bio-scaffolds derived from natural products exhibit significant variations across diverse governmental bodies and geographical areas. It is imperative to understand that regulations are susceptible to modification, and it is preferable to check out the most recent guidelines provided by pertinent regulatory bodies. The Food and Drug Administration (FDA) is responsible for overseeing the regulation of natural product-based bio-scaffolds in the United States. The Food and Drug Administration (FDA) has implemented regulatory frameworks in order to guarantee the safety and efficacy of medical devices, namely, those pertaining to biomaterials and bio-scaffolds (Jeevanandam et al. 2018). Medical devices, such as bio-scaffolds, are classified according to their intended purpose and the level of danger they pose. The Food and Drug Administration (FDA) utilises a classification framework that spans from Class I, denoting minimal risk, to Class III, indicating a significant danger. Bio-scaffolds are expected to be categorised under Class II or III. Manufacturers commonly initiate the process of obtaining permission for their bio-scaffold by submitting a 510(k) application, which serves to establish the significant equivalence of their product to an already legally marketed device. For products with a higher risk, a PMA application may be needed, which includes more information. Bio-scaffolds must meet strict biocompatibility standards because they come in touch with living tissues. To do this, possible toxicity, immunogenicity and other cellular effects must be assessed. The safety of bio-scaffold materials, whether natural or manufactured, requires thorough analysis. Natural product dangers must be identified and managed (Ghelich et al. 2022).

The European Medicines Agency (EMA) is responsible for regulating bio-scaffolds in the European Union (EU), including those based on natural products (Isasi et al. 2018). Since rules and regulations can change, it's best to check the EMA's website for the most up-to-date information on the official guidelines and regulations. Good Manufacturing Practice (GMP) regulations must be followed by companies making bio-scaffolds. GMP makes sure that the goods are always made and checked to make sure they meet quality standards. As part of the marketing authorisation procedure, a comprehensive quality and safety assessment is conducted on pharmaceuticals. This evaluation includes a review of the production method, composition and quality control. The continuous monitoring of bio-scaffolds is crucial for ensuring their safety and performance. Manufacturers have a responsibility to incorporate pharmacovigilance and post-market surveillance systems in order to identify and document any untoward incidents (Kingston et al. 2021; Laurent et al. 2021). Biological products are regulated in China by the National Medical Products Administration (NMPA), and the approval procedure includes evaluations of the product's quality, safety and efficacy. The approval of bio-scaffolds that are based on natural products is subject to China's very own unique set of rules (Tian et al. 2022). The PMDA in Japan oversees the approval process for biological goods, which includes checks for quality, safety and efficacy (Shin and Kim 2022). Harmonisation attempts, like those led by the International Council for Harmonisation of Technical Requirements for Pharmaceuticals for Human Use (ICH), try to make rules the same in all areas. However, different countries may have slightly different rules. There may be special considerations for bio-scaffolds based on natural products used in alternative medicine. In many nations, the practice of traditional medicine is governed by particular norms or guidelines, which take into account its cultural and historical context. During the first phases of research and development, initiatives involving bio-scaffolds may be required to comply with ethical norms and local rules pertaining to research (Molzon et al. 2011; Pazhouhnia et al. 2022).

11.6 Regulatory Aspects of Natural Product-Based Bio-Scaffold in India

In India, the Central Drugs Standard Control Organisation (CDSCO), which is part of the Indian Ministry of Health and Family Welfare, is in charge of regulating the environment in which natural product-based bio-scaffolds can be used in the country. The regulatory framework is characterised by its dynamic nature and susceptibility to modifications. Therefore, it is recommended to refer to the most recent guidelines and regulations in order to obtain the most up-to-date information (Patil and Das 2023). The following is an outline of the common procedures that are followed in India in order to regulate bio-scaffolds:

Bio-scaffolds made from natural products can be categorised as either medical devices or pharmaceuticals, depending on their intended purpose and other characteristics. The regulatory pathway is determined by classification (Dromel and Singh 2021). The bio-scaffold may fall under the purview of the pharmaceutical industry if its primary function is therapeutic. The New Drug Application (NDA) process may be necessary for approval. The implementation of rigorous quality control and assurance systems is of utmost importance. This encompasses the examination and evaluation of the bio-scaffold in order to ascertain its compliance with predetermined quality criteria. Clinical trials may be necessary if the bio-scaffold is intended for therapeutic purposes. The data obtained from these trials will serve as an essential element in the regulatory submission. The New Drug Administration (NDA) process is frequently followed when it comes to pharmaceutical items. This necessitates the provision of exhaustive information regarding the product's viability, efficiency and standard of excellence (Vishal et al. 2014; Tsintou et al. 2021). Continuous post-market surveillance is required in order to effectively monitor the safety of bio-scaffolds. It is anticipated that manufacturers will be required to establish and execute pharmacovigilance processes, as well as promptly notify any instances of adverse events (Kumar 2017; Mironska et al. 2019). During the product development process, developers and manufacturers are urged to work with the CDSCO. Getting scientific information and holding meetings before submitting can help you figure out what the rules are. Sometimes rules and regulations change, so it's important to know about any new or updated rules or directions from the CDSCO (Pawar et al. 2023). In a nutshell, the regulatory framework governing natural product-derived bio-scaffolds in India encompasses categorisation according to their intended application (medical device or pharmaceutical), compliance with established quality criteria, provision of complete data and implementation of post-market monitoring. Active involvement with the Central Drugs Standard Control Organisation (CDSCO) during the entirety of the product development process is crucial in order to guarantee adherence to regulatory obligations. It is advisable to consistently consult the most recent rules and regulations provided by the Central Drugs Standard Control Organisation (CDSCO) in order to obtain the most precise and up-to-date information.

11.7 Current Understanding and Future Prospects

The potential future implications of utilising bio-scaffolds derived from natural products are highly promising and expected to witness breakthroughs and widespread usage in diverse disciplines (Fig. 11.3). The potential of bio-scaffolds derived from natural products for use in regenerative medicine is significant. These materials have the ability to be customised in order to facilitate the process of tissue regeneration and repair, rendering them highly advantageous in several fields like bone regeneration, wound healing and organ transplantation. Bio-scaffolds have the potential to be used as drug delivery vehicles. Drug distribution, bioavailability and

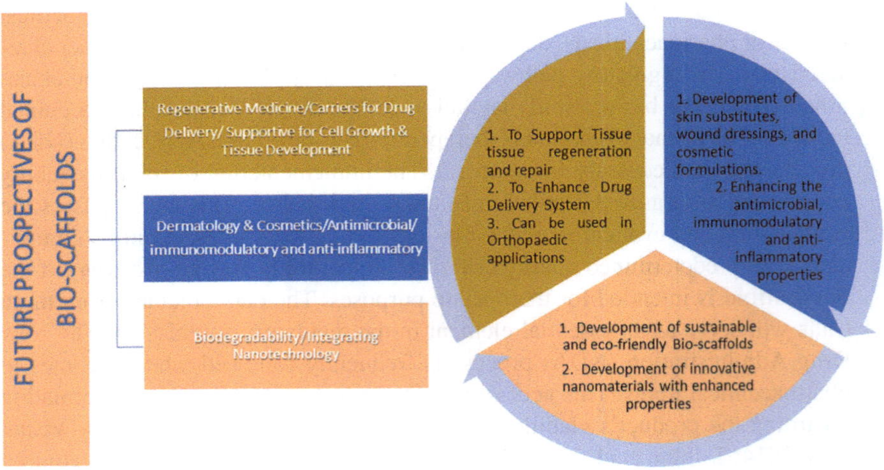

Fig. 11.3 Futuristic approach-based design for natural product-based bio-scaffold

controlled release of medicinal substances can all be improved by including natural ingredients into these scaffolds. The discipline of tissue engineering encompasses the development of functional tissues and organs with the purpose of transplantation. Bio-scaffolds made from natural products can be very helpful in creating an ideal environment for cells to grow and tissues to form. Orthopaedic uses for bio-scaffolds generated from natural sources include the creation of implants for joint and bone healing. Because of their biocompatibility, these materials have a better chance of fusing with the host's tissue. Bio-scaffolds made from natural products can be used in skincare and cosmetics. These substances can be used to make skin replacements, wound dressings and cosmetics that help the skin grow back faster. Natural products are widely recognised for their inherent antibacterial characteristics. The integration of these characteristics into biological scaffolds has the potential to facilitate the creation of materials that possess inherent resistance to infections, rendering them appropriate for use in medical implants and wound management. Immunomodulatory and anti-inflammatory effects can frequently be found in naturally occurring substances. The use of bio-scaffolds constructed with these natural materials has the potential to modify immunological responses, which could lead to a reduction in inflammation and an acceleration of the healing process in a variety of medical settings. The increasing emphasis placed on sustainable and environmentally friendly materials is compatible with the biodegradability of natural product-based bio-scaffolds. The creation of biodegradable scaffolds has the potential to lessen the influence that humans have on the environment over the long run. The progression of bio-scaffold technology has the potential to make significant contributions to the advancement of personalised medical methodologies. Customising scaffolds to meet the specific requirements of individual patients has the potential to improve treatment outcomes, particularly in disciplines such as tissue engineering and regenerative medicine. Combination therapy using natural product-based bio-scaffolds could have synergistic benefits. This method may improve medical treatments. The integration of sophisticated imaging technologies into bio-scaffolds has the potential to provide the continuous observation of tissue regeneration processes in real time. The assessment of treatment success and subsequent changes may be contingent upon this factor. The combination of nanotechnology and bio-scaffolds derived from natural products has the potential to result in the production of novel materials that have improved characteristics. It may be possible to include nanoparticles obtained from natural products in order to provide targeted drug administration and imaging.

11.8 Conclusions

In summary, bio-scaffolds derived from natural products hold a significant potential in the domains of biotechnology and medicine. The bio-scaffolds, which are produced from naturally occurring substances, provide a vast array of intricate and diverse molecular structures that possess inherent biological activity. The use of natural products as scaffolds establishes a fundamental basis for the advancement of innovative pharmaceuticals, medicinal agents and biomaterials. One of the primary benefits associated with bio-scaffolds derived from natural products is their extensive track record of safety and effectiveness in traditional medicinal practises. Nature has created complex chemical structures that often have very specific and strong biological effects. If these scaffolds are used in drug research and development, it might be possible to make medicines with better pharmacological profiles and fewer side effects. The investigation and application of bio-scaffolds derived from natural products are regarded as a rapidly evolving and very promising area within the fields of bioengineering and drug development. Through ongoing research, innovation and interdisciplinary collaboration, these bio-scaffolds possess the capacity to make substantial contributions to the advancement of next-generation therapies and biomedical technology.

References

Abasalizadeh F, Moghaddam SV, Alizadeh E, Akbari E, Kashani E, Fazljou SMB, Torbati M, Akbarzadeh A (2020) Alginate-based hydrogels as drug delivery vehicles in cancer treatment and their applications in wound dressing and 3D bioprinting. J Bio Eng 14(8):1–22

Abdellatif AAH, Mohammed HA, Khan RA, Singh V, Bouazzaoui A, Yusuf M, Akhtar N, Khan M, Al-Subaiyel A, Mohammed SAA, Al-Omar MS (2021) Nano-scale delivery: a comprehensive review of nano-structured devices, preparative techniques, site-specificity designs, biomedical applications, commercial products, and references to safety, cellular uptake, and organ toxicity. Nanotechnol Rev 10:1493–1559. https://doi.org/10.1515/ntrev-2021-0096

Abrass CK, Hansen KM, Patton BL (2010) Laminin α4-null mutant mice develop chronic kidney disease with persistent overexpression of platelet-derived growth factor. Am J Pathol 172(2):839–849. https://doi.org/10.2353/ajpath.2010.090570

Alamán-Díez P, García-Gareta E, Napal PF, Arruebo M, Pérez MA (2022) In vitro hydrolytic degradation of polyester-based scaffolds under static and dynamic conditions in a customized perfusion bioreactor. Materials 15(7):1–18. https://doi.org/10.3390/ma15072572

Alamo L, Ware JS, Pinto A, Gillilan RE, Seidman JG, Seidman CE, Padron R (2017) Effects of myosin variants on interacting-heads motif explain distinct hypertrophic and dilated cardiomyopathy phenotypes. Elife 6:e24634. https://doi.org/10.7554/eLife.24634

Aljohani W, Ullah MW, Zhang X, Yang G (2018) Bioprinting and its applications in tissue engineering and regenerative medicine. Int J Biol Macromol 107(Pt A):261–275. https://doi.org/10.1016/j.ijbiomac.2017.08.171

Anitua E, Nurden P, Prado R, Nurden AT, Padilla S (2019) Autologous fibrin scaffolds: when platelet-and plasma-derived biomolecules meet fibrin. Biomaterials 192:440–460. https://doi.org/10.1016/j.biomaterials.2018.11.029

Arora D, Pant P, Sharma PK (2021) Trends in functional biomaterials in tissue engineering and regenerative medicine. In: Bhaskar B, Sreenivasa Rao P, Kasoju N, Nagarjuna V, Baadhe RR (eds) Biomaterials in tissue engineering and regenerative medicine. Springer, Singapore, pp 215–269. https://doi.org/10.1007/978-981-16-0002-9

Barnes EC, Kumar R, Davis RA (2016) The use of isolated natural products as scaffolds for the generation of chemically diverse screening libraries for drug discovery. Nat Prod Rep 33(3):372–381. https://doi.org/10.1039/c5np00121h

Bettini S, Bonfrate V, Valli L, Giancane G (2020) Paramagnetic functionalization of biocompatible scaffolds for biomedical applications: a perspective. Bioengineering 7(4):153. https://doi.org/10.3390/bioengineering7040153

Bhadra B, Sakpal A, Patil S, Patil S, Date A, Prasad V, Dasgupts S (2021) A guide to collagen sources, applications and current advancements. Sys Bios Eng 1(2):67–87. https://doi.org/10.37256/sbe.1220211043

Bisht B, Hope A, Mukherjee A, Paul MK (2021) Advances in the fabrication of scaffold and 3D printing of biomimetic bone graft. Ann Biomed Eng 49(4):1128–1150. https://doi.org/10.1007/s10439-021-02752-9

Bitting S, Derme T, Lee J, Mele TV, Dillenburger B, Block P (2022) Challenges and opportunities in scaling up architectural applications of mycelium-based materials with digital fabrication. Biomimetics 7(2):44. https://doi.org/10.3390/biomimetics7020044

Boyetey MJB, Torgbo S, Sukyai P (2023) Bio-scaffold for bone tissue engineering with focus on bacterial cellulose, biological materials for hydroxyapatite synthesis and growth factors. Eur Polym J 194:112168. https://doi.org/10.1016/j.europolymj.2023.112168

Braga AS, Catirse ABCBE, Vaz LG, Spadaro ACC (2005) Quantitative analysis of potentially toxic metals in alginates for dental use. Rev Ciênc Farm Básica Apl 26(2):125–130

Brovold M, Almeida JI, Pla-Palacín I, Sainz-Arnal P, Sánchez-Romero N, Rivas JJ, Almeida H, Dachary PR, Serrano-Aulló T, Soker S, Baptista PM (2018) Naturally-derived biomaterials for tissue engineering applications. In: Chun H, Park K, Kim CH, Khang G (eds) Novel biomaterials for regenerative medicine. Adv Exp Med Biol, vol 1077. Springer, Singapore, pp 421–449. https://doi.org/10.1007/978-981-13-0947-2

Campiglio CE, Ponzini S, De Stefano P, Ortoleva G, Vignati L, Draghi L (2020) Cross-linking optimization for electrospun gelatin: challenge of preserving fiber topography. Polymers 12(11):2472. https://doi.org/10.3390/polym12112472

Cerda C, Bruguera M, Peres A (2013) Hepatotoxicity associated with glusamine and chondroitin sulfate in patients with liver disease. World J Gastroenterol 19(32):5381–5384. https://doi.org/10.3748/wjg.v19.i32.538

Chen K, Li Y, Pan W, Tan G (2023a) Silk fibroin combined with electrospinning as a promising strategy for tissue regeneration. Macomol Biosci 23(2):e2200380. https://doi.org/10.1002/mabi.202200380

Chen W, Ming Y, Wang M, Huang M, Liu H, Huang Y, Huang Z, Qing L, Wang Q, Jia B (2023b) Nanocomposite hydrogels in regenerative medicine: applications and challenges. Macromol Rapid Commun 44(15):e2300128. https://doi.org/10.1002/marc.202300128

Chintanaphol M, Orgil BO, Alberson NR, Towbin JA, Purevjav E (2022) Restrictive cardiomyopathy: from genetics and clinical overview to animal modeling. Rev Cardiovasc Med 23(3):108. https://doi.org/10.31083/j.rcm2303108

Chung HJ, Park TG (2007) Surface engineered and drug releasing pre-fabricated scaffolds for tissue engineering. Adv Drug Deliv Rev 59(4–5):249–262. https://doi.org/10.1016/j.addr.2007.03.015

Chung L, Maestas DR Jr, Housseau F, Elisseeff JH (2017) Key players in the immune response to biomaterial scaffolds for regenerative medicine. Adv Drug Deliv Rev 114:184–192. https://doi.org/10.1016/j.addr.2017.07.006

Ciolacu DE, Nicu R, Ciolacu F (2022) Natural polymers in heart valve tissue engineering: strategies, advances and challenges. Biomedicines 10(5):1095. https://doi.org/10.3390/biomedicines10051095

Collins MN, Ren G, Young K, Pina S, Reis RL, Oliveira JM (2021) Scaffold fabrication technologies and structure/function properties in bone tissue engineering. Adv Funct Mater 31(21):1–22

Cong Z, Shi Y, Wang Y, Wang Y, Chen N, Niu J, Xue H (2018) A novel controlled drug delivery system based on alginate hydrogel/chitosan micelle composites. Int J Biol Macromol 107(Pt A):855–864. https://doi.org/10.1016/j.ijbiomac.2017.09.065

Cornwell KG, Lei P, Andreadis ST, Pins GD (2007) Crosslinking of discrete self-assembled collagen threads: effects on mechanical strength and cell–matrix interactions. J Biomed Mater Res A 80(2):362–371. https://doi.org/10.1002/jbm.a.30893

Dewar DH, Amato M, Ellis HJ, Pollock EL, Gonzalez-Cinca N, Wieser H, Ciclitira PJ (2006) The toxicity of high molecular weight glutenin subunits of wheat to patients with coeliac disease. Eur J Gastroenterol Hepatol 18(5):483–491. https://doi.org/10.1097/00042737-200605000-00005

Ding YW, Zhang XW, Mi CH, Qi XY, Zhou J, Wei DX (2023) Recent advances in hyaluronic acid-based hydrogels for 3D bioprinting in tissue engineering applications. Smart Mater Med 4:59–68. https://doi.org/10.1016/j.smain.2022.07.003

Dong C, Lv Y (2016) Application of collagen scaffold in tissue engineering: recent advances and new perspectives. Polymers 8(2):42. https://doi.org/10.3390/polym8020042

Dromel PC, Singh D (2021) Biomanufacturing. In: 3D printing in medicine and surgery. Woodhead Publishing, Cambridge, MA, pp 137–170

Dzionek A, Wojcieszyńska D, Guzik U (2022) Use of xanthen gum for whole cell immobilization and its impact in bioremediation—a review. Bioresour Technol 351:126918. https://doi.org/10.1016/j.biotech.2022.126918

Fatimi A, Okoro OV, Podstawczyk D, Siminska-Stanny J, Shavandi A (2022) Natural hydrogel-based bio-inks for 3D bioprinting in tissue engineering: a review. Gels 8(3):179. https://doi.org/10.3390/gels8030179

Federer C, Kurpiers M, Bernkop-Schurch A (2021) Thiolated chitosans: a multi-talented class of polymers for various applications. Biomacromolecules 22(1):24–56. https://doi.org/10.1021/acs.biomac.0c00663

Feng B, LaPerle JL, Chang G, Verma MVS (2010) Renal clearance in drug discovery and development: molecular descriptors, drug transporters and disease state. Expert Opin Drug Metab Toxicol 6(8):939–952. https://doi.org/10.1517/17425255.2010.482930

Gentile P, Chiono V, Carmagnola I, Hatton PV (2014) An overview of poly(lactic-co-glycolic) acid (PLGA)-based biomaterials for bone tissue engineering. Int J Mol Sci 15:3640–3659. https://doi.org/10.3390/ijms15033640

Ghelich P, Kazemzadeh-Narbat M, Najafabadi AH, Samandari M, Memić A, Tamayol A (2022) (Bio)manufactured solutions for treatment of bone defects with an emphasis on US-FDA regulatory science perspective. Adv Nanobiomed Res 2(4):1–26. https://doi.org/10.1002/anbr.202100073

Gholap AD, Rojekar S, Kapare HS, Vishwakarma N, Raikwar S, Garkal A, Mehta TA, Jadhav H, Prajapat MK, Annapure U (2023) Chitosan scaffolds: expanding horizons in biomedical applications. Carbohydr Polym 323:121394. https://doi.org/10.1016/j.carbpal.2023.121394

Glowacki J, Mizuno S (2008) Collagen scaffolds for tissue engineering. Biopolymers 89(5):338–344

Goonoo N, Bhaw-Luximon A, Bowlin GL, Jhurry D (2013) An assessment of biopolymer- and synthetic polymer-based scaffolds for bone and vascular tissue engineering. Polym Int 62(4):523–533. https://doi.org/10.1002/pi.4474

Graciela CQ, Juan ECJ, Gieraldin CL, Alejandra PMX, Gabriel AA (2023) Hyaluronic acid—extraction methods, sources and applications. Polymers 15(16):3473. https://doi.org/10.3390/polym15163473

Gurib-Fakim A (2006) Medicinal plants: traditions of yesterday and drugs of tomorrow. Mol Asp Med 27:1–93. https://doi.org/10.1016/j.mam.2005.07.008

Helenius G, Bäckdahl H, Bodin A, Nannmark U, Gatenholm P, Risberg B (2006) In vivo biocompatibility of bacterial cellulose. J Biomed Mater Res A 76(2):431–438. https://doi.org/10.1002/jbm.a.30570

Huang J, Best SM (2022) Ceramic biomaterials for tissue engineering. In: Tissue engineering using ceramics and polymers, 3rd edn. Woodhead Publishing Series in Biomaterials, pp 3–40. https://doi.org/10.1016/B978-0-12-820508-2.00007-6

Inchingolo AD, Inchingolo AM, Malcangi G, Avantario P, Azzollini D, Buongiorno S, Viapiano F, Campanelli M, Ciocia AM, De Leonardis N, de Ruvo E, Ferrara I, Garofoli G, Montenegro V, Netti A, Palmieri G, Moncini A, Patano A, Piras F, Marinelli G, Pede CD, Laudadio C, Rapone B, Hazballa D, Corriero A, Fatone MC, Palermo A, Lorusso F, Scarano A, Bordea IR, Venere DD, Inchingo F, Dipalma G (2022) Effects of resveratrol, curcumin and quercetin supplementation on bone metabolism—a systematic review. Nutrients 14(7):3519. https://doi.org/10.3390/nu14173519

Iorio V, Troughton LD, Hamill KJ (2015) Laminins: roles and utility in wound repair. Adv Wound Care (New Rochelle) 4:250–263

Isasi R, Knoppers B, Borry P, Shabani M, Nys H (2018) Umbilical cord blood banking, research and clinical applications: report and recommendations. In: Governing stem cells regenerative medicine in Europe: the vision and recommendations from the EUCelLEX project, pp 60–75

Jansen K, Van Der Werff JFA, Van Wachem PB, Nicolai J-PA, De Leij LFMH, Van Luyn MJA (2004) A hyaluronan-based nerve guide: in vitro cytotoxicity, subcutaneous tissue reactions, and degradation in the rat. Biomaterials 25(3):483–489. https://doi.org/10.1016/s0142-9612(03)00544-1

Jeevanandam J, Barhoum A, Chan YS, Dufresne A, Danquah MK (2018) Review on nanoparticles and nanostructured materials: history, sources, toxicity and regulations. Beilstein J Nanotechnol 9:1050–1074. https://doi.org/10.3762/bjnano.9.98

Ji H, Li Y, Su B, Zhao W, Kizhakkedathu JN, Zhao C (2023) Advances in enhancing hemocompatibility of hemodialysis hollow-fiber membranes. Adv Fiber Mater 1–43. https://doi.org/10.1007/s42765-023-00277-5

Kasravi M, Ahmadi A, Babajani A, Mazloomnejad R, Hatamnejad MR, Shariatzadeh S, Bahrami S, Niknejad H (2023) Immunogenicity of decellularized extracellular matrix scaffolds: a bottleneck in tissue engineering and regenerative medicine. Biomater Res 27(10):1–24. https://doi.org/10.1186/s40824-023-00348-z

Khanna A, Zamani M, Huang NF (2021) Extracellular matrix-based biomaterials for cardiovascular tissue engineering. J Cardiovasc Dev Dis 8(11):137. https://doi.org/10.3390/jcdd8110137

Kim SM, Gwak EJ, Jeong SH, Lee SM, Sim WJ, Kim JS (2019) Toxicity evaluation of cellulose nanofibers (Cnfs) for cosmetic industry application. J Toxicol Risk Assess 5(2):1–6. https://doi.org/10.23937/2572-4061.1510029

Kingston R, Sioris K, Gualtieri J, Brutlag A, Droege W, Osimitz TG (2021) Post-market surveillance of consumer products: framework for adverse event management. Regul Toxicol Pharmacol 126:1–8. https://doi.org/10.1016/j.yrtph.2021.105028

Krishani M, Shin WY, Suhaimi H, Sambudi NS (2023) Development of scaffolds from bio-based natural materials for tissue regeneration applications: a review. Gels 9(2):100. https://doi.org/10.3390/gels9020100

Kumar A (2017) Pharmacovigilance: importance, concepts, and processes. Am J Health Syst Pharm 74(8):606–612. https://doi.org/10.2146/ajhp151031

Kumar S, Kang HJ, Berthiaume F (2019) Scaffolds for epidermal tissue engineering. In: Handbook of tissue engineering scaffolds, vol 2. Woodhead Publishing, pp 173–191

Kumari A, Singh K (2022) P463 toxicological evaluation of *Aureobasidum pullulans* var. *pullulans* induced mycotoxicosis in amurine model. Med Mycol 60(Suppl 1):myac072P463. https://doi.org/10.1093/mmy/myac072.P463

Laurent A, Abdel-Sayed P, Grognuz A, Scaletta C, Hirt-Burri N, Michetti M, de Buys Roessingh AS, Raffoul W, Kronen P, Nuss K, von Rechenberg B, Applegate LN, Darwiche SE (2021) Industrial development of standardized fetal progenitor cell therapy for tendon regenerative medicine: preliminary safety in xenogeneic transplantation. Biomedicines 9:380. https://doi.org/10.3390/biomedicines9040380

Liang Y, Zhang T, Tang M (2022) Toxicity of quantum dots on target organs and immune system. J Appl Toxicol 42(1):17–40. https://doi.org/10.1002/jat.4180

Litowczenko J, Woźniak-Budych MJ, Staszak K, Wieszczycka K, Jurga S, Tylkowski B (2022) Milestones and current achievements in development of multifunctional bioscaffolds for medical application. Bioact Mater 6(8):2412–2438. https://doi.org/10.1016/j.bioactmat.2021.01.007

Liu S, Yu JM, Gan YC, Qiu XZ, Gao ZC, Wang H, Chen SX, Xiong Y, Liu GH, Lin SE, McCarthy A, John JV, Wei DX, Hou H (2023) Biomimetric natural biomaterials for tissue engineering and regenerative medicine: new biosynthesis methods, recent advances, and emerging applications. Mil Med Res 10(1):1–30. https://doi.org/10.1186/s40779-023-00448-w

Lopes SV, Collins MN, Reis RL, Oliveira JM, Silva-Correia J (2021) Vascularization approaches in tissue engineering: recent developments on evaluation tests and modulation. ACS Appl Bio Mater 4(4):2941–2956. https://doi.org/10.1021/acsabm.1c00051

Lu RXZ, Lai BFL, Rafatian N, Gustafson D, Campbell SB, Banerjee A, Kozak R, Mossman K, Mubareka S, Howe KL, Fish JE, Radisic M (2022) Vasculature-on-a-chip platform with innate immunity enables identification of angiopoietin-1 derived peptide as a therapeutic for SARS-CoV-2 induced inflammation. Lab Chip 22(6):1171–1186. https://doi.org/10.1039/d1lc00817j

Madduri S, Gander B (2012) Growth factor delivery systems and repair strategies for damaged peripheral nerves. J Control Release 161(2):274–282. https://doi.org/10.1016/j.jconrel.2011.11.036

Magaz A, Spencer BF, Hardy JG, Li X, Gough JE, Blaker JJ (2020) Modulation of neuronal cell affinity on PEDOT–PSS nonwoven silk scaffolds for neural tissue engineering. ACS Biomater Sci Eng 6(12):6906–6916. https://doi.org/10.1021/acsbiomaterials.0c01239

Makadia HK, Siegel S (2011) Poly lactic-co-glycolic acid (PLGA) as biodegradable controlled drug delivery carrier. Polymers 3(3):1377–1397. https://doi.org/10.3390/polym3031377

Mansur HS, Costa HS (2008) Nanostructured poly (vinyl alcohol)/bioactive glass and poly (vinyl alcohol)/chitosan/bioactive glass hybrid scaffolds for biomedical applications. Chem Eng J 137(1):72–83. https://doi.org/10.1016/j.cej.2007.09.036

Matica A, Menghui G, Ostafe V (2017) Toxicity of chitosan based products. New Front Chem 26(1):65–74

Mironska E, Chapple C, MacNeil S (2019) Recent advances in pelvic floor repair. F1000Research 8:1–11. https://doi.org/10.12688/f1000research.15046.1

Moghadam ET, Yazdanian M, Alam M, Tebyanian H, Tafazoli A, Tahmasebi E, Ranjbar R, Yazdanian A, Seifalian A (2021) Current natural bioactive materials in bone and tooth regeneration in dentistry: a comprehensive overview. J Mater Res Technol 13:2078–2114. https://doi.org/10.1016/j.jmrt.2021.05.089

Mohanty AK, Wu F, Mincheva R, Hakkarainen M, Raquez JM, Mielewski DF, Narayan R, Netravali AN, Misra M (2022) Sustainable polymers. Nat Rev Methods Primers 2(1):46

Molzon JA, Giaquinto A, Lindstrom L, Tominaga T, Ward M, Doerr P, Hunt L, Rago L (2011) The value and benefits of the International Conference on Harmonisation to drug regulatory authorities: advancing harmonization for better public health. Clin Pharmacol Ther 89(4):503–512. https://doi.org/10.1038/clpt.2011.10

Mortensen A, Aguilar F, Crebelli R, Di Domenico A, Frutos MJ, Galtier P, Gott D, Gundert-Remy U, Lambré C, Leblanc JC (2017) Re-evaluation of guar gum (E412) as a food additive. EFSA 15(2):e04669. https://doi.org/10.2903/j.efsa.2017.4669

Naserzadeh P, Mortazavi SA, Ashtari K, Seydi E, Pourahmad J (2018) Evaluation of the toxicity effects of silk fibroin on isolated fibroblast and Huvec cells. Iran J Pharm Res 17(Suppl 2):134–145

Nikolova MP, Chavali MS (2019) Recent advances in biomaterials for 3D scaffolds: a review. Bioact Mater 4:271–292. https://doi.org/10.1016/j.bioactmat.2019.10.005

Noori A, Ashrafi SJ, Vaez-Ghaemi R, Hatamian-Zaremi A, Webster TJ (2017) A review of fibrin and fibrin composites for bone tissue engineering. Int J Nanomedicine 12:4937–4961. https://doi.org/10.2147/IJN.S124671

Okolie O, Kumar A, Edwards C, Lawton LA, Oke A, McDonald S, Thakur VK, Nijuguna J (2023) Bio-based sustainable polymers and materials: from processing to biodegradation. J Compos Sci 7(6):1–54. https://doi.org/10.3390/jcs7060231

Oryan A, Kamali A, Moshiri A, Baharvand H, Daemi H (2018) Chemical crosslinking of biopolymeric scaffolds: current knowledge and future directions of crosslinked engineered bone scaffolds. Int J Biol Macromol 107:678–688. https://doi.org/10.1016/j.ijbiomac.2017.08.184

Ozdil D, Aydin HM (2014) Polymers for medical and tissue engineering applications. J Chem Technol Biotechnol 89(12):1793–1810. https://doi.org/10.1002/jctb.4505

Pacurari M, Lowe K, Tchounwou PB, Kafoury R (2016) A review on the respiratory system toxicity of carbon nanoparticles. Int J Environ Res Public Health 13(3):325. https://doi.org/10.3390/ijerph13030325

Paladini F, Pollini M (2022) Novel approaches and biomaterials for bone tissue engineering: a focus on silk fibroin. Materials 15(19):6952. https://doi.org/10.3390/ma15196952

Pankajakshan D, Agrawal DK (2010) Scaffolds in tissue engineering of blood vessels. Can J Physiol Pharmacol 88(9):855–873. https://doi.org/10.1139/y10-073

Pastorello AE, Farioli L, Pravettoni V, Scibilia J (2009) Maize food allergy: lipid-transfer proteins, endochitinases, and alpha-zein precursor are relevant maize allergens in double-blind placebo-controlled maize-challenge-positive patients. Anal Bioanal Chem 395(1):93–102. https://doi.org/10.1007/s00216-009-2945-z

Patil US, Das S (2023) Clinical trials in India: towards effective regulation a critical analysis of the regulatory system in India. Lat Am J Pharm 42(2):242–250

Pawar KN, Gore RT, Palekar SR (2023) A review on approval process and regulation of medical devices as per US FDA and CDSCO. Int J Drug Regul Affairs 11(1):61–70. https://doi.org/10.22270/ijdra.v11.586

Pazhouhnia Z, Beheshtizadeh N, Namini MS, Lotfibabakhshaiesh N (2022) Portable hand-held bioprinters promote in situ tissue regeneration. Bioeng Transl Med 7(3):e10307. https://doi.org/10.1002/btm2.10307

Pentšuk N, van der Laan JW (2009) An interspecies comparison of placental antibody transfer: new insights into developmental toxicity testing of monoclonal antibodies. Birth Defects Res B Dev Reprod Toxicol 86(4):328–344. https://doi.org/10.1002/bdrb.20201

Petre A, Bjarnadottir A (2019) What is gellan gum? Uses, benefits, and safety. Healthline

Piccinin E, Villani G, Moschetta A (2019) Metabolic aspects in NAFLD, NASH and hepatocellular carcinoma: the role of PGC1 coactivators. Nat Rev Gastroenterol Hepatol 16(3):160–174. https://doi.org/10.1038/s41575-018-0089-3

Prasathkumar M, Sadhasivam S (2021) Chitosan/hyaluronic acid/alginate and an assorted polymers loaded with honey, plant, and marine compounds for progressive wound healing—know-how. Int J Biol Macromol 186:656–685. https://doi.org/10.1016/j.ijbiomac.2021.07.067

Rai V, Dilisio MF, Dietz NE, Agrawal DK (2017) Recent strategies in cartilage repair: a systemic review of the scaffold development and tissue engineering. J Biomed Mater Res A 105(8):2343–2354. https://doi.org/10.1002/jbm.a.36087

Rampersad SN (2012) Multiple applications of Alamar Blue as an indicator of metabolic function and cellular health in cell viability bioassays. Sensors 12(9):12347–12360. https://doi.org/10.3390/s120912347

Rasouli M, Rahimi A, Soleimani M, Keshel SH (2021) The interplay between extracellular matrix and progenitor/stem cells during wound healing: opportunities and future directions. Acta Histochem 123(7):151785. https://doi.org/10.1016/j.acthis.2021.151785

Reina G, González-Domínguez JM, Criado A, Vázquez E, Bianco A, Prato M (2017) Promises, facts and challenges for graphene in biomedical applications. Chem Soc Rev 46(15):4400–4416. https://doi.org/10.1039/C7CS00363C

Roberts IV, Bukhary D, Valdivieso CYLV, Tirelli N (2020) Fibrin matrices as (injectable) biomaterials: formation, clinical use, and molecular engineering. Macromol Biosci 20:e1900283. https://doi.org/10.1002/mabi.201900283

Robscheit-Robbins FS, Miller LL, Whipple GH (1944) Gelatin—its usefulness and toxicity: blood protein production impaired by continued gelatin by vein. J Exp Med 80(2):145–164. https://doi.org/10.1084/jem.80.2.145

Rodriguez-Vazquez M, Vega-Ruiz B, Quinones-Olvera LF (2015) Chitosan and its potential use as a scaffold for tissue engineering in regenerative medicine. Biomed Res Int 2015:821279. https://doi.org/10.1155/2015/821279

Ruiz EL, Gonzalez GJ, de Cienfuegos LA, Acedo AC, Sabata R, Marchal JA, Galvez P (2019) Advances of hyaluronic acid in stem cell therapy and tissue engineering, including current clinical trials. Eur Cell Mater 37:186–213. https://doi.org/10.22203/eCM.v037a12

Sahranavard M, Zamanian A, Ghorbani F, Shahrezaee MH (2020) A critical review on three dimensional-printed chitosan hydrogels for development of tissue engineering. Bioprinting 17:e00063. https://doi.org/10.1016/j.bprint.2019.e00063

Saiz E, Zimmermann EA, Lee JS, Wegst UG, Tomsia AP (2013) Perspectives on the role of nanotechnology in bone tissue engineering. Dent Mater 29(1):103–115. https://doi.org/10.1016/j.dental.2012.08.001

Sarasati A, Syahruddin MH, Nuryanti A, Ana ID, Barlian A, Wijaya CH, Ratnadewi D, Wungu TDK, Takemori H (2023) Plant-derived exosome-like nanoparticles for biomedical applications and regenerative therapy. Biomedicines 11(4):1–27. https://doi.org/10.3390/biomedicines11041053

Schilrreff P, Alexiev U (2022) Chronic inflammation in non-healing skin wounds and promising natural bioactive compounds treatment. Int J Mol Sci 23(9):1–31. https://doi.org/10.3390/ijms23094928

Shin H, Kim E (2022) Analysis of GMP for marketing authorization of ATMPs: comparison in the US, the EU, Japan and South Korea. Regen Med 17(5):283–297. https://doi.org/10.2217/rme-2021-0153

Singh M, Morris CP, Ellis RJ, Detamore MS, Berkland C (2008) Microsphere-based seamless scaffolds containing macroscopic gradients of encapsulated factors for tissue engineering. Tissue Eng Part C Methods 14(4):299–309. https://doi.org/10.1089/ten.tec.2008.0167

Singh N, Sahoo SK, Kumar R (2020) Hemolysis tendency of anticancer nanoparticles changes with type of blood group antigen: an insight into blood nanoparticle interactions. Mater Sci Eng C 109:1–10. https://doi.org/10.1016/j.msec.2020.110645

Song P, Wang H (2020) High-performance polymeric materials through hydrogen-bond cross-linking. Adv Mater 32(18):e1901244. https://doi.org/10.1002/adma.201901244

Swapan MA (2018) Polyurethanes from vegetable oils and applications: a review. J Polym Res 25:1–15

Tian J, Song X, Wang Y, Cheng M, Lu S, Xu W, Gao G, Sun L, Tang Z, Wang M, Zhang X (2022) Regulatory perspectives of combination products. Bioact Mater 10:492–503. https://doi.org/10.1016/j.bioactmat.2021.09.002

Tsintou M, Dalamagkas K, Moore TL, Rathi Y, Kubicki M, Rosene DL, Makris N (2021) The use of hydrogel-delivered extracellular vesicles in recovery of motor function in stroke: a testable experimental hypothesis for clinical translation including behavioral and neuroimaging assessment approaches. Neural Regen Res 16(4):605–613. https://doi.org/10.4103/1673-5374.295269

Tupone MG, d'Angelo M, Castelli V, Catanesi M, Benedetti E, Cimini A (2021) A state-of-the-art of functional scaffolds for 3D nervous tissue regeneration. Front Bioeng Biotechnol 9:639765. https://doi.org/10.3389/fbioe.2021.639765

Uddin MS, Hall C, Murphy P (2015) Surface treatments for controlling corrosion rate of biodegradable Mg and Mg-based alloy implants. Sci Technol Adv Mat 16(5):1–24. https://doi.org/10.1088/1468-6996/16/5/053501

Unal AZ, West JL (2020) Synthetic ECM: bioactive synthetic hydrogels for 3D tissue engineering. Bioconjug Chem 31(10):2253–2271. https://doi.org/10.1021/acs.bioconjchem.0c00270

Varma RS, Iravani S (2022) Cellulose-based composites as scaffolds for tissue engineering: recent advances. Molecules 27(24):8830. https://doi.org/10.3390/molecules27248830

Vieira WT, de Farias MB, Spaolonzi MP, da Silva MGC, Vieira MGA (2021) Endocrine-disrupting compounds: occurrence, detection methods, effects and promising treatment pathways—a critical review. J Environ Chem Eng 9(1):104558. https://doi.org/10.1016/j.jece.2020.104558

Vishal P, Rahulgiri G, Pratik M, Badjatya JK (2014) A review on drug approval process for US, Europe and India. Int J Drug Regul Affairs 2(1):1–11. https://doi.org/10.22270/ijdra.v2i.7

Wang W, Xue C, Mao X (2020) Chitosan: structural modification, biological activity and application. Int J Biol Macromol 164:4532–4546. https://doi.org/10.1016/j.ijbiomac.2020.09.042

Weber M, Steinle H, Golombek S, Hann L, Schlensak C, Wendel HP, Avci-Adali M (2018) Blood-contacting biomaterials: in vitro evaluation of the hemocompatibility. Front Bioeng Biotechnol 6:1–11. https://doi.org/10.3389/fbioe.2018.00099

Winkler HC, Notter T, Meyer U, Naegeli H (2018) Critical review of the safety assessment of titanium dioxide additives in food. J Nanobiotechnol 16(51):1–19

Yamauchi K, Maniwa M, Mori T (1998) Cultivation of fibroblast cells on keratin-coated substrata. J Biomater Sci Polym Ed 9(3):259–270. https://doi.org/10.1163/156856298X00640

Yang D, Xiao J, Wang B, Li L, Kong X, Liao J (2019) The immune reaction and degradation fate of scaffold in cartilage/bone tissue engineering. Mater Sci Eng C Mater Biol Appl 104:1–17. https://doi.org/10.1016/j.msec.2019.109927

Yao S, Cui J, Chen S, Zhou X, Li J, Zhang K (2022) Extracellular matrix coatings on cardiovascular materials—a review. Coatings 12(8):1–20. https://doi.org/10.3390/coatings12081039

Yazdimamaghani M, Razavi M, Vashae D, Moharamzadeh K, Boccaccini RA, Tayebi L (2017) Porous magnesium-based scaffolds for tissue engineering. Mater Sci Eng C 71:1253–1266. https://doi.org/10.1016/j.msec.2016.11.027

You J, Zhang Y, Zhou Y (2022) Strontium functionalized in biomaterials for bone tissue engineering: a prominent role in osteoimmunomodulation. Front Bioeng Biotechnol 10:1–22. https://doi.org/10.3389/fbioe.2022.928799

Zeugolis DI, Paul GR, Attenburrow G (2009) Cross-linking of extruded collagen fibers—a biomimetic three-dimensional scaffold for tissue engineering applications. J Biomed Mater Res A 89(4):895–908. https://doi.org/10.1002/jbm.a.32031

Zhang Y, Tang J, Fang W, Zhao Q, Lei X, Zhang J, Chen J, Li Y, Zuo Y (2023) Synergetic effect of electrical and topographical cues in aniline trimer-based polyurethane fibrous scaffolds on tissue regeneration. J Funct Biomater 14(4):185. https://doi.org/10.3390/jfb14040185

Zhou X, Liu C, Han Y, Li C, Liu S, Li X, Zhao G, Jiang Y (2022) An antibacterial chitosan-based hydrogel as a potential degradable bio-scaffold for alveolar ridge preservation. R Soc Chem 12:32219–32229. https://doi.org/10.1039/D2RA05151F

Correction to: Natural Gum (Xanthan, Gellan, Arabic, Guar, Ghatti Gum, Etc.)-Based Bio-Scaffold and Their Application in Tissue Engineering

Swati Gajbhiye, Sanjay Dhoble, and Desmond Tobin

Correction to:
Chapter 4 in: M. Kumar et al. (eds.), *Natural Product Inspired Scaffolds*, https://doi.org/10.1007/978-981-97-3111-4_4

The original version of this chapter was inadvertently published with the incorrect affiliation of the author Dr. Desmond Tobin. The correct affiliation should read as:

Dr. Desmond Tobin

The Charles Institute of Dermatology, School of Medicine, The Conway Institute, University College Dublin, Belfield, Dublin 4, Ireland

The affiliation of the chapter author has been updated in the original version with this erratum.

The updated version of this chapter can be found at
https://doi.org/10.1007/978-981-97-3111-4_4

M. Kumar et al. (eds.), *Natural Product Inspired Scaffolds*,
https://doi.org/10.1007/978-981-97-3111-4_12

FSC
www.fsc.org
MIX
Papier aus verantwortungsvollen Quellen
Paper from responsible sources
FSC® C105338

If you have any concerns about our products,
you can contact us on
ProductSafety@springernature.com

In case Publisher is established outside the EU,
the EU authorized representative is:
Springer Nature Customer Service Center GmbH
Europaplatz 3, 69115 Heidelberg, Germany

Printed by Libri Plureos GmbH
in Hamburg, Germany